MW01504588

OPTICAL WIRELESS COMMUNICATIONS

OTHER TELECOMMUNICATIONS BOOKS FROM AUERBACH

Active and Programmable Networks for Adaptive Architectures and Services
Syed Asad Hussain
ISBN: 0-8493-8214-9

Ad Hoc Mobile Wireless Networks: Principles, Protocols and Applications
Subir Kumar Sarkar, T.G. Basavaraju, and C. Puttamadappa
ISBN: 1-4200-6221-2

Comprehensive Glossary of Telecom Abbreviations and Acronyms
Ali Akbar Arabi
ISBN: 1-4200-5866-5

Contemporary Coding Techniques and Applications for Mobile Communications
Onur Osman and Osman Nuri Ucan
ISBN: 1-4200-5461-9

Context-Aware Pervasive Systems: Architectures for a New Breed of Applications
Seng Loke
ISBN: 0-8493-7255-0

Data-driven Block Ciphers for Fast Telecommunication Systems
Nikolai Moldovyan and Alexander A. Moldovyan
ISBN: 1-4200-5411-2

Distributed Antenna Systems: Open Architecture for Future Wireless Communications
Honglin Hu, Yan Zhang, and Jijun Luo
ISBN: 1-4200-4288-2

Encyclopedia of Wireless and Mobile Communications
Borko Furht
ISBN: 1-4200-4326-9

Handbook of Mobile Broadcasting: DVB-H, DMB, ISDB-T, AND MEDIAFLO
Borko Furht and Syed A. Ahson
ISBN: 1-4200-5386-8

The Handbook of Mobile Middleware
Paolo Bellavista and Antonio Corradi
ISBN: 0-8493-3833-6

The Internet of Things: From RFID to the Next-Generation Pervasive Networked Systems
Lu Yan, Yan Zhang, Laurence T. Yang, and Huansheng Ning
ISBN: 1-4200-5281-0

Introduction to Mobile Communications: Technology, Services, Markets
Tony Wakefield, Dave McNally, David Bowler, and Alan Mayne
ISBN: 1-4200-4653-5

Millimeter Wave Technology in Wireless PAN, LAN, and MAN
Shao-Qiu Xiao, Ming-Tuo Zhou, and Yan Zhang
ISBN: 0-8493-8227-0

Mobile WiMAX: Toward Broadband Wireless Metropolitan Area Networks
Yan Zhang and Hsiao-Hwa Chen
ISBN: 0-8493-2624-9

Optical Wireless Communications: IR for Wireless Connectivity
Roberto Ramirez-Iniguez, Sevia M. Idrus, and Ziran Sun
ISBN: 0-8493-7209-7

Performance Optimization of Digital Communications Systems
Vladimir Mitlin
ISBN: 0-8493-6896-0

Physical Principles of Wireless Communications
Victor L. Granatstein
ISBN: 0-8493-3259-1

Principles of Mobile Computing and Communications
Mazliza Othman
ISBN: 1-4200-6158-5

Resource, Mobility, and Security Management in Wireless Networks and Mobile Communications
Yan Zhang, Honglin Hu, and Masayuki Fujise
ISBN: 0-8493-8036-7

Security in Wireless Mesh Networks
Yan Zhang, Jun Zheng, and Honglin Hu
ISBN: 0-8493-8250-5

Wireless Ad Hoc Networking: Personal-Area, Local-Area, and the Sensory-Area Networks
Shih-Lin Wu and Yu-Chee Tseng
ISBN: 0-8493-9254-3

Wireless Mesh Networking: Architectures, Protocols and Standards
Yan Zhang, Jijun Luo, and Honglin Hu
ISBN: 0-8493-7399-9

AUERBACH PUBLICATIONS
www.auerbach-publications.com
To Order Call: 1-800-272-7737 • Fax: 1-800-374-3401
E-mail: orders@crcpress.com

OPTICAL WIRELESS COMMUNICATIONS

IR for Wireless Connectivity

Roberto Ramirez-Iniguez

Sevia M. Idrus

Ziran Sun

CRC Press
Taylor & Francis Group
Boca Raton   London   New York

CRC Press is an imprint of the
Taylor & Francis Group, an **informa** business
AN AUERBACH BOOK

Auerbach Publications
Taylor & Francis Group
6000 Broken Sound Parkway NW, Suite 300
Boca Raton, FL 33487-2742

© 2008 by Taylor & Francis Group, LLC
Auerbach is an imprint of Taylor & Francis Group, an Informa business

Library of Congress Cataloging-in-Publication Data

Ramirez-Iniguez, Roberto.
    Optical wireless communications : IR for wireless connectivity / Roberto
Ramirez-Iniguez, Sevia M. Idrus.
        p. cm.
    Includes bibliographical references and index.
    ISBN 978-0-8493-7209-4 (hardback : alk. paper) 1. Wireless communication
    systems. 2. Optical communications. 3. Infrared technology. I. Idrus, Sevia M.
    II. Title.

TK5103.2.R347 2007
621.384--dc22   2007029822

Visit the Taylor & Francis Web site at
http://www.taylorandfrancis.com

and the Auerbach Web site at
http://www.auerbach-publications.com

Contents

List of Figures

List of Tables

Preface

Optical wireless communications systems have begun to permeate all areas of our daily life. Devices such as mobile phones, PDAs, and computer peripherals, to name just a few, are increasingly incorporating infrared transceivers that provide an alternative to radio for wireless connectivity. This book examines some of the most important characteristics of optical wireless communications systems, presenting an up-to-date review of the features and techniques employed by wireless IR systems for indoor and outdoor use.

The book has been organized in ten chapters: the first chapter presents an introduction to optical wireless communication technology, covering aspects such as system topologies, evolution of wireless IR communication systems, and design fundamentals. Here, the benefits and limitations of IR as a medium to convey wireless information are examined and the advantages and disadvantages of infrared are compared to radio systems.

Chapters 2 and 3 describe atmospheric and other types of data transmission limitations. Eye safety issues are discussed in Chapter 3, together with techniques to minimize the risk posed by some optical sources. In addition, some of the most important characteristics of LEDs and LDs are introduced and special considerations for outdoor systems are presented.

Techniques to compensate for some of the data transmission limitations suffered by wireless IR systems are explained in Chapters 4 and 5. The former introduces fundamentals of optical concentration employed in Chapter 5 to design and describe a variety of imaging and nonimaging concentrators. In addition, different optical filtering techniques employed to reduce the noise introduced by background illumination are also analyzed.

Chapters 6 and 7 are devoted to the design of optical wireless transmitters and receivers. Chapter 6 covers topics such as optical sources, driver circuit design, predriver, data retiming, automatic control, and transmitter linearization techniques. Chapter 7 presents topics such as detector types, receiver noise considerations, bit error rate and sensitivity, bandwidth, and signal amplifications techniques.

Different modulation techniques employed in optical wireless communication systems are explained in Chapter 8. This chapter also includes comparisons of different modulation techniques in terms of power and bandwidth requirements, as well as comparisons of the modulation schemes in the presence of noise and multipath distortion. Multiple access techniques are also presented.

Topics concerning wireless IR protocols and wireless IR networking are described in Chapters 9 and 10, respectively. The characteristics of the IrDA protocol, for instance, are introduced in Chapter 9, while important networking issues such as network architecture, optical wireless network specifications, ad hoc networking, and QoS are covered in Chapter 10.

A section entitled summary and conclusions has been incorporated at the end of most chapters to highlight the main points covered and to discuss the implications of the material presented. The aim of this book is to serve as an introduction to wireless IR communications for undergraduate students interested in the topic. It is also intended as a reference for postgraduate students, researchers, and designers working in this discipline.

About the Authors

Dr. Roberto Ramirez-Iniguez received a B.Eng. degree in Electrical and Electronic Engineering from the Universidad Nacional Autónoma de México (UNAM) in 1996, and an M.Sc. in Communications and Real-Time Electronic Systems from the University of Bradford in 1998. After obtaining his first degree, he worked for a short period of time for Tektronix Mexico in the area of Tools and Business Units. In 1998, he started postgraduate studies at the University of Warwick, from which he received his Ph.D. in 2002. Shortly afterwards, he began working as a researcher for Optical Antenna Solutions within the Communications and Signal Processing group in the Photonics and Communications Laboratory of Warwick University. He currently is a lecturer in Communications and Embedded Systems at Glasgow Caledonian University.

His areas of interest include optical wireless communications, optical antennae design, and research on directive elements for wireless IR transmitters and receivers.

Dr. Sevia M. Idrus is Deputy Director of the Photonics Technology Centre in the Universiti Teknologi Malaysia. She received a B.Eng. and a Masters degree in Electrical Engineering both from the Universiti Teknologi Malaysia. She obtained a Ph.D. in Optical Communication Systems from the University of Warwick in 2004. She has worked at the Universiti Teknologi Malaysia since 1998, both as academic and administrative staff. Her main areas of research include the characterization, modeling, and design of opto-electronic devices, radio-over-fiber systems, optical transceiver design, and optical wireless communication technology.

Dr. Ziran Sun received a B.Eng. degree in Automation Control in 1995 and an M.Sc. in Communication and Electronics in 1998 in China. In 1999, she was awarded an ORS scholarship from the British Government and a Studentship from the University of Warwick for doctoral studies. She was awarded a Ph.D. by the University of Warwick in 2005. In 2002, she worked as a Researcher in the University of Bradford and later moved to the University of Cambridge (in 2004). She joined the telecommunications industry in 2005, where she currently works as a software engineer. Her areas of interest include optical wireless communications, communications protocol design for mobile hosts, and wireless networks.

List of Abbreviations

ADR	Angle diversity receiver
AEL	Allowable exposure limit
AGC	Automatic gain control
Air	Advanced infrared
AM	Amplitude modulation
ANSI	American National Standards Institute
AODV	Ad Hoc On-Demand Distance Vector Routing Protocol
AP	Access point
APC	Automatic power control
APD	Avalanche photodetector
AR	Antireflection
ARQ	Automatic repeat request
ASK	Amplitude-shift keying
AWGN	Additive white Gaussian noise
BER	Bit error rate
BJT	Bipolar junction transistor
BPSK	Binary phase-shift keying
BSA	Basic service area
BTA	Bootstrapped transimpedance amplifier
BW	Bandwidth
CA	Collision avoidance
CAS	Collision avoidance slots
CDMA	Code division multiple access
CDR	Clock and data recovery
CENELEC	European Committee for Electrotechnical Standardisation
CGH	Computer-generated hologram
CGSR	Clustered gateway switch routing
CTS	Clear to send
CMU	Clock multiplication unit
CNR	Carrier-to-noise ratio
CPC	Compound parabolic concentrator

CRZ	Chirped return-to-zero
CSMA/CA	Carrier sense multiple access with collision avoidance
CSMA/CD	Carrier sense multiple access with collision detection
CSRZ	Carrier suppressed return-to-zero
CVSD	Continuous variable slope delta
CW	Continuous wave (also contention window)
DBF	Distributed feedback
DBIF	Directed beam infrared
DCF	Distributed coordination function
DD	Direct detection
DFE	Decision-feedback equalizer
DFIR	Diffuse infrared
DHPIM	Dual header-pulse interval modulation
DMUX	Demultiplexer
DPIM	Digital pulse interval modulation
DPPM	Differential pulse position modulation
DSDV	Dynamic destination sequenced distance vector
DSR	Dynamic Source Routing Protocol
DSSS	Direct Sequence Spread Spectrum
DTIRC	Dielectric totally internally reflecting concentrator
EAM	Electro-absorption modulator
ECL	Emitter-coupled logic
EDCF	Enhanced distributed coordination function
EGC	Equal-gain combining
EMI	Electromagnetic interference
EOB	End-of-burst
EOBC	End-of-burst confirmation
ER	Extinction ratio
FCC	Federal Communications Commission (USA)
FDMA	Frequency-division multiple access
FEC	Forward error correction
FET	Field effect transistor
FFE	Feed-forward equalizer
FFP	Far-field pattern
FHSS	Frequency Hopping Spread Spectrum
FIR	Fast infrared (also finite impulse response)
FM	Frequency modulation
FOV	Field-of-view
FSK	Frequency-shift keying
FSR	Fisheye state routing
FSO	Free-space optics
GA	Genetic algorithm
GBN	Go-back-N

GBW	Gain bandwidth
GO	Geometrical optics
GSR	Global state routing
HCF	Hybrid coordination function
HDLC-ABM	High-level data link control asynchronous balanced mode
HDTV	High-definition television
HSR	Hierarchical state routing
HTTP	HyperText Transport Protocol
IAS	Information access service
IDA	Instantaneous digital adaption
IEC	International Electrotechnical Commission
IM	Intensity modulation
IOEC	Integrated opto-electronics
IR	Infrared
IrCOMM	Infrared communication
IrDA	Infrared Data Association
ISI	Intersymbol interference
LA	Limiting amplifier
LAN	Local area network
LAP	Link Access Protocol
LAR	Location aided routing
LC	Link control
LD	Laser diode
LED	Light emitting diode
LM	Link manager
LMP	Link Management Protocol
LOS	Line-of-sight
L-PAM	Pulse amplitude modulation with L levels
LPF	Low-pass filter
L-PPM	Pulse position modulation with L levels
LSAP-Sel	Logical service access point selector
LSD	Loss of signal detector
LSMS	Line-strip multi-beam systems
LUT	Look-up table
MAC	Media Access Control
MIR	Medium infrared
MLSD	Maximum likelihood sequence detection
MPE	Maximum permissible exposure
MRC	Maximum ratio combining
MSDS	Multi-spot diffusing system
MSM	Metal-semiconductor-metal
MUX	Multiplexer
MZM	Mach-Zehender modulation

NAV	Network allocation vector
NFB	Negative feedback
NIST	National Institute of Standards and Technology
NRZ	Non-return-to-zero
OA	Optical antenna
OBEX	Object exchange
OfCom	Office of Communications (United Kingdom)
OOK	On-off-keying
OLSR	Optimized link state routing
OPPM	Overlapping pulse position modulation
OSHA	Occupational Safety and Health Administration
OSI	Open Systems Interconnection
PAM	Pulse amplitude modulation
PDA	Personal digital assistant
PDF	Probability density function
PDM	Pulse duration modulation
PHY	Physical layer
PIN	Positive-intrinsic-negative
PLL	Phase-locked loop
PPM	Pulse position modulation
PSK	Phase-shift keying
QDIR	Quasi-diffuse infrared
QoS	Quality-of-Service
QPSK	Quadrature phase-shift keying
RF	Radio frequency
RIN	Relative intensity noise
RLL	Run-length limited
RMA	Receiver main amplifier
RTS	Request to send
RZ	Return-to-zero
SA	Simulated annealing
SBR	Solar background radiation
SCFL	Source-coupled FET logic
SCM	Subcarrier multiplexing
SD	Selection diversity
SDMA	Space division multiple access
SDT	Spot-diffusing transmitters
SIFS	Short inter-frame space
SIG	Special interest group
SIP	Serial infrared interaction pulse
SIR	Standard infrared
SNDR	Signal-to-noise and distortion ratio
SNR	Signal-to-noise ratio

SONET	Synchronous Optical Network
SW	Stop and wait
TBP	Ticket-based probing
TDD	Time division duplex
TDM	Time division multiplexing
TDMA	Time division multiple access
TE	Transverse electric
TIR	Total internal reflection
TM	Transverse magnetic
TORA	Temporarily ordered routing algorithm
UART	Universal asynchronous receiver transmitter
UV	Ultraviolet
VCCS	Voltage control current source
VCSEL	Vertical cavity surface emitting laser
VCO	Voltage-controlled oscillator
VFIR	Very fast infrared
VGA	Variable gain amplifier
VPN	Virtual private network
WDMA	Wavelength division multiple access
WEP	Wired Equivalent Privacy
WGN	White Gaussian noise
WiFi	Wireless Fidelity
WLAN	Wireless local area network
WRP	Wireless Routing Protocol
WSK	Wavelength shift keying
ZRP	Zone Routing Protocol

Acknowledgments

We would like to thank the Universidad Nacional Autónoma de México (UNAM), the University of Warwick, the University of Glasgow Caledonian, and the Universiti Teknologi Malaysia for their encouragement and support (in one way or another) to produce this book. We are particularly grateful to Professor Roger J. Green from the University of Warwick for reading the manuscript and for making useful comments about parts of the book that could be corrected or improved. Thanks are also due to Optical Antenna Solutions Ltd. for their permission to use part of the material presented in Chapter 5.

Finally, we would like to thank our respective families for their patience, understanding and support during the preparation of the manuscript.

R. Ramirez-Iniguez, S. M. Idrus, and Z. Sun

Chapter 1

Introduction

1.1 Technology Overview

In the past few decades, an unprecedented demand for wireless technologies has been taking place. Both industrial and private customers are demanding products — for a wide range of applications — that incorporate wireless features, which allow them to exchange, receive, or transmit information without the inconvenience of having to be fixed to any particular location. Laptops, personal digital assistants (PDAs), and mobile phones, to name just a few examples, are becoming part of the everyday life of a growing number of people, and all these devices are increasingly incorporating technologies that allow them to work without cables.

The benefits of wireless technologies are not limited to user convenience — in terms of mobility — and flexibility in the placement of terminals. Significant reductions in cost and time also can be achieved, in a number of applications, using wireless solutions. Reconfiguring computer terminals or microcontroller systems (in places such as laboratories, conference rooms, offices, hospitals, production floors, or educational institutions), for instance, can be done relatively cheaply and quickly with wireless networks. Maintaining and reconfiguring wired networks, on the other hand, is usually carried out in more expensive, time-consuming, and complicated ways (especially in situations where cables are grounded or installed in inaccessible places). Furthermore, cables are susceptible to damage, which means potential disruption to the network operation.

Radio and infrared (IR) are currently the main parts of the electromagnetic spectrum used to transmit information wirelessly. In this book the term "radio" refers to the radiofrequency and microwave parts of the spectrum, and "IR" refers to the near-infrared part of it. Despite the fact that the most commonly used medium for wireless communications thus far is radio, IR is becoming more popular every day; and it is being preferred (due to its inherent advantages) over its radio counterpart for a number of applications. From a spectrum management point of view, for example, IR offers potentially huge bandwidths that are currently unregulated worldwide. The radio part of the spectrum, on the other hand, gets more congested every year, and the allocation of radio frequencies is increasingly difficult and expensive. Moreover, due the fact that the authorities that regulate the allocation of radio frequencies vary from one country to another (for example, *OfCom* — the *Office of Communications* — in the United Kingdom [1, 2], and the FCC — the *Federal Communications Commission* — in the United States [3], to name only two), there is a potential risk of system or product incompatibility in different geographical locations.

Another advantage of IR over radio is its immunity to electromagnetic interference (EMI). This makes IR the preferred option in environments where interference must be minimized or eliminated. In addition, IR does not interfere with and is not affected by radio frequencies, which is particularly relevant in hospitals, as explained in a number of published articles in the area [4–6]. Hagihira et al. [4], for example, indicate that while radio frequency local area networks (LANs) transporting and managing patient information interfere with medical devices such as pacemakers and infusion pumps — with potentially fatal consequences —, infrared LANs offer a safe and effective alternative for the same application.

IR also presents advantages over radio in terms of security. Because IR radiation behaves like visible light, it does not penetrate walls, which means that the room where the energy is generated encloses the emitted signal completely (assuming there are no windows or transparent barriers between rooms). This prevents the transmitted information from being detected outside and implies intrinsic security against eavesdropping [7, 8]. In addition, IR offers the possibility of rapid wireless deployment and the flexibility of establishing temporary communication links.

Further advantages of IR over radio include the low cost, the small size, and the limited power consumption of IR components. This is explained by the fact that wireless IR communication systems make use of the same opto-electronic devices that have been developed and improved over the past decades for optical fiber communications and other applications. One such component is the light-emitting diode (LED), which, due to its now faster response times, high radiant output power, and improved efficiency, is becoming the preferred option for short-distance optical wireless applications. Other components that have been remarkably improved over the past decades include the laser diode (LD) and the positive-intrinsic-negative (PIN) detector, which are becoming faster and are making higher data rates possible.

Despite the advantages presented by the infrared medium, IR is not without its drawbacks. Optical wireless links are susceptible to blocking from persons and objects, which can result in the attenuation of the received signal or in the disruption of the link (depending on the configuration of the system). In addition, wireless IR systems generally operate in environments where other sources of illumination are present. This background illumination has part of its energy in the spectral region used by wireless IR transmitters and receivers, and introduces noise in the photodetector, which limits the range of the system. Moreover, optical wireless systems are also affected by the high attenuation suffered by the IR signal when transmitted through air, and by atmospheric phenomena such as fog and snow that further reduce the range of the system and deteriorate the quality of the transmission when operating outdoors (cf. Chapter 2).

Some of the drawbacks presented by IR (such as attenuation and background illumination noise) can be compensated to some extent by increasing the optical power level at the transmitter. Unfortunately, due to the fact that a high emission power from some emitters can be potentially dangerous to the retina (which has motivated the creation of eye safety regulations), and because of power budget considerations, there is a limit to the optical power that can be safely and efficiently emitted by wireless IR transmitters (cf. Chapter 3).

Taking into account both the advantages and disadvantages of IR, it is questionable that it will replace radio as the only medium to transmit information wirelessly. It is more likely that radio and IR will continue operating in a complementary manner, with one being preferred over the other, depending on the specific application. IR will probably continue being favored for short-distance systems where security, low cost, and immunity to radio interference are required. Radio, on the other hand, will very likely continue being used for transmission over longer distances, in situations where high mobility is necessary or for systems operating in environments where the atmospheric conditions favor it over IR.

The idea of using IR as a medium to transmit information between computer terminals without cables was first proposed in the late 1970s by Gfeller et al. [9–11]. They described a way to interconnect a group of computer terminals at low speed over short distances (up to 50 m) using near-IR radiation. In their description, the terminals, which were located within the same room, exchanged information through an electro-optical satellite located on the ceiling. This optical satellite transferred information using diffusely scattered IR radiation as the carrier to convey information to and from the computer terminals. Its controller was connected to a host via a network ring, such as the one illustrated in Figure 1.1; and each satellite, as well as the computer terminals, contained an IR transceiver conformed by an array of LEDs and photodiodes aimed in different directions.

The system proposed by Gfeller and colleagues included automatic gain control features at the receivers, which allowed them to compensate for variations in the signal level caused by a potential obstruction of the link. In addition, by making use of the reflective properties of walls and ceilings, the system guaranteed maximum coverage, avoiding the need to have line-of-sight between the terminals and the optical satellite. Collision between the up- and down-links was avoided using different carrier frequencies — with phase-shift keying (PSK) or frequency

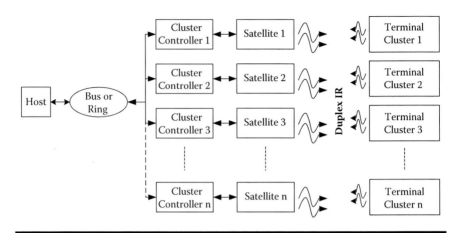

Figure 1.1   Possible structure of a wireless IR LAN. (*Source:* Adapted from [9, 10].)

modulation (FM) — or using emitting sources at different wavelengths that could be separated at the receiver using optical filters. They also described basic duplex transmission between the controller and the terminals, which, by combining broadcast and dedicated channels, eliminated collisions from different terminals.

From the late 1970s to date, there has been an exponential growth of research in the field of optical wireless communications, and a number of commercial products based on this technology have become a reality (see [12–16] for some examples).

The creation of standards for IR and radio communication (the Infrared Data Association [IrDA] [17] for IR — which is explained in more detail in Chapter 9 — and Bluetooth [18] for radio, for example) and the increasing demand for wireless applications indicate that both technologies will continue to develop and expand. This explains why researchers and manufacturers are trying to find better, more efficient, and more reliable techniques to provide wireless communication.

1.2 System Configurations

Optical wireless systems for indoor and outdoor use can be arranged in a number of configurations depending on the specific requirements of a system. In general, the topologies used for indoor optical wireless communication systems are classified according to two parameters: (1) the existence of an unobstructed path between the transmitter and the receiver (LOS – non-LOS), and (2) the degree of directionality of the transmitter, the receiver, or both (directed, non-directed, or hybrid) [19–23]. In this book, this classification is extended to outdoor systems when it applies.

The different configurations of optical wireless systems are illustrated in Figure 1.2. In the line-of-sight (LOS) topologies, the transmitter and the receiver are in direct view of each other, without any object obstructing the path between them. Non-LOS systems, on the other hand, may have obstacles blocking the direct

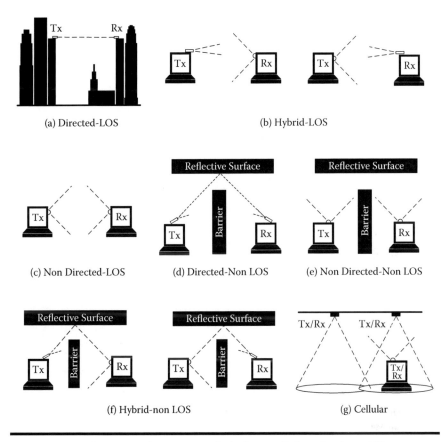

(a) Directed-LOS

(b) Hybrid-LOS

(c) Non Directed-LOS

(d) Directed-Non LOS

(e) Non Directed-Non LOS

(f) Hybrid-non LOS

(g) Cellular

Figure 1.2 Different configurations of wireless IR links. The dotted lines represent the different FOVs. (*Source:* Adapted from [47].)

path between the transmitter and the receiver. Therefore, these configurations rely on the use of reflective surfaces to create an alternative path for the link.

With regard to directionality, when the emitted beam from the transmitter and the field-of-view (FOV) of the receiver are narrow, the system is classified as "directed." If only one of them is narrow (directive), the system is denominated "hybrid;" and when contrary to directed systems, both the receiver and the transmitter have a wide FOV and a wide emission beam, respectively, the system is classified as "non-directed."

LOS configurations improve power efficiency and minimize multipath distortion. Unfortunately, due to the fact that alignment of the transmitter, the receiver, or both is sometimes required, the flexibility of some LOS-based systems is restricted. Non-LOS topologies, on the other hand, increase link robustness because they allow the system to operate even when obstacles are blocking the direct path between the transmitter and the receiver [21]. The problem of systems based on non-LOS topologies is that, as they rely on the use of reflective surfaces to create alternative paths between the transmitter and the receiver, they also suffer from multipath distortion.

The directed-LOS topology, also called "directed beam infrared" (DBIR) by some authors [24], is the configuration that provides maximum power efficiency because the transmitted power is concentrated into a narrow emission cone. This allows the use of narrower FOV receivers that not only improve the link's power budget, but also minimize the probability of having undesired energy (from a reflection of the information signal or from unwanted energy at a different wavelength) impinging on the detector, which reduces the problems of multipath distortion and noise. This also means that, compared to other configurations and for a given distance, a directed-LOS system requires a lower transmission power (due to the fact that the divergence of the emitted beam is small, which increases the power flux density at the receiver). Outdoor free space optics (FSO) links, used to exchange information between buildings, are good examples of this topology.

The disadvantages of the directed-LOS configuration include its susceptibility to blocking, its restricted mobility (unless the system contains a tracking mechanism), and the fact that it requires careful aiming of the transmitter and the receiver. The narrow emitted beam required by these systems (to achieve relatively long distances) is generally provided by a laser source, which can emit beams as narrow as 1° [25]. Directed-LOS systems also employ very narrow FOV receivers that incorporate imaging or non-imaging high-gain concentrators that allow the use of smaller photodetectors with reduced capacitance (which increases the speed of the system). Unfortunately, alignment of such directive elements is usually complicated.

Tracked systems are a special case of the directed-LOS configuration. Here, the directive transmitter or receiver is made to "move" by mechanical or electro-optical means to maintain a continuous LOS while providing mobility. This offers maximum power efficiency, high bit rates, and wide coverage. The transmitted power required for a tracked system, for example, is several orders of magnitude smaller than the power required for a diffuse system for a specific data rate and range. This is illustrated in Table 1.1 where the transmitted power required for a system operated under the diffuse and the tracked configurations (for different data rates) is shown. These values were obtained by Smith et al. [25] for a 4 × 4-m room. In the case of the LOS topology, the system emitted a 200-mm beam at the transmitter and used a 1-mm^2 detector and a 1-cm^2 lens at the receiver. In the diffuse case, the system used incorporated 1-cm^2 Si PIN detectors.

Another special case of the directed-LOS configuration is the cellular topology. Here, semi-directive LOS transmitters create spots or cells of energy — with minimum overlap between each other — that provide mobility while maintaining a reasonable power efficiency. The emission angle of the transmitter of a cellular system is typically around 40°, while the FOV of cellular receivers is around 40° [25]. Computer-generated holograms (CGHs) have been proposed for this type of configuration to give designers control over the shape of the emitted beam (cf. Chapter 3).

Table 1.1   Values of Transmitted Output Power for Diffuse and Tracked Topologies (Data obtained from [25])

Data Bit Rate(Mbps)	Optical Power (Diffuse System)	Optical Power (Tracked System)	Optical Power (Cellular System)
0.1	3×10^{-2} W	6×10^{-7} W	3×10^{-3} W
0.5	6×10^{-2} W	8×10^{-7} W	5.8×10^{-3} W
1	8×10^{-2} W	1×10^{-6} W	8×10^{-3} W
5	1×10^{-1} W	3×10^{-6} W	1×10^{-2} W
10	2.5×10^{-1} W	6×10^{-6} W	2×10^{-2} W
50	6×10^{-1} W	9×10^{-6} W	3×10^{-2} W

Despite the advantages presented by cellular systems, which include high mobility, improved power budget, and minimum multipath distortion (compared to the diffuse configuration), they are still susceptible to blocking from persons and objects. This situation can be alleviated to some extent by having some degree of overlapping between the cells (and cooperation between the optical satellites) in such a way that if one of the cell paths is blocked, the path with another cell can be used. An example of a system based on this configuration can be found in Natarajan et al. [26]. There, the authors considered the conceptual design of an optical wireless LAN that made use of existing wired infrastructure as the backbone of the optical satellites, which could communicate with groups of mobile terminals through wide beams of energy that defined optical cells.

A comparison of the power required at the transmitter to achieve different data rates (in a 4 ×4-m room with an optical satellite attached to the ceiling at 3 m over the receivers) for the cellular and the tracked configurations is also shown in Table 1.1. The cellular system used for this comparison was based on a receiver that contained a bare detector with capacitance $C = 70$ pF, while the tracked system used a receiver that contained a 1-mm^2 detector with capacitance $C = 2$ pF and an optical concentrator of 1 cm^2. The transmitter of the tracked system generated a 200-mm beam spot. Both systems used bootstrapped transimpedance preamplifiers (BTAs) and non-return-to-zero on-off-keying (NRZ-OOK) modulation. In addition, both systems were exposed to background illumination that introduced a photocurrent density of 10^4 A/cm^2.

These results show that the transmitted power required for a cellular system (to support a bit error rate (BER) of 10^{-9} and data rates between 100 kbps and 1 Gbps) is more than three orders of magnitude higher than the power required for a tracked system.

A non-directed-LOS system can be created using arrays of directive emitters and detectors orientated in different directions, which provide a wide emission beam and an overall large FOV. If the FOV of each independent emitter or detector is known, their respective FOVs can be accommodated in such a way that they barely overlap, forming a continuous total emission or acceptance angle that is the sum of each independent FOV. The system described by Pauluzzi et al. [27] is a good example of this configuration.

In hybrid LOS systems, either the transmitter or the receiver has a wide FOV, while the other element has a narrow FOV. A typical hybrid LOS system might consist of a very directive transmitter that optimizes the power budget and the distance of the link by emitting a very narrow beam aimed at the receiver, and a wide FOV receiver that relaxes the alignment constraints and allows the reception of signals from other transmitters within its FOV. An example of this configuration can be found in the quasi-diffuse infrared (QDIR) array proposed by Singh et al. [24]. They described a system consisting of a base station (based on an active or a passive reflector) attached to the ceiling that provided wide coverage to a number of terminals within a room. The information transmitted from a terminal to the base station could be reflected by the latter in such a way that other terminals within the reflector's range could collect the transmitted energy. In one of the versions described for this system, the beam angle of the terminal's transmitter was narrow while the FOV of the receiver was wide, which allowed the terminal to receive the information reflected from the base station from different points within the room. In another version, the transmitter of the terminal had a wide emission angle but the terminal's receiver had a narrow FOV needing to be pointed at the base station.

Directed non-LOS systems incorporate transmitters and receivers (with very narrow emission angles and FOVs) that, rather than being pointed at each other, are aimed at a reflective surface that allows the link to overcome a barrier. The main advantage of this topology is that, in addition to the possibility of overcoming an obstacle, the information signal is received after a single reflection, which minimizes multipath dispersion. Unfortunately, due to the very directive nature of the transmitter and the receiver, alignment is potentially problematic.

Hybrid non-LOS configurations have either a very directive transmitter or receiver, while the other element of the transceiver has a wide FOV. This relaxes the need for careful alignment between the transmitter and the receiver, but allows the system to benefit from either an optimized emitted power at the transmitter or from a better sensitivity at the receiver (which can be achieved using a narrow FOV high-gain optical concentrator). While hybrid non-LOS systems do not suffer from blocking problems, they are affected by multipath distortion, which increases with room area.

One of the most attractive configurations is the non-directed-non-LOS, or diffuse [28, 29]. Systems working under this configuration (also called "diffuse infrared" or DFIR by some authors [24]) use wide emission beam transmitters and wide FOV receivers that do not require line-of-sight, or alignment between them. Here, the IR energy emitted by the transmitters is spread as uniformly as possible within a room by making use of the reflective properties of its walls and ceiling. This is possible due to the fact that a number of materials commonly used for offices and buildings present a high reflectivity and a low absorption, which means that the losses from each reflection are low. Systems working under the diffuse configuration present maximum mobility and can operate even when there are barriers between the transmitter and the receiver, which makes it the most robust and flexible topology.

Despite the advantages presented by DFIR systems, this configuration is not without drawbacks. The fact that the transmitted optical signal can reach the detector after one or several reflections from surrounding surfaces and objects means that temporal dispersion (that limits the bit rate of the system) is introduced. The larger the room, the higher the multipath dispersion and the lower the bit rates achieved. The data rate limit related to intersymbol interference for this type of topology is 260 Mb·m·s^{-1} [10]. In addition, as different optical power levels reach the detector (depending on the position of the receiver within the diffusely illuminated room), the receiver needs to have a large dynamic range, which increases the complexity of the system. Furthermore, the larger the size of the room, the higher the required optical power at the transmitter.

One way to overcome the disadvantages presented by the diffuse configuration while maintaining wide coverage is to use quasi-diffuse transmitters and angle diversity receivers. A quasi-diffuse transmitter consists of multiple narrow beam transmitters orientated in different directions and simulating a wide emission angle transmitter. An angle diversity receiver, on the other hand, provides wide coverage by using multiple receiving elements orientated in different directions. The receiving elements incorporate imaging or non-imaging concentrators with reduced FOV. This means that they provide high gain while allowing the use of small and fast photodetectors. The main advantages of the quasi-diffuse configuration when compared to the diffuse topology are reduced power consumption and reduced multipath distortion. Figure 1.3 presents an illustration of the quasi-diffuse configuration.

The majority of the manufacturers of indoor and outdoor IR systems base their designs on the directed-LOS and the hybrid-LOS configurations. The Canobeam point-to-point system manufactured by Canon [30] is a good example of the directed-LOS topology (this is probably the most popular configuration for commercial systems). Other examples include those of Terescope [31] and OPTICOMM [32].

Some manufacturers of indoor systems have continued developing Gfeller's original idea of a wireless IR network; but due to the benefits offered by the directed- and the hybrid-LOS topologies, they have opted for the latter over the diffuse configuration. An example of a LAN based on the hybrid-LOS topology is the VIPSLAN system for indoor use manufactured by JVC [33]. This system is

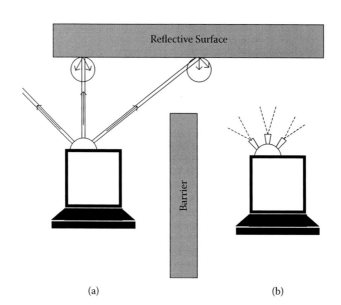

Figure 1.3   Quasi-diffuse configuration: (a) quasi-diffuse transmitter, and (b) angle diversity receiver. The dotted lines represent the FOV of each receiver. (*Source:* Adapted from [24].)

based on the idea of a group of portable computers within a room that have access to a host computer via a base station located on the ceiling.

Commercial diffuse systems generally use a large number of LEDs at the transmitter to achieve maximum coverage. The average emitted optical power at the transmitter is in the range of 100 to 500 mW (at wavelengths between 850 and 950 nm) [21]. It is not uncommon to find that they also incorporate several photodetectors (orientated in different directions) at the receiver with the same intention (optimizing the coverage of the system).

Table 1.2 shows the main characteristics of the different configurations presented in this section.

1.3 Evolution of Infrared Communication Systems

Optical wireless communication systems have experienced a huge development since the late 1970s when IR was first proposed as an alternative way (to radio) to connect computer networks without cables. Since then, a growing number of companies and research institutions have invested increasing resources to develop systems that exploit the advantages of the IR medium and to overcome the drawbacks from which it suffers. This section presents a chronological review of research carried out in this field, both in educational institutions and companies, for indoor

Table 1.2 Characteristics of the Different Configurations of Wireless IR Communication Systems

	Line-of-Sight (LOS)	*Non-Line-of-Sight (Non-LOS)*
Directed	Direct view between the transmitter and the receiver Collimated emission beam at the transmitter Narrow FOV at the receiver Maximum bit rate Restricted mobility Improved power efficiency High susceptibility to blocking Minimum multipath distortion	No direct view between the transmitter and the receiver Makes use of reflective surfaces Narrow FOV at the receiver Collimated emission beam at the transmitter Restricted mobility
Non-directed	Direct view between the transmitter and the receiver Wide emission beam at the transmitter Wide FOV at the receiver Maximum mobility Low susceptibility to blocking	No direct view between the transmitter and the receiver Wide emission beam at the transmitter Wide FOV at the receiver Maximum mobility Minimum susceptibility to blocking Maximum multipath distortion Maximum flexibility and robustness
Hybrid	Direct view between the transmitter and the receiver Collimated emission beam at the transmitter and wide FOV at the receiver or wide emission beam at the transmitter and narrow FOV at the receiver Relative mobility Relative susceptibility to blocking	No direct view between the transmitter and the receiver Collimated emission beam at the transmitter and wide FOV at the receiver or wide emission beam at the transmitter and narrow FOV at the receiver Relative mobility Relative susceptibility to blocking

and outdoor applications. It explains the characteristics of the proposed systems as well as the maximum speeds achieved at every stage.

IBM was one of the first organizations to work on wireless IR networks. The first reports on IBM's experimental work were published between 1978 and 1981 [9–11]. Here, as explained in Section 1.1, they described a duplex IR link that achieved a bit rate of 64 kbps using PSK and a carrier frequency of 256 kHz. The system consisted of an optical satellite containing 40 GaAs LEDs, with emission powers of 16 mW each (at 950 nm), and terminal stations equipped with nine photodetectors and ten LEDs each. In this system, a voltage-controlled oscillator (VCO) at the transmitter provided the carrier that generated the PSK signal used to modulate the LEDs' output power. Clock extraction and synchronization at the receiver was achieved by transforming the transmitted digital non-return-to-zero (NRZ) information into return-to-zero (RZ) code. In addition, two phase-locked loops (PLLs) provided signal detection and clock extraction, respectively.

In 1983, Minami et al. from Fujitsu [34] described a full-duplex LOS system that operated under the same principles as the network described by Gfeller. That system consisted of an optical satellite attached to the ceiling and connected to a network node via a cable, and of a number of computer terminals that communicated to the server via the optical satellite. It operated at 19.2 kbps (over 10 m) with an error rate of 10^{-6} when working under fluorescent illumination. The wavelength of operation was 880 nm. The transmitter sources were based on LEDs (nine for the satellite and five for each terminal) with emission angles of ± 60°; and the receivers were based on silicon PIN photodetectors (four for the satellite and two for each terminal) with 1-cm² areas. To optimize the power consumption, the total emission angle of the transmitters was reduced through a rectangular pipe to ±30°. The modulation scheme used was frequency-shift keying (FSK) over intensity modulation (IM); and optical filters were used at the receivers to minimize the noise introduced by ambient illumination.

By 1985, the Fujitsu team had managed to improve the data rate of its system to 48 kbps, as reported by Takahashi and Touge [35]. This newer version of the system could operate as a hybrid-LOS configuration in a point-to-multipoint fashion or as a directed-non-LOS topology with point-to-point connection. These configurations were preferred over the non-directed-non-LOS topology to avoid the problems related to diffuse propagation. The downlink of this system presented a wide emission angle (120°), and the uplink had a medium emission angle of between ±10 and ±30°. The preferred wavelength continued to be within the range of 800 to 900 nm. With regard to the modulation scheme used, this time they employed PSK.

In their report they presented an expression to calculate the link distance r achievable from direct line propagation:

$$r = \sqrt{\frac{P_s A_r L_1 L_2}{2\pi P_{rm}(1 - \cos\phi)}} \qquad (1.1)$$

Here, P_s represents the optical output power from the transmitter (in mW), A_r is the active area of the photodetector, L_1 is the transmittivity of the transmitter filter, L_2 is the transmittivity of the filter at the receiver, P_{rm} is the optical power required (in mW) to obtain a specific carrier-to-noise ratio at the receiver, and □ is the half angle of the energy radiated by the optical source. From this expression, they calculated achievable distances (depending on the FOV), which in their case covered a range of between 10 and 20 m.

In the same year (1985), researchers from two other companies (Hitachi and HP Labs) presented their own work in the area of wireless IR communications. In the case of Hitachi, Nakata et al. [36] reported a directed-LOS network system that replaced the optical satellite on the ceiling with an optical reflector. This system achieved a data rate of up to 1 Mbps with a BER of less than 10^{-7} for a distance of 5 m. This was achieved by using arrays of LEDs that provided a total emission power of 300 mW (both the terminal units and the reflector contained arrays of LEDs). In the case of the terminal units, the emitted beam was concentrated through a lens to narrow down the angle to 10° and to focus the energy on the reflector. The reflector's angle of emission, on the other hand, was 120°. For the detectors at the transceiver terminals and the reflector, the system employed PIN photodiodes. The transceiver's receiver also incorporated a Fresnel lens that provided optical gain. The use of such an optical element was possible due to the fact that the incoming energy arrived at the receiver only from the direction of the reflector, which meant that the focal point shift originated by angular variations at the entrance of the lens was minimal. To facilitate alignment between the reflector and the terminals (especially if the terminals were moved to a new position), the transceiver of each terminal had the optical unit separated from the main body and connected to it through a flexible arm. The modulation scheme used for this system was the same as the one employed for the first optical wireless network version of Fujitsu: FSK.

With regard to the IR networking system developed by Hewlett-Packard, Yen and Crawford [37] published their results in 1985. This system (similarly to the ones reported by Fujitsu and IBM) was based on the idea of an optical satellite — which could be attached to the ceiling — that communicated with a number of terminal units. The configuration used in this case was the directed-LOS (proposed as a way of minimizing the optical power requirement when compared to the diffuse configuration), and the data rate of this system was 1 Mbps, with a reported range of 50 m achieved through arrays of LEDs that emitted with a maximum power of 165 mW. The BER achieved was 10^{-9}.

The wavelength employed for the HP optical satellite transmitter was 660 nm and the satellite's receiver wavelength was around 880 nm. This constitutes a rather unusual selection of wavelength for the downlink, taking into account that 660 nm corresponds to the visible part of the electromagnetic spectrum, which means that any transmission from the satellite could be seen. Moreover, the noise contribution from some sources of ambient illumination is stronger at visible wavelengths. This noise was minimized through an optical filter with a cut-off wavelength of 880 nm. The filter was also used to avoid interference from the signals of the downlink transmission. The number of LEDs used for the transmitter of the optical satellite was 108, and they were arranged in a number of clusters controlled by different drivers so as to have some control over the emitted radiation pattern. They were positioned at the focal plane of a transmitter lens that focused the light within a 3° emission angle.

The receiver part of the satellite used a PIN photodetector of 1 cm² connected to a wide dynamic range amplifier. This was done with the idea of accommodating the different power levels originated by variable distances between the satellite and the terminals.

With regard to the network terminals, they were equipped with a transmitter that incorporated a single LED transmitting information at a wavelength of 880 nm and emitting an optical power of 5 mW over a beam with divergence equal to 2° (which was achieved through an optical lens). The terminals' receiver contained a PIN photodetector, an optical filter (with cut-off wavelength at 600 nm), and a molded glass optical concentrator.

In 1986, another company reported research in the field of wireless IR communications. This time, Motorola Incorporated, through its researchers Kotzin and Van den Heuvel [38], reported a diffuse, full-duplex portable telephone for indoor use that communicated with a fixed transceiver through nine LEDs emitting at a wavelength of 950 nm with an optical power of 16 mW each. The LEDs on the terminal unit were positioned as a radial array on top of a rod, providing an emission field of around 360° on one plane, which achieved 95 percent coverage within a room of 12 × 12 × 4 m. The photodetector used was a 7.6-mm² silicon PIN diode that incorporated a wide FOV *fish-eye* type lens of 1 cm diameter and a gelatin long-pass optical filter. The photodetector was connected to an amplifier-filter (low-pass and high-pass) combination that eliminated the low-frequency components created by the electronic ballasts of fluorescent lamps. The multiplexing and modulation techniques used by the telephone system from Motorola were time division multiplexing (TDM) and continuously variable slope delta modulation (CVSD).

In 1987, AT&T Bell presented their work on optical wireless communications. They reported a directed-LOS system that operated at 45 Mbps over a wavelength of 800 nm [39]. Shortly afterward, in 1988, Fuji and Kikawa [40] from Matshushita Electronic Components Corporation presented a full-duplex, free space optical transmission module that operated at 19.2 kbps over a maximum distance of 10 m with a BER of 10^{-6}. The idea in this case too was to exchange information between computer terminals and an optical satellite attached to the ceiling via IR radiation. This system used LEDs for the transmitter and PIN photodetectors for the receiver (both operating at a wavelength of 880 nm). The modulation scheme used was FSK-IM. Their system also incorporated band-pass filters to reduce background illumination noise.

In 1992, an alternative to the optical telephony system presented by Motorola in 1986 was demonstrated by Poulin et al. from MPR Teltech [41]. The transmitter end of their system contained 900-nm, high-power LEDs, and the receiver contained PIN photodiodes with active areas of 20 mm² and 7 mm² incorporating optical filters and optical concentrators. It operated at a data rate of 230.4 kbps over a 20-m range. The wavelength of operation of the LEDs was in the range 800 to 950 nm, and the optical sources were located at the end of a 6-inch antenna (to avoid shadowing from the head) on the telephone unit as an array that provided an omni-directional emission pattern. The handset also contained a low-noise amplifier. They selected time division multiple access, time division duplex (TDMA-TDD) as the multiple access method (with Manchester coding). For the outbound part of MPR's system, the transmitter used a total of 36 LEDs to flood the area with as much near-infrared radiation as possible. The total power emitted by the inbound and outbound parts of the system was 50 mW into a 180° full-angle field-of-view.

A year later (1993), BT Labs reported a cellular wireless IR system that operated at 50 Mbps [20, 25]. Something characteristic from the systems developed by BT is the use of computer-generated holograms to increase the eye safety of the transmitters while controlling the coverage of the emitted beams (this topic is reviewed in more detail in Chapter 3). The original system's transmitter included a single laser and a hologram, and emitted an optical power of +14 dBm at a wavelength of 850 nm. The receiver included a silicon PIN diode of 7×7 mm incorporating a thin-film optical filter with passband $\Delta\lambda = 50$ nm, and a lens that provided an optical gain of +4 dB. The receiver sensitivity was –33 dBm. The system provided full coverage in a room of $2.2 \times 2.2 \times 3$ m.

Shortly afterward, in 1994, BT managed to increase the data bit rate of this system to 155 Mbps. This newer version of the system employed 40-mW extended sources and 5-mm² active area avalanche photodetectors. A further modification, a change of the system topology from cellular to tracking, allowed them to increase the speed of the system to 1 Gbps [20]. Tracking in this system was achieved using steerable optics. They also reported exploring the option of using a solid-state array tracking system in which arrays of transmitters or receivers followed or preceded by imaging lenses could produce different emission angles or FOVs. The advantage of the second option is that the use of mobile parts is avoided, which simplifies the design of the system and increases its robustness.

In the early 1990s, while some commercial companies were investigating ways to exploit the advantages offered by the IR medium, a number of academic institutions were also actively trying to find better and more efficient ways of transmitting information wirelessly. A good example of these efforts is reflected in the research reports produced by the University of Berkeley in 1994 and 1996 [42, 43]. During this period, Marsh and Kahn reported a 50-Mbps diffuse FSO system that used OOK with decision-feedback equalization to provide wireless IR communication indoors. The BER of this system in the presence of background illumination was 10^{-6} at a distance of 2.9 m. Their demonstrator used an optical source (operating at 806 nm) pointed at a white painted ceiling in order to use the reflective properties of the ceiling and the walls (which were also painted white) to diffuse the energy and distribute it as uniformly as possible within a room. The average power of this source was 475 mW. The receiver of the system incorporated a hemispherical optical concentrator of 2-cm radius and refractive index of 1.78 mounted on an antireflection-coated, 1-cm^2 silicon PIN photodiode. This concentrator provided an overall FOV of 78° and a gain of 4.5 dB. The receiver also incorporated a multilayer thin-film optical band-pass filter with central wavelength at 815 nm and passband $\Delta\lambda$ = 30 nm. This filter (whose peak transmission was 68 percent) was bonded to the hemispherical surface of the concentrator.

The PIN photodiode of the receiver was connected to a 10-kΩ hybrid high-impedance preamplifier, followed by low-pass and high-pass electronic filters. The preferred transmission/detection technique for their system was intensity modulation with direct detection (DD) due to its simplicity and its efficient spatial diversity. The system used an array of eight laser diodes behind a Plexiglas diffuser that provided a nearly Lambertian emission pattern to obtain the right amount of IR energy necessary to illuminate an entire room (of dimensions 8.4 × 6.3 × 3.9 m). This array of IR sources provided an on-axis irradiance of 97.2 W·sr^{-1}·m^{-2}. Timing recovery was provided by a second-order phase-locked loop (PLL) [44].

In 1995, Smyth et al. [25] presented an interesting comparison between the BT cellular system operating at 50 Mbps and the diffuse system from Berkeley University operating at the same data rate. Their report shows that while the demonstrator presented by Berkeley achieves a longer range and has a wider FOV at the receiver, the system from BT requires a lower transmitter power.

The work produced by Oxford University is another good example of academic research in the area of optical wireless communications. Faulkner et al. [44] reported a system that used an optical wireless satellite (in their case called the "base station") attached to the ceiling to exchange information with computer terminals via IR radiation. For their demonstrator, they produced a 155-Mbps system operating at a wavelength of 980 nm, which used Manchester coding. Here, the transmission from the base station to the computer terminals was considered the downlink and that from the computer terminals to the base station the uplink. The system included tracking capabilities (using solid-state arrays) and achieved a distance between the base station and the terminal in the range of 2 to 4 m.

The base station incorporated two arrays of semiconductor flip-chip sources, which in this case were resonant cavity LEDs (RCLEDs) emitting at 980 nm with a 20- to –30-nm on-axis linewidth, bonded to arrays of driver electronics. The sources were followed by optical imaging elements that minimized power consumption by collimating the emitted energy. These optical elements were used to optimize coverage by redirecting the emitted beams with particular angles. Power consumption was also minimized by activating the specific sources of the array required for a particular terminal only. Each terminal incorporated arrays of low-capacitance and narrow-FOV flip-chip detectors, combined with an imaging lens that focused the incoming beam into a different photodetector depending on the angle of incidence. This system was rendered as tracking because, through this arrangement of detectors plus imaging lens, the receiver could identify the angle of incidence of the incoming beam and could therefore activate the appropriate transmitter. The fact that the detectors had a narrow FOV facilitated a higher rejection of unwanted background illumination; and the low capacitance of the receivers' photodetectors allowed higher data rates.

With regard to the detectors at the receiver, the system incorporated an array of hexagonal flip-chip InGaAs/InP PIN diodes. These devices had a capacitance of 24 pF/mm^2 and were sensitive to wavelengths from 980 nm to around 1700 nm. Each detector was connected to a transimpedance preamplifier followed by a post-amplifier. They also incorporated an optical filter layer that minimized ambient illumination noise. The detectors were preceded by four imaging optical elements in cascade that redirected the incoming beam to the detector array. The size of this imaging optics array was about 20 mm.

More recently, Showa Electric [45] reported a 100-Mbps short-range IR wireless transceiver that operated over a maximum range of 20 m and used LEDs for the transmitter and avalanche photodetectors (APDs) for the receiver. Another system, proposed by Singh et al. in 2004 [24], was based on the idea of a base station attached to the ceiling and connected to the network via a backbone. The proposed network operated at 100 Mbps and was based on DPPM with carrier sense multiple access with collision detection (CSMA/CD) for the Media Access Control (MAC) protocol. The proposed transmitter in this case consisted of a diffractive element preceded by an array of vertical-cavity surface-emitting lasers transmitting at a wavelength above 1.4 μm; and the proposed receiver comprised an imaging lens followed by an array of InGaAs detectors with narrow FOVs that operated at around 1.5 μm.

Figure 1.4 shows the chronology of research published by some companies' laboratories and some educational institutions in the field of optical wireless communications. It indicates the topology, the maximum bit rates, and some other characteristics of different wireless IR systems.

1.4 The Optical Wireless Channel

As explained above, a great number of applications use intensity modulation/direct detection (IM/DD) as the transmission-reception technique due to its simplicity of implementation. Intensity modulation is easily obtained through variations on the bias current of the transmitter device, which modulates the desired signal (the information) onto the instantaneous power of the infrared carrier. On the down-conversion side, direct detection is easily obtained because the photodetector generates a current proportional to the received optical power.

For communication systems using intensity modulation, the selection of the channel model depends on the background illumination conditions. In environments where background illumination is low, the received signal is generally modeled as a Poisson process with rate $\lambda_s(t) + \lambda_n$ [46]. Here, λ_n is proportional to the power of the background illumination and $\lambda_s(t)$ is proportional to the instantaneous optical power of the received signal. In environments where background

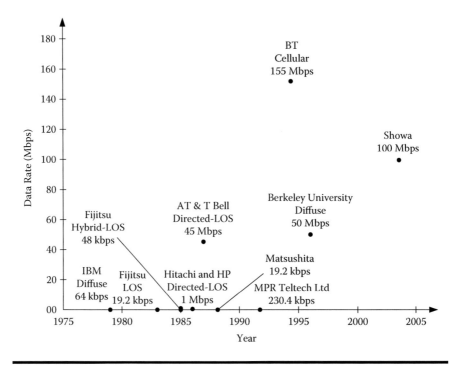

Figure 1.4   Chronology of indoor optical wireless communication research.

illumination is high and the receiver employs a wideband receiver, on the other hand, the shot noise at the photodetector can be modeled as an additive white Gaussian noise (AWGN) plus a DC offset.

For indoor applications using the diffuse configuration, it is possible to model the baseband channel as shown in Figure 1.5 [21], where $p(t)$ represents the transmitted optical power of a signal in a wireless IR link, and $i(t)$ represents the photocurrent produced by the photodetector. Here, the current $i(t)$, generated by the photodetector, is proportional to the integral, over the photodiode active area, of the squared electric field. In a diffuse configuration, the multipath propagation produces dispersion of the pulses, which is modeled in Figure 1.5 by the impulse response $rh(t)$. In this case, r is the responsivity of the photodetector in (A/W); $h(t)$ is time invariant and linear if the transmitter and the receiver are stationary with respect to fixed reflecting surfaces. If the receiver moves, the impulse response varies marginally with time.

The electric thermal noise generated by the preamplifier is generally negligible when compared to the very intense white Gaussian noise introduced in the photodiode by different sources of ambient illumination. The latter is consequence of the high optical power emitted by the sun, and by fluorescent and incandescent lamps, which can in some cases exceed that of the information signal by up to 25 dB [47]. If the system were operated in an environment where no background illumination

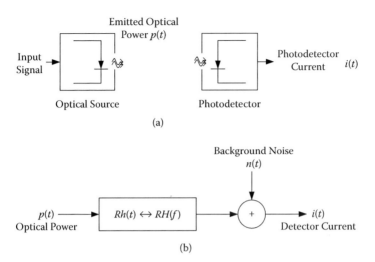

Figure 1.5   Channel model from transmitted signal power to generated photo-current: (a) intensity modulation and direct detection (IM/DD); and (b) block diagram of the optical wireless channel and the IM/DD process. (*Source:* Adapted from [21].)

existed, the dominant source of noise in the receiver would be the receiver pream-plifier noise; but as this is not normally the case, the noise is modeled as white and signal independent. Therefore, the background illumination noise is represented by $n(t)$ in the above model.

According to Park et al. [46, 48], the baseband channel model for optical wire-less communication using (IM/DD) can be described as:

$$y(t) = \int_{-\infty}^{\infty} x(\tau)h(t-\tau)d\tau + n(t) \qquad (1.2)$$

where $y(t)$ represents the instantaneous current generated by the photodetector at the receiver, $x(t)$ is the instantaneous optical power of the transmitter, $n(t)$ represents the white Gaussian noise, and $h(t)$ represents the multipath impulse response.

Kahn et al. [21] have also described the channel in terms of its frequency response (the Fourier transform of $h(t)$) as follows:

$$H(f) = \int_{-\infty}^{\infty} h(t)e^{-j2\pi ft} dt \qquad (1.3)$$

It must be said that the relation between $h(t)$ and $H(f)$ is linear due to the many electromagnetic modes of the received signal. In addition, the channel is modeled as fixed because the channel model only changes when significant movements of the transmitter or the receiver happen.

The primary difference between the model of the IR channel and that of a conventional linear Gaussian noise channel is that the input $p(t)$ in the former does not represent amplitude, but power; and, while the limitation to in the power of the channel input of the conventional channel is given by $\int p^2(t)dt$, the average value of the transmitted signal in the IR model must not exceed a predetermined value P and be positive [47]:

$$\lim_{T \to \infty} \frac{1}{2T} \int_{-T}^{T} p(t)dt \le P \tag{1.4}$$

1.5 Design Fundamentals

As explained in Section 1.1, optical fiber technology has undergone major developments in the past decades; and, thanks to the fact that wireless IR communication systems use some of the same components employed in optical fiber systems, wireless IR systems benefit from mature and efficient devices that are available at a relatively low cost.

The selection of the opto-electronic components for the transmitter and the receiver is generally done according to the configuration desired for a system. Directed topologies require directed sources and detectors, while non-directed links require wide emission beams and wide FOVs. One of the things that can be observed from the information on different systems developed for wireless IR communications presented in Section 1.3 is that the two optical sources most commonly used for wireless IR transmitters are (1) light emitting diodes and (2) laser diodes. LEDs present wider emission beams than LDs, which makes them the preferred option of the indoor non-directed and the hybrid configurations. In addition, they are generally considered as eye safe, which means that they can be used at higher emission powers than LDs. This, plus the fact that they are more difficult to damage and cheaper than LDs, favors their use for indoor applications. Other important features of LEDs include their lower sensitivity to temperature variations (compared to LDs) and the simplicity of the driver associated with them. LDs, on the other hand, require more complex driver circuits and are more sensitive to temperature fluctuations. Despite these limitations, LDs can be modulated at higher speeds than LEDs, which makes them the only option in applications that require a very high data rate. Moreover, the fact that their emission beams are very narrow means that they can be used over longer distances, which favors their use in directed-LOS links outdoors.

With regard to the opto-electronic devices used at the receiver, the two detectors generally preferred for wireless IR communication systems are (1) the PIN diode and (2) the APD. PIN diodes are generally cheaper than APDs. In addition, they present a lower bias-voltage requirement and lower temperature sensitivity. Unfortunately, they are also about 10 to 15 dB less sensitive than APDs.

1.6 Power Budget Considerations

The power budget is one of the most important considerations when designing a wireless communication system because it defines the battery size and the operation time of portable units. Power consumption is determined by a number of factors, such as the electronic and the optical components used, the modulation scheme, the topology, and the emitted power of a wireless system. For this reason, several researchers have investigated ways of improving the power consumption of a system by analyzing every aspect of it. Examples of research carried out to address the problem of power efficiency include the efforts of Yu et al. [49] and Otte et al. [50]. Yu et al. [49], for example, investigated power-efficient, multiple subcarrier modulation schemes, while Otte et al. [50] analyzed different aspects of a wireless IR system and their requirements with regard to power consumption.

The type of technology used also affects power consumption. IR transceivers, for example, present a lower power requirement than their RF counterparts. According to Smyth et al. [25], an optical wireless transceiver operating at 1 Mbps consumes 150 mW, while a radio LAN transceiver consumes 1.5 W, which corresponds to a 25 percent extra drain on the power supply of a laptop.

The power consumption of a system is strongly affected by the power emitted by the transmitter. This power should be high enough to cover the desired range of a particular system, as well as to supply the receiver with sufficient energy to allow the system to operate at the bit rate required under different conditions of background illumination. The power at the receiver is determined by the range of the link, the topology used, the geometry of the room where the system is operating, and the reflective properties of its walls and ceiling. From the topology point of view, for example, applications using portable units and requiring minimum power consumption benefit from the use of the directed-LOS configuration, as it optimizes power efficiency through the concentrated transmitted power achieved by using a narrow emission cone. For this reason, the use of a collimated source (such as a laser) can be advantageous. In addition, the use of an optical collimating element can minimize the power consumption at the transmitter by transforming an extended source into a concentrated source with narrow emission angles. When this is the case, care must be taken to comply with eye safety regulations. The use of collimated sources also allows the use of narrower FOV receivers, which, due to their directive nature, can present high optical gain (when using an optical concentrator, as explained below), increasing the sensitivity of the receiver and reducing the need for a high transmitted power for a given distance. The use of angle-diversity receivers and multi-spot transmitters also helps to reduce power consumption while maintaining wide coverage.

Another way of improving power consumption is through the use of an optical concentrator at the receiver. This is possible due to the fact that an optical concentrator improves the sensitivity of the receiver, which means that a lower emitted power may be required at the transmitter (for a given range) compared to the same system without a concentrator. Further information on optical concentrators is presented in Chapters 4 and 5.

To optimize the power consumption, it is also important to transmit only the relevant information, to use an effective signal coding, and to perform the required signal processing at low power if possible. With regard to the modulation used, different schemes for wireless IR communications have been analyzed in the past. From their analysis and comparisons it has been concluded that one of the modulation schemes that presents better power efficiency is L-PPM. Unfortunately, this is done at the expense of reducing the bandwidth efficiency, which is another very important parameter in a number of applications. A detailed analysis of different modulation schemes (in terms of bandwidth and power requirements) is presented in Chapter 8.

1.7 Summary and Conclusions

This chapter presented an introductory discussion to optical wireless communication systems. The reasons why IR is preferred over radio for a number of applications include:

- The very high bandwidth offered by IR, which is unregulated worldwide
- The fact that IR is inherently secure because IR radiation does not penetrate walls and the energy is confined within the room where it was generated
- Its immunity to radio and electromagnetic interference
- Its immunity to interference from other IR systems operating in neighboring rooms
- The limited cost, size, and power consumption of IR components

The disadvantages of IR technology also were discussed with the aim of having a complete overview of the technology and the possibility to compare it to radio as a medium for wireless communication. The drawbacks of IR include the:

- Possibility of an obstruction in the communication path
- Fact that noise is introduced in the detector by sources of background illumination
- High attenuation suffered by the IR signal when transmitted through air
- Restrictions to the emitted optical power due to eye safety

It is not probable that either IR will replace radio or that radio will replace IR as the only media to transmit information wirelessly. What has been happening in the past few years is that both technologies have been operating in a complementary way; and one is being preferred over the other depending on the specific requirements of a particular system.

Different system configurations have been presented, and their advantages and limitations explained. IR systems can be classified according to their degree of directionality as directed, non-directed, or hybrid; and according to the existence of a line-of-sight between the transmitter and the receiver as LOS or non-LOS. Due to their simplicity, low cost, and low multipath distortion, the majority of commercial optical wireless systems are based on the directed-LOS or the hybrid-LOS configurations. Indoor wireless IR systems generally incorporate LEDs at the transmitter, due to the fact that they are inherently safe and can make use of a simple driver. Outdoor systems, on the other hand, generally employ LDs, which can transmit information at high data rates and, thanks to the small divergence of the emitted beam, can reach longer distances. At the receiver end, the preferred photodetector is the PIN photodiode, which is usually combined with an optical element that provides concentration and an optical filter that rejects unwanted radiation.

A number of wireless IR networks have been produced for academic or commercial purposes following the model first presented by Gfeller at al. [10]: see, for example, [26, 29]) For outdoor systems, optical wireless links operate at distances of up to 1 Km and at up to 1.25 Gbps; but for indoor systems, distances are typically up to 30 m and transmission speeds up to a maximum of 155 Mbps (for diffuse configurations). Therefore, increasing efforts are being made to improve the optical and electronic components involved in the system, to achieve higher signal-to-noise ratios, to reduce the system power consumption, and to maximize the range of the link. Researchers and manufacturers are also trying to find ways to improve the modulation techniques and the data bit rates offered by current systems, while maintaining the costs as low as possible. Furthermore, novel optical wireless system protocols and architectures have been proposed, and a number of experiments and analyses of the wireless IR channel have been carried out.

Chapter 2

Atmospheric Transmission Limitations

2.1 Introduction to Atmospheric Propagation

As Chapter 1 explained, its inherent characteristics make IR (infrared) an attractive medium for indoor and outdoor applications. When used for outdoor systems, IR offers — in addition to the availability of huge unregulated bandwidth, high security, and immunity to EMI (electromagnetic interference) — the simplicity of deployment (of permanent and temporary links). Unfortunately, outdoor optical wireless links also present additional challenges when compared to their indoor counterparts. This means that, in addition to the free-space loss encountered in directed-LOS systems, FSO (free space optics) service providers need to take into account the quality of service that can be offered under adverse atmospheric conditions, as phenomena such as fog, rain, and snow affect the performance of wireless IR links to different degrees. This is why FSO service providers and manufacturers are trying to find ways to compensate for the unwanted effects introduced by different atmospheric phenomena and by intense solar illumination. An example of one of the solutions proposed is the use of backup links.

A good understanding of atmospheric phenomena such as fog, haze, mist, and snow, and their effect on the performance of a wireless IR link is of great importance when designing the transmitter and the receiver of an optical wireless system. By understanding weather parameters such as humidity, visibility, and temperature, together with the features of an FSO system and its deployment characteristics, it is possible to model atmospheric propagation. This chapter introduces a number of definitions that are important for a better comprehension of these atmospheric phenomena and their effect on the transmission of an optical signal. A variety of atmospheric phenomena are presented and their impact on the propagation of IR radiation is explained. The origin and the effects of scintillation as well as its main characteristics are introduced in Section 2.5. Finally, this chapter presents a summary and conclusions in Section 2.6.

2.2 Important Definitions

Before proceeding to explain the characteristics and effects of the different atmospheric phenomena to which outdoor wireless IR links are exposed, it is important to define a few concepts that are relevant to them. The knowledge of these concepts — which include atmosphere, aerosol, absorption, scattering, and radiance — facilitates the comprehension of the next sections of this chapter. Absorption and scattering, for example, are related to the loss and redirection of the energy of an electromagnetic beam when intercepted by suspended particles in the atmosphere. Their combined effect is known as *extinction*.

2.2.1 Atmosphere

Atmosphere is defined as the gaseous mass or envelope surrounding a celestial body, which in this case is the Earth. The atmosphere reaches over 560 kilometers (km) from the Earth's surface and it is retained by the Earth's gravitational force. Free space optical (FSO) links operate at the lower part of the atmosphere (which is the one closest to the Earth's surface and is called the troposphere) where it behaves as a fluid in hydrostatic equilibrium and its density is higher. An understanding of atmospheric physics helps to comprehend the formation of clouds and fog, which are particularly detrimental to the propagation of an optical beam.

The atmosphere is primarily composed of nitrogen (N_2, 78 percent), oxygen (O_2, 21 percent), and argon (Ar, 1 percent), but there are also a number of other elements, such as water (H_2O, 0 to 7 percent) and carbon dioxide (CO_2, 0.01 to 0.1 percent), present in smaller amounts. There are also small particles that contribute to the composition of the atmosphere; these include particles generated by combustion, dust, debris, and soil. The combination of all the elements comprising the atmosphere contributes to its absorption and its behavior.

The density of the air increases with water (due to the high density of the latter). The combination of water and dry air in the atmosphere is determined by parameters such as water vapor saturation (defined as the maximum point — temperature dependent — at which the air can hold no more moisture), relative humidity* of

---

* The ratio, defined as a percentage, of absolute humidity to saturation, where absolute humidity is the mass of water vapor per unit volume of air.

the current mass of water per volume of gas and the mass per volume of a saturated gas, partial pressure of dry air, partial pressure of water vapor, and total atmospheric pressure. From these, the total atmospheric pressure P can be calculated as [51]:

$$P = P_d + e \tag{2.1}$$

where P_d is the total pressure of dry air and e is the partial pressure of water vapor. The relative humidity can be calculated as [51]:

$$HR = \frac{e}{e_s} \tag{2.2}$$

Here, e_s is the pressure with respect to liquid water. Water vapor H_2O and carbon dioxide CO_2 molecules have a direct impact on the propagation of IR radiation. Unfortunately, this propagation is difficult to predict due to the variable concentration of CO_2 and H_2O vapor with geographic location, altitude, and season. The CO_2 concentration, for example, varies between 0.03 and 0.04 percent, with lower concentrations during the months of November and December and higher concentrations during the months of April and May.

2.2.2 Aerosol

An aerosol is a suspension of solid or liquid particles in a gaseous medium. There are not clearly defined upper and lower limits to the size of the particles composing the dispersed phase in an aerosol, but their size range is commonly considered to be — according to Veck [52] — between 0.01 and 10 μm in radius. Other authors such as Achour [51] consider the size range of aerosol particles to be only from 0.1 to 1 μm. The larger the drop size, the smaller the concentration of drops in an aerosol. Haze particles, clouds, and fog may be considered aerosols. Table 2.1 shows the concentration and size parameters of some aerosol particles and hydrometeors. From this table, it can be seen that hydrometeors (which are the product of precipitation, such as rain and snow formed from the condensation of water vapor in the atmosphere) are larger than aerosol particles. The particle concentration is highly dependent on the region, with much higher concentrations in maritime locations and lower concentrations in continental areas. Some values of aerosol particle concentrations based on location and on particle size are presented in Table 2.2. The most significant differences in this case are observed in particles of very small radius. The distribution of these particles is obtained through different mechanisms, depending on their size. The distribution of particles above 0.5 μm, for example, can be obtained through light scattering. Particle collection through electrical precipitation can be used to obtain the distribution of particles between 0.1 and 0.5 μm, and smaller particles distributions can be derived from the diffusion parameter [51].

Table 2.1 Size and Concentration of Some Atmospheric Particles

Particle Type	Concentration (cm⁻³)	Radius (μm)
Raindrop	10^{-3}–10^{-2}	10^{2}–10^{4}
Fog droplet	10–100	1–10
Cloud droplet	10–300	1–10
Haze particle	10–10^{3}	10^{-2}–1
Aitken nucleus	10^{2}–10^{4}	10^{-3}–10^{-2}
Air molecule	10^{19}	10^{-4}

Source: From [52, 55].

Table 2.2 Concentration of Aerosol Particles Depending on Location and Particle Size

Particle Radius	Particle Concentration in Continental Regions	Particle Concentration in Maritime Regions
<0.01 μm	1600	3
0.01–0.032 μm	6800	83
0.032–0.1 μm	5800	105
0.1–0.32 μm	940	14
0.32–1 μm	29	2
1–3.2 μm	0.94	0.47
>3.2 μm	0.29	0.029

Source: From [51].

The growth of an aerosol particle radius depends on the equilibrium of the droplet with its surrounding medium, and on the relative humidity of the aerosol. For water vapor condensation to grow continuously, it is necessary that its particles are in a supersaturated and unstable state with its surroundings. The supersaturation level of particles is of great importance to fog and cloud formation because it defines, together with the size of the particles, which particles grow. Small particles, for example, require a high supersaturation state to grow, while larger particles need a lower supersaturation state to grow.

Aerosol suspended droplets or particles can absorb and scatter incoming light. They, together with fog droplets, are the most important atmospheric elements that contribute to the attenuation of an optical signal (through Mie scattering, cf. Section 2.2.5). The loss of light by scattering increases the atmospheric extinction, which, as explained above, is the sum of actual absorption and scattering. Sunlight scattering is also responsible for the hazy appearance of air. The concentration of aerosol particles is high close to the Earth's surface and low at high altitudes.

A concept closely related to that of aerosols is haze. Haze is an atmospheric aerosol of sufficient concentration to be visible. Its particles cannot be seen individually due to their small size; but when they are seen in a sufficiently large concentration, they reduce the visual range.

2.2.3 Attenuation

Atmospheric attenuation is defined as the process whereby some or all of the energy of an electromagnetic wave is lost (absorbed and/or scattered) when traversing the atmosphere [53]. For the typical wavelengths employed by wireless IR links (850 nm, 1550 nm), the attenuation contribution from the atmosphere is relatively low compared to the attenuation contribution from weather conditions. Gebhart et al. [54], for example, mention attenuation values of around 0.2 dB/km for clear atmospheric conditions, and 10 dB/km in urban regions (due to dust). Attenuation due to heavy fog, on the other hand, can reach values greater than 300 dB/km.

2.2.4 Absorption

Absorption, in the context of electromagnetic waves and light, is defined as the process of conversion of the energy of a photon to internal energy, when electromagnetic radiation is captured by matter. When particles in the atmosphere absorb light, this absorption provokes a transition (or excitation) in the particle's molecules from a lower energy level to a higher one. The only light that can be absorbed is the one from energy that can create transitions from one energy level to another. The molecules go back to their original unexcited states through discrete emissions of radiation.

Whenever a medium is irradiated with electromagnetic energy or with a light beam, the wavelength at which the attenuation of the propagated light is maximum (or at which a particular medium absorbs more power) is defined as the *absorption peak*. In the context of optical wireless communications, the absorption peak refers to the specific wavelength at which most power is absorbed by a particular impurity in a specific medium.

The atmospheric absorption is wavelength dependent. The atmospheric windows due to absorption are created by atmospheric gases, but neither nitrogen nor oxygen, which are two of the most abundant, contribute to absorption in the infrared part of the spectrum.

2.2.5 Scattering

Scattering is defined as the dispersal of a beam of particles or of radiation into a range of directions as a result of physical interactions. When a particle intercepts an electromagnetic wave, part of the wave's energy is removed by the particle and re-radiated into a solid angle centered at it. The scattered light is polarized, and of the same wavelength as the incident wavelength, which means that there is no loss of energy to the particle.

The behavior of the re-radiation (scattering) depends on the characteristics of the particle: its size in relation to the wavelength of the intercepted energy, its index of refraction, and its isotropy (the property of molecules and materials of having identical physical characteristics in all directions). If all these parameters are known, the scattering pattern of the particle can be predicted. The size of the particle, for example, defines the type of symmetry of the scattered energy with regard to the direction of propagation of the incident energy. If the size of the particle is equal to the wavelength of the incident light, the scattering by the particle presents a large forward lobe and small side lobes that start to appear. As the size of the particle becomes smaller, the backward lobe becomes larger and the side lobes disappear. When the size of the particle is 10 percent the size of the wavelength of the incident beam, the backward lobe is symmetrical with the forward lobe. Table 2.1 shows a number of atmospheric particles with their respective radius and concentrations [52, 55].

There are three main types of scattering: (1) Rayleigh, (2) Mie, and (3) non-selective. Rayleigh scattering refers to the scattering of light by particles whose size is small compared to the wavelength of the electromagnetic radiation incident upon them. Despite the fact that this type of scattering is mainly observed in gases, it may also occur in transparent solids and liquids. The scattering by particles with diameters larger than one tenth of the incident light wavelength up to a diameter equal to the wavelength of the incident light is known as Mie scattering [23]. Finally, non-selective scattering occurs to particles whose size is larger than the incident wavelength. In the context of wireless IR communications, Rayleigh and Mie are probably the most relevant types of scattering, as they refer to the scattering of light from molecules of air, and to the scattering of light from other atmospheric particles such as fog, mist, and haze [56]. As mentioned above, neither oxygen nor nitrogen contribute to absorption in the IR part of the spectrum. However, they do contribute to Rayleigh scattering.

Mie scattering is wavelength dependent. Its theory was developed from Rayleigh scattering, for which the exponential absorption coefficient is given by:

$$\sigma_R = \frac{1.04x10(n-1)^2}{\lambda^4} \tag{2.3}$$

where n is the refractive index of the particle, and λ is the wavelength of the incident energy in microns (the use of this expression can be extended to molecules up to about 10 percent the size of the wavelength of the incident light).

The effects of scattering and absorption can be taken into account to evaluate the total extinction coefficient, which is required to calculate the attenuation of radiation from atmospheric particles. The total extinction coefficient β_T is given by [52]:

$$\beta_T = \beta_S + \beta_a \tag{2.4}$$

where

$$\beta_S = \beta_m + \beta_p \tag{2.5}$$

Here, β_S is the extinction coefficient due to scattering, which is the result of the addition of the effects of scattering due to molecules β_m and scattering due to particles β_p. The contribution from absorption is represented by β_a.

The emergent power per unit area at a distance d along the radiation path can be calculated by using the extinction coefficient as follows [52]:

$$E_p = \frac{I_0^{(-\beta_T d)}}{d^2} \tag{2.6}$$

where E_p is the emergent power, I_0 is the intensity of the incident radiation, and β_T is the total extinction coefficient. This expression was derived from Bouguer's exponential law of attenuation, which describes the exponential attenuation of light with distance.

According to Veck [52], the extinction coefficient can be expressed in terms of its angular coefficient $\beta(\theta)$ as:

$$\beta = \int_0^{4\pi} \beta(\theta) d\omega \tag{2.7}$$

where $\beta(\theta)$ is the angular coefficient, and ω is a solid angle in radians. Other important parameters that can be calculated include the optical thickness Γ, which can be obtained from the molecular or particle scattering in a homogeneous path as $\Gamma = \beta_x$, and the transmittance β, which can be calculated as the ratio of the emergent power E_x over the incident power E_0.

The attenuation due to Mie scattering can reach values of hundreds of dB/km [56, 57] (with the highest contribution arising from fog). This scattering can be so strong as to disrupt a communication link even for distances of less than 100 meters. The attenuation due to fog can reach values as high as 300 dB/km, which is well beyond the attenuation values of rain for microwave links [23]. The effect of fog in FSO links is explored in more detail in Section 2.4.

An alternative way to calculate the attenuation due to scattering has been presented by Gebhart et al. [54]. Their expression is based on visibility rather than on

Table 2.3 Variation in Atmospheric Attenuation due to Scattering Based on Visibility (Data obtained from [54])

Visibility S (Line of Sight) (km)	$\lambda = 800$ nm (dB/km)	$\lambda = 1500$ nm (dB/km)
0.5	32.5	30.8
0.7	23	21
0.9	18	16
1.1	14.5	12.5
1.3	12	10
1.5	10	8.33

the properties of the particles, which may not always be known. They define the visibility range — in relation to the contrast resolution of the human eye — as the range over which an attenuation of approximately 17 dB occur for a wavelength of 555 nm. From this visibility parameter, which can be obtained from meteorological stations or airports, they estimate that the attenuation a_{scatt} in decibel per kilometer can be calculated as follows:

$$a_{scatt} = \frac{17}{S} \left(\frac{555}{\lambda} \right)^{0.195 S} \tag{2.8}$$

where S is the visibility of the human eye in kilometers, and λ is the wavelength of the transmitted light (in nanometers). Some values of atmospheric attenuation due to scattering based on visibility are presented in Table 2.3

2.2.6 Radiance

Radiance is a physical quantity that represents the amount of electromagnetic radiation leaving a given point of a real or imaginary surface, in a given direction. That is, radiance indicates the rate at which light radiation is emitted in a given direction per unit of projected surface area, where this projected area can be calculated by multiplying the cosine of the angle of the emitted energy with respect to the surface normal ($\cos \theta_R$) by the surface area. It is then the projection of the surface on to the plane orthogonal to the direction in which light is being emitted. Thus, radiance is the power per unit solid angle per unit projected area of a source. It is generally used in applications such as radiometry to quantify the intensity of a beam of light; and its unit are watts per square meter per steradian per unit wavelength ($Wm^{-2}sr^{-1}\mu m^{-1}$).

Radiance is present in the atmosphere through scattering, where the atmosphere has a radiance of its own in the ultraviolet and visible parts of the electromagnetic spectrum. In addition, due to its temperature (which ranges between 250 and 300K), the atmosphere radiates microwave and thermal infrared energy with Planckian distribution. This radiation has an intensity value on the order of 8 to 9 $Wm^{-2}sr^{-1}\mu m^{-1}$ (for a vertical path through the entire atmosphere) and peaks at approximately 10 μm [52].

The atmospheric radiance characteristics depend on a number of factors, including wavelength, polarization effects, global latitude, angle of view with respect to the sun, solar zenith angle, and haze conditions.

2.3 Atmospheric Transmission

Due to their relevance to wireless IR communications and other applications, atmospheric transmittance and radiance have been investigated by a number of researchers. Veck [52], for example, obtained values of transmittance (and radiance) in the range of 280 nm to 28.57 μm. He calculated these values using transmittance curves in combination with a program that estimated the equivalent amount of a particular atmospheric absorber at sea level. His calculations took into account parameters such as geometrical latitude, time of the year, attenuation due to cirrus, extinction due to haze, and radiation or advection of fog. The program used also took into account geometrical angle and altitude in order to estimate the path through the atmosphere allowing for refraction. Figure 2.1 shows the space-ground atmospheric transmission in the visible and near-IR regions based on these calculations. This type of graph is useful not only to evaluate the amount of sunlight that reaches the surface of the Earth at different wavelengths (relevant of the evaluation solar illumination noise at the receiver), but also as an indicator or the atmospheric attenuation or absorption windows that must be avoided when selecting the operation wavelength of a wireless IR link. A similar graph, showing the same absorption windows, was presented by Smyth et al. [25] for transmission over a 1000-foot horizontal air path at sea level.

One of the things observed in Figure 2.1 is that there are transmission and absorption windows along the spectral regions corresponding to visible and near-IR radiation. The four absorption bands along the visible and near-IR regions are approximately: 900 to 980 nm, 1.1 to 1.16 μm, 1.3 to 1.5 μm, and 1.8 to 2 μm. This is one of the reasons why the operation wavelengths of outdoor optical wireless systems are generally between 780 and 900 nm, between 1.2 and 1.3 μm, and between 1.5 and 1.7 μm, as reported by Eardley et al. [56].

While the effect of latitude on the transmission characteristics of the atmosphere in the visible part of the spectrum is small in general, its effect is more considerable in the IR region. In the tropics, for example, the transmission characteristics change due to warm air. The transmission characteristics of the atmosphere also

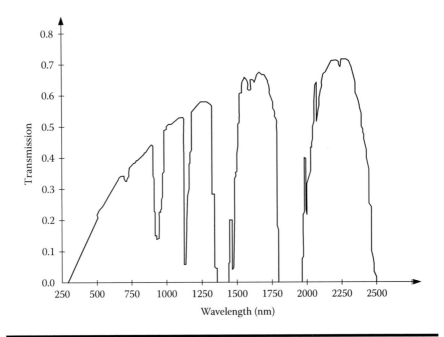

Figure 2.1 **Atmospheric transmission from space to ground at mid-latitude (summer time) in the visible and near IR regions. (*Source:* Adapted from [52].)**

change, depending on geographical location. Urban areas present a more reduced visibility than maritime and rural areas. This variation derives mainly from the existence of different types of aerosols (which create haze of different characteristics) in the atmosphere. It must be mentioned that the near-IR region is the least affected by geographical variations.

2.4 Effect of Rain, Fog, and Mist

The surface of the Earth is the main generator of water vapor in the atmosphere. The concentration of condensed water is higher at the lower part of the atmosphere, where it is present in the form of clouds or fog. Water is also present in the atmosphere in the form of raindrops and snow.

There are two main factors that contribute to the growth of droplets. One of them is the collection of small droplets (in this case, larger drops are formed from the collision of smaller drops). The other is the heterogeneous nucleation of aerosol particles that attracts water vapor molecules to their surface at relative humidity levels below 100 percent. This generates cloud drops and fog. Aerosol particles can grow to form fog, clouds, or ice droplets with an increase of relative humidity.

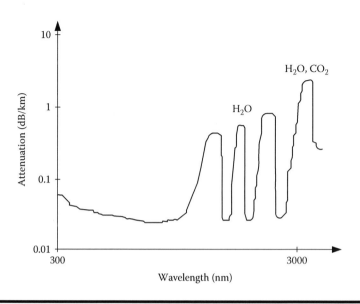

Figure 2.2 Mean attenuation by atmospheric gases (at 20°C) in the visible and near IR regions. (*Source:* Adapted from [52].)

Several models have been developed to describe the phenomenon related to the growth of droplets. One of these models is the stochastic collection model that describes the probability related to drop collision and their combination. This model assumes that some droplets are statistically favored for growth.

In general terms, fog can be considered a stratus cloud at ground level that does not produce precipitation, but that can originate drizzle. It is created by the cooling of the evaporated moisture from the Earth's surface, which generates condensation. It is well known that in the visible and near-IR parts of the spectrum, fog, mist, and clouds can reduce the performance of an optical wireless link and, in some cases, disrupt the communication completely. Thick fog (created by a high density of very small water droplets in the atmosphere), for example, can reduce visibility down to a few meters; and maritime mist and clouds can affect visibility in the same way. Attenuation due to water vapor in the visible and near-IR regions is proportional to the percentage of water droplets in the atmosphere: the larger the percentage of water droplets, the higher the attenuation. This is illustrated in Figure 2.2, where attenuation due to atmospheric gases is shown for visible and near-IR wavelengths.

The three main parameters used to calculate attenuation due to fog and haze are (1) visibility, (2) relative humidity, and (3) temperature. Among them, the most important is visibility (also known as *meteorological visual range*), which is broadly defined as the range at which an object can be distinguished from its background. This parameter is generally measured using the lowest level of 2 percent contrast that the eye can detect. Therefore, the meteorological visual range can be defined as the maximum distance at which the object contrast against its background at 550 nm is 2 percent. The reason why the wavelength of 550 nm is used for the attenuation calculations is because this is the wavelength at which the human eye is more sensitive. The meteorological visual range can be obtained from its extinction coefficient β at the reference wavelength (500 nm) as follows [51]:

$$V = \frac{|\ln(0.02)|}{\beta(\lambda = 550nm)} = \frac{3.91}{\beta(\lambda = 550nm)} \quad [\text{km}] \quad (2.9)$$

The atmospheric attenuation due to fog and clouds can be calculated through the coefficients of scattering β_{scat} and absorption β_{abs} as [51]:

$$\tau_a = e^{-(\beta_{abs}+\beta_{scat})R} \quad (2.10)$$

where R is optical depth. This expression is known as Beer's law.

The scattering coefficient for an aerosol can be calculated from the visibility value V obtained from Equation (2.9) as follows [51]:

$$\beta_{scat}(\lambda) = \frac{3.91}{V}\left(\frac{\lambda}{0.55}\right)^{-\delta} \quad [\text{km}^{-1}] \quad (2.11)$$

where δ is a parameter used to describe the quality of the visibility. Good visibility (visibility above 50 km), for example, is related to a value of δ = 1.6. Medium quality visibility (visibility range between 6 and 50 km) is assigned a value δ = 1.3, while poor visibility (visibility below 6 km) is related to a value of [51]:

$$\delta_{poor} = 0.585\sqrt[3]{V} \quad (2.12)$$

Unfortunately, neither this expression nor the relationship between attenuation and visibility in Equation (2.12) takes into account the size of the droplets. This has motivated the development of techniques that do take the droplet size into account. An example of software based on an alternative model is Simulight™, which takes into account the aerosol size distribution for its calculations. Table 2.4 shows attenuation due to scattering losses obtained by Clay et al. using Simulight™. The scattering loss value is given in dB units, and the results are also compared to experimental data and to results obtained with Equation (2.11).

The stable fog field in Table 2.4 refers to stable fog and low clouds, which generally present visibility up to 1.5 km. Evolving fog, on the other hand, refers to variable conditions between light haze, dense haze, and stable fog, where the visibility usually ranges between 1 and 4 km. It can be observed here that the results

Table 2.4 Attenuation of Electromagnetic Radiation in dB when Transmitted through Fog (Data obtained from [51] and [64])

Fog model	Wavelength Range (655–750 nm)				Wavelength (1.23 μm)				Wavelength (10.1 μm)			
	Evolving	Stable	Stable		Evolving	Stable	Stable		Evolving	Stable	Stable	
Visibility (km)	1	0.075	0.3		1	0.075	0.3		1	0.075	0.3	
Mie loss - Simulight™ (dB/km)	17.09	227.06	57.72		18.8	239.2	65.78		0.93	213.65	12.67	
Experimental loss(dB/km)	15.2	226.2	62.9		17.37	247.33	68.4		7.3	263.4	18.45	
Equation loss(dB/km)	14.16	209.73	50.13		10.16	185.64	41.3		3.1	110.42	18.1	

obtained with Equation (2.13) differ more with the experimental data than the data obtained through the program (that takes into account the droplet size).

Attenuation of near-IR radiation due to rainfall is less critical than attenuation due to other particles (which is more critical at longer wavelengths). The impact of particles whose diameter ranges between 0.1 and 5 mm is more likely to affect radio communication links than optical links [54]. Snow particles, which are of irregular shape, present a variable attenuation of near-IR radiation. Particles in the range of 2 to 25 mm absorb light according to the relation between their size and that of the receiving optical area.

The attenuation due to rain can be calculated as a function of rainfall rate. Its value has been calculated as approximately 6 dB/km during heavy rain (10 mm/hr) [23, 56]. This value was calculated based on the model presented by Rensch and Long [58]. Other reports — presented by Bramson [56, 57] — indicate that the attenuation due to rain can reach values of up to 17 dB/km, while the attenuation due to snow can be as high as 60 dB/km.

2.5 Scintillation

With regard to communications, scintillation is defined as a random fluctuation on the received field strength caused by irregular changes in the transmission path over time. In the specific case of optical wireless communications, this term refers to the strength variation of an optical signal as it travels through air, and it derives from small fluctuations in the index of refraction along the optical path. These fluctuations, called *optical turbulence*, originate from atmospheric turbulence that creates thermal inhomogeneities along the path of the transmitted optical signal. The fluctuations originate from the energy conversion of solar energy, which becomes manifest as wind energy as the solar power heats the atmosphere. This wind energy is eventually reconverted into heat but, in the process, different cells in the atmosphere exhibit different temperatures. This reflects as variations in the index of refraction of the atmosphere that create fluctuations in the amplitude of the received optical signal with a frequency spectrum between 0.01 and 200 Hz [23]. This is due to the fact that light transmission in a medium occurs according to the principle that light traveling from one point to another follows the shortest optical path (Fermat's principle), and this depends not only on the geometrical distance, but also on the optical characteristics of the medium, from which one of the most important ones is the index of refraction. The index of refraction value in the atmosphere depends on temperature, pressure, and humidity of air and on the wavelength used for the transmission.

After fog, low clouds, and direct sunlight, scintillation is the factor that causes most significant performance deterioration of an optical wireless link for distances greater than 500 m. This problem is especially relevant at distances over 4 km (where scintillations can reach values greater than 20 dB). Scintillation also appears to be more significant when the receiver (of an optical wireless system) has a small aperture [59].

Scintillation has been analyzed by a number of authors due to its effect on the performance of wireless IR systems. These authors have created a variety of theoretical models to characterize its behavior [60–62]. The losses and quality of a wireless IR link related to scintillation, for example, can be expressed as probability of fade and mean fade time, which are probabilistic terms. This is due to the fact that optical turbulence is not a deterministic, but rather a probabilistic process. The intensity fluctuation of a laser beam due to atmospheric turbulence can be expressed as a probability density function (PDF), and a number of researchers have developed different PDF to calculate these intensity fluctuations. Two of these PDF types are (1) the lognormal PDF and (2) the gamma-gamma PDF.

One of the most useful parameters when modeling and characterizing scintillation is the Ryotov variance, which is used to calculate the strength of the optical turbulence. This variance is used to define the PDF of the intensity fluctuation due to atmospheric turbulence. The Ryotov variance σ_l^2 is calculated as [59]:

$$\sigma_l^2 = 1.23 C_n^2 \left(\sqrt[6]{k^7 L^{11}} \right) \tag{2.13}$$

where C_n^2 is the refractive index structure parameter, L is the distance of propagation, and k is the wavenumber. The refractive index structure parameter can be measured to calculate the Ryotov variance and the PDF. Other parameters that can be measured to calculate the PDF include the optical power under different scintillation conditions and the meteorological parameters.

Experimental results on scintillation were reported by Gebhart et al. [54] for a very long wireless IR link (of over 61 km). Their results show amplitude variations (at the receiver) originated by air turbulence during clear sky conditions of up to 30 dB. Their results also indicate that this problem becomes more significant as the distance between the transmitter and the receiver increases. Results obtained from another experiment (performed over a link distance of 2.7 km and using a slightly larger beam divergence) show that there is a clear dependence of scintillation on the time of the day. This is due to the fact that parameters such as humidity, temperature, and sunlight change during the day. Daytime and night, for example, present the highest variations (peak values of up to 10 dB), while sunrise and sunset present more stable air conditions (with variations on the order of 0.6 dB). From this experiment it was also observed that the duration of fade during daytime was in the range of 40 to 60 ms, while the duration of fade at night was between 10 and 150 ms.

2.6 Summary and Conclusions

Optical wireless communication systems are affected by a variety of atmospheric phenomena that limit their data transmission rate, range, and reliability. In the case of outdoor systems, atmospheric phenomena such as haze, fog, rain, and scintillation have a detrimental effect on their performance. From these, fog and haze constitute the most important atmospheric scatterers. Their attenuation, which can reach values of over 300 dB/km (corresponding to very thick fog), can affect the performance of a wireless IR link for distances as small as 100 m. A similar effect occurs when smoke is within the communication path.

Rain and snow do not introduce a significant attenuation in wireless IR links. This is due to the fact that scattering is wavelength dependent, and raindrops affect mainly radio and microwave systems that transmit energy at longer wavelengths. A heavy rain of 10 mm/hr, for example, produces an attenuation of around 6 dB/km, which is well below the maximum attenuation values presented by fog [25]. An attenuation of approximately 60 dB/km is presented by snow and dust [57, 63]. Their effect is generally neglected for short distance links.

The atmosphere originates losses even when atmospheric scatterers are not present (that is, under clear air conditions). This phenomenon is denominated by clear air absorption, and it is equivalent to the absorption that happens in optical fibers. For this reason, atmospheric windows must be taken into account when selecting the wavelength of operation of a system to minimize the attenuation.

Air turbulence does not produce significant variations in the received optical power of a system when it operates over a short distance. For longer distances, however, it may become a problem. Scintillation, which refers to variation in the index of refraction of air due to heating, depends on a number of parameters, such as time of the year and the time of the day. Some of the solutions proposed by Gebhart et al. [54] to compensate these effects include:

■ The use of adaptive optics that vary the optical power reaching the detector and therefore compensate for the power fluctuations due to turbulence
■ The use of coding technologies to overcome fade
■ Equalization through different emitters that transmit the same information over different paths

Results presented by Gebhart et al. [54] suggest that, with regard to the bit error rate (BER), the average BER in a year under different atmospheric conditions is low. Their measurements, performed in Graz (Austria) for distances over 2.5 km (with a beam divergence of between 2 and 6 mrad) indicate a BER of 10^{-8} under bad weather conditions and 10^{-12} under good ones. Their results also show that there is a strong relationship between link availability and visibility, with an average availability of around 94 percent (per year) for a system working over 2.7 km with a link margin of 7 dB per kilometer. Haze and fog during specific day times during autumn and winter (December being the most problematic month) appeared to be the main cause of link disruption, which occurred mainly during the evening or early morning.

Chapter 3

Data Transmission Limitations and Eye Safety

3.1 Data Transmission Limitations

There are three major factors that limit the data rate of an optical wireless system: (1) ambient light, (2) multipath distortion (in the case of non-LOS links), and (3) the response time of the opto-electronic components used for the transmitter and the receiver. As explained in previous chapters, background illumination introduces noise in the detector, which reduces the SNR (signal-to-noise ratio) and the range of the system and limits its data rate. Indoor systems, for example, are exposed to illumination from fluorescent and incandescent lamps, while outdoor systems must be able to operate under intense solar illumination, which makes the design of the receiver particularly challenging. The solar power reaching the photodetector of an optical wireless system can be several orders of magnitude larger than the maximum power emitted by the transmitter and saturate the detector. Moreover, the laser diodes (LDs) used at the transmitter rely in many cases on a relatively large area monitor photodiode that controls and stabilizes the LD output power through its current. These LDs can have their operation affected if their monitor photodiode (which is sometimes located at the back of the laser diode) is exposed to solar energy.*

---

* This is due to the fact that sunlight may reach the lens at the output of the emitter, which means that its energy may be amplified — by the lens — and detected by the monitor photodiode, affecting the performance of the transmitter by increasing the current of the photodiode and reducing the LD output power.

Multipath dispersion originates from pulses of light arriving at the receiver at slightly different times. This data transmission limitation affects mainly indoor systems based on non-LOS (non-line-of-sight) configurations. Outdoor systems, on the other hand, are usually based on the directed-LOS topology to optimize optical power and maximize the range of the system. This makes them practically immune to multipath distortion. However, despite the fact that outdoor systems rarely suffer from multipath dispersion, they are affected by atmospheric phenomena such as fog, haze, and snow that affect the propagation of the transmitted energy and attenuate the signal (cf. Chapter 2).

This chapter describes different sources of background illumination noise and their effect on an optical wireless link. Multipath dispersion is explained and different issues related to eye safety are introduced. The classification of optical sources as extended or collimated is presented in Section 3.3. In addition, a description of holographic diffusers is introduced in Section 3.4, the main differences between LEDs and LDs are presented in Section 3.5, and special considerations for outdoor systems are presented in Section 3.6. Finally, a summary and conclusions regarding data transmission limitations are presented in Section 3.7.

3.1.1 Ambient Illumination Noise

In most indoor and outdoor environments, the photodetector of a wireless infrared (IR) receiver is not only exposed to the near-IR radiation emitted by the transmitter, but also to ambient illumination from lamps and from the sun. These sources of background illumination have a fraction of energy in the IR part of the spectrum, which introduces noise in the detector, thus reducing the SNR of the system.

The SNR for any optical detection system is defined by Kotzin and Van den Heuvel [38] as:

$$\frac{S}{N} = \frac{\left(nP_s\right)^2}{2enB\left(P_s + P_b + P_o\right) + \dfrac{4kTB}{R}} \tag{3.1}$$

where n is the conversion efficiency of the photodetector in A/W, P_s is the signal power reaching the detector (in watts), B is the signal bandwidth in hertz, P_b is the background illumination optical power reaching the detector (in watts), P_o is the detector's dark current equivalent power (in watts), and R is the detector's load resistance.

In the case of indirect wireless IR links, the noise contribution from background illumination dominates over all the other sources of noise, which reduces the SNR expression to [38]:

$$\frac{S}{N} = \frac{nP_o^2}{2eBP_b} \tag{3.2}$$

From Equations (3.1) and (3.2) it can be seen that the sensitivity of a wireless IR system is determined by the amount of unwanted background illumination that it is able to reject. Therefore, a good understanding of the different sources of background illumination and of the techniques available to reduce or eliminate this unwanted radiation is important for the design of wireless IR systems.

The three most common sources of ambient light found in indoor environments are (1) fluorescent lamps, (2) incandescent lamps, and (3) sunlight. Their normalized power densities per unit wavelength are illustrated in Figure 3.1 [9], where it is seen that the spectral power density of fluorescent lamps is minimum at the near-IR wavelengths at which the majority of near-IR detectors are most sensitive (typically around 850 and 950 nm). Daylight and incandescent light, on the other hand, present a broadband spectrum and a higher amount of energy in the near-IR part of the spectrum. Tungsten lamps, for example, present a high power spectral density in the 800- to 1400-nm wavelength range.

The noise generated by sunlight illumination affects indoor optical wireless systems particularly when the receiver is located near windows. For outdoor systems, however, daylight constitutes a more acute problem, which can only be reduced by appropriate optical or electronic filtering (and by reducing the FOV [field of view] of the receiver). Direct and reflected solar illumination is discussed in detail in Section 3.1.1.1.

Moonlight and stellar illumination also introduce unwanted radiation in wireless IR receivers, but their effect can be considered negligible due to the fact that the moon is only a reflector of solar energy and does not emit radiation of its own (which means that even during a full moon, the radiation it re-emits is of much lower intensity) and the stars are at such a long distance that the illumination they produce is also extremely low (the overall starlight intensity in the visible part of the spectrum is even lower than the intensity produced by a full moon). Therefore, both types of ambient illumination do not introduce any significant noise in the receiver. In addition, depending on the phase of the moon, the re-emitted energy intensity may be even lower, as just a fraction of the moon surface may be reflecting solar energy at a specific time [52].

As the moon's surface reflects all the wavelengths received from the sun, the spectrum of the energy re-emitted by the moon is very similar to that of the sun. However, this re-emission is not completely identical. This is due to the fact that the surface of the moon is composed of materials that absorb and reflect just a fraction of the energy received from the sun at different wavelengths, which means that the intensity of some reflected wavelengths may be low. The spectral distribution of the energy emitted by the stars, on the other hand, is different from that of the sun because they are not just reflectors, but emitters of their own energy. For example, due to the fact that the majority of the stars in the Earth's part of the galaxy are cooler, the intensity of stellar illumination tends to be higher in the near-IR part of the spectrum. Sunlight, on the other hand, presents higher intensities in the visible part of the spectrum (cf. Figure 3.1).

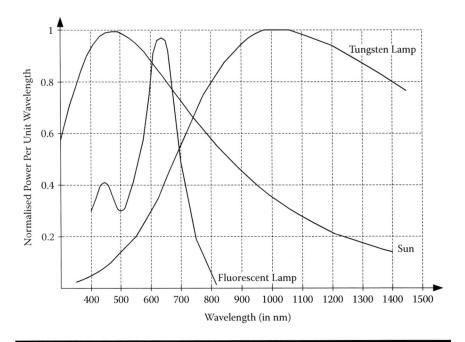

Figure 3.1 Spectral power densities of three typical sources of background illumination. (*Source:* From [9].)

Comparing the flux of the sun and the moon at the Earth's surface helps us understand the different effects that they could have on an optical wireless link. While the flux from the sun at Earth's surface is 1.36 kW/m², the flux of a full moon (at earth surface), taking into account that the moon's reflectivity is around 0.067, is only 6.9 × 10⁻⁶ kW/m². This means that the lunar flux is about 197,000 times smaller than the solar flux [52].

As mentioned above, the overall radiation flux of the stars is even smaller than that produced by a full moon. The lunar radiation flux between 760 and 900 nm (during sub-arctic winter, with maritime haze and full moon at the zenith), for example, is 5 × 10⁻⁴ kW/m² while the radiation flux of the stars at the same wavelengths is only 40% of that value at around 2 × 10⁻⁴ kW/m².

The majority of offices and indoor environments where optical wireless systems are used employ, rather than incandescent lamps, fluorescent lamps that introduce less noise into the photodetector. Unfortunately, the light emitted by fluorescent lamps presents rapid fluctuations that create spectral lines in the resulting photodetector current at multiples of the line frequency [65]. The frequency range of the harmonics generated by fluorescent illumination depend on the lamp model, with older models having harmonics up to around 500 kHz, and newer models (mainly employed in the United States and Japan) having harmonics up to several megahertz [20]. The interference generated from these spectral lines can be avoided by modulating the signal onto a subcarrier with a frequency well over 100 kHz, which corresponds to the

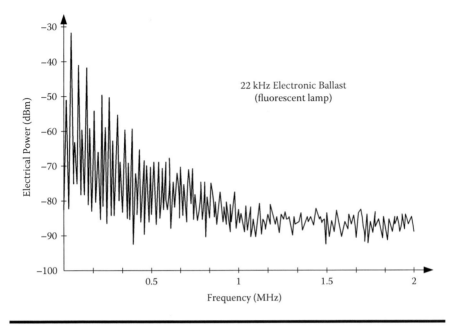

Figure 3.2   Variation of electrical power with frequency of a fluorescent lamp. (*Source:* From [65].)

frequency of the highest significant harmonic of the line in the old lamp models [66]. Alternatively, a data coding scheme with a suppressed spectrum at low frequencies can be used [29]. It must be noted that some of these lamps (especially the newer models) also generate higher levels of noise for several minutes at the moment they are switched on (until they become stable). The detected power spectrum of IR emission from a fluorescent lamp driven by a 22-kHz electronic ballast is shown in Figure 3.2.

There are a number of techniques that can be used to reduce the noise introduced by background illumination. One of these techniques, for instance, consists of restricting the field of view of the receiver. Alternatively, optical filters can be used before detection by the photodetector. The filters can be either bandpass or longpass, and they must match the linewidth of the source in order to minimize the unwanted noise without attenuating the desired signal. Longpass filters (which can be made of colored plastic or glass), for example, allow the pass of energy at wavelengths beyond their cut-off wavelength; and when combined with a silicon photodiode, the filter-detector combination performs jointly as a bandpass filter [9]. The fact that the transmission characteristics of colored longpass filters are primarily independent of the angle of incidence of the incoming energy makes them the most commonly used type of filter in commercial IR systems. Unfortunately, the overall passband of the combined filter-detector combination is generally wide. This is illustrated in Figure 3.3, where the spectral sensitivity of a silicon photodiode (notice that its responsivity rolls off just below 1100 nm) and the transmittance of a longpass colored optical filter are shown superimposed on the same graph [9].

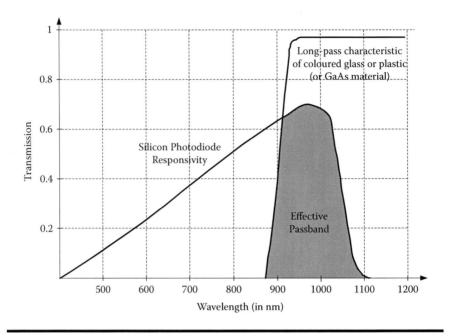

Figure 3.3   Spectral sensitivity of a Si photodiode and transmittance of a material with longpass characteristics. (*Source:* From [10].)

An alternative filtering technique presented by Gfeller et al. [9] to reduce unwanted background illumination, consists of developing unexposed color film and using it as a filter at the entrance of the photodetector. They calculated that, using this technique, fluorescent light can be reduced by a factor of 50, while sunlight can be reduced by a factor of approximately 3.

Thin-film optical bandpass filters are another option to reduce ambient light in wireless IR receivers. These filters consist of several layers of dielectric slabs that can be combined to form very narrow optical bandwidths, therefore providing maximum background illumination reduction. It has been demonstrated that a thin-film filter with a bandpass of around 60 nm can reduce fluorescent light by a factor of 300, and sunlight by a factor of 10 [9]. The main disadvantage of thin-film optical filters is that their transmission characteristics depend significantly on the angle of incidence of the received energy, which makes the design of the optical front-end of a wireless IR receiver quite challenging. Designers of systems employing optical concentrators in combination with thin-film optical filters need to take into account the angular response of the filter to accommodate the angles of the rays at the exit of the concentrator (when the filter is sandwiched between a non-imaging concentrator and the detector). According to Kahn et al. [21], the average optical power P_n created by ambient illumination reaching a receiver consisting of an optical concentrator, an optical filter, and a photodetector is given by:

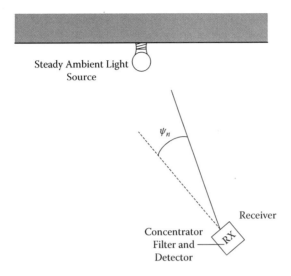

Figure 3.4 Source of illumination introducing noise in a wireless IR receiver.

$$P_n = p_n \Delta\lambda_n T_0 A g(\psi_n) \cos\psi_n \tag{3.3}$$

where p_n is the spectral irradiance (independent of wavelength) of the ambient light noise in W/cm² nm, $\Delta\lambda$ is the passband of the optical filter, T_0 is the peak transmission of the filter, A is the area of the detector, and $g(\psi_n)$ is the gain of the concentrator. This expression assumes that the light originates from a localized source forming an angle ψ_n with the normal of the receiver as illustrated by Figure 3.4.

If the light comes from an isotropic source and the receiver employs an ideal optical concentrator with constant gain over its entire FOV, the received optical power from the ambient illumination source can be calculated as [21]:

$$P_{ni} = p_n \Delta\lambda T_0 A_n^2 \tag{3.4}$$

In this case, n^2 represents the gain of the concentrator over the entire FOV. From these expressions (Equations (3.3) and (3.4)) it can be seen that by narrowing the bandpass of the optical filter, the ambient illumination power reaching the detector can be significantly reduced.

3.1.1.1 Direct and Reflected Sunlight

As explained, the performance of outdoor FSO (free space optics) links is significantly affected by sunlight, which can reach the photodetector directly or through reflections from surrounding objects, generating additional current (in the detector)

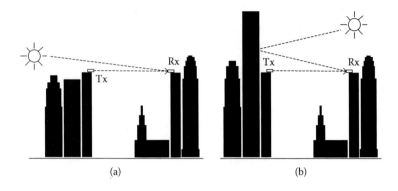

Figure 3.5 Solar illumination on a wireless IR link: (a) direct solar conjunction and (b) reflected solar conjunction.

and creating white Gaussian noise (WGN) in the receiver. This WGN reduces the SNR of the system, limiting its distance and its transmission speed. It also introduces problems such as detector impedance reduction and flicker noise (which creates irregular disturbances and affects the throughput of long data packets) [54]. Moreover, when the photodetector is exposed to direct solar radiation, the high optical power of the sun can potentially saturate the photodetector, interrupting the communication completely (despite the use of optical filters). For these reasons, a good understanding of the power contribution from solar illumination is necessary because it helps in designing and evaluating the performance of wireless IR communication links under intense background illumination conditions. By knowing the amount and type (direct or reflected) of solar optical power reaching the detector, for example, it is possible to calculate the optical power required at the transmitter to achieve a given bit error rate (BER).

As discussed in Section 3.1.1, the noise contribution from solar illumination can be reduced by restricting the FOV of the receiver and by using optical and electronic filters. However, in situations where the sun is positioned behind the transmitter and within the FOV of the receiver, it is very difficult to reduce this noise.

The effect of having the sunlight within the FOV of the receiver is commonly known as *solar conjunction*. As explained by Rollins et al. [67], this effect does not only occur when the solar disk (which from the earth appears to have an angle of around 9.31 mrad in diameter and a relative movement of 0.0728 mrad/s⁻¹) is within the field-of-view of the receiver, but also when its diffraction corona (which is explained in more detail below) is within the receiver FOV, as illustrated in Figures 3.5 and 3.7. Here it can be seen that, despite the fact that either transceiver in a wireless IR link can suffer from solar conjunction, the ones with positive elevation (or elevation close to 0°) are more prone to experience it. This is explained by the fact that receivers with positive elevation point to potential areas of direct solar illumination and that most reflecting surfaces in outdoor environments are vertical, which reduces the probability of reflections from underneath the receiver.

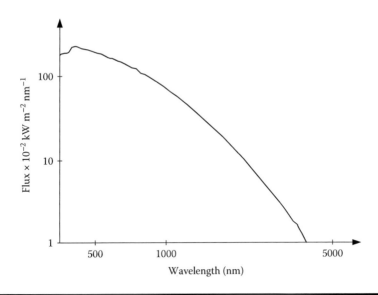

Figure 3.6   Spectrum of the solar radiation outside the Earth's atmosphere in the visible and near-IR regions. (*Source:* Adapted from [52].)

Despite the fact that characterizing the background illumination conditions of an FSO link is not always straightforward — as this characterization depends on a number of parameters such as environmental conditions and system geometry — a number of researchers have investigated ways to calculate the optical power at the receiver under different solar illumination scenarios. Some of these expressions are presented in this section.

Some of the models used to represent solar emission (which reaches its peak at around 460 nm) assume that the radiation from a black body at a temperature of 5770K (the temperature of the photosphere of the sun) reaching the earth is equal to 1360 W/m², while the radiation at ground level is only around 1000 W/m² (during clear sky conditions) due to atmospheric absorption, which as mentioned in Chapter 2 is wavelength dependent. Figure 3.6 shows the solar radiation spectrum in the visible and near-IR regions outside the Earth's atmosphere [52]. Here it can be seen that, in agreement with the information presented previously in this chapter, solar radiation is stronger at visible and near-IR wavelengths close to visible light, and weaker at wavelengths in the medium-IR part of the electromagnetic spectrum.

The solar radiation incident upon a specific point in the atmosphere or the ground consists of two components. One of these components is direct energy from the sun, and the other is solar diffuse energy originating from scattering. This scattered energy reaches the point from different directions. Direct sunlight accounts for more than 90 percent of the energy reaching the Earth, while less than 10 percent corresponds to solar energy that loses directivity and is scattered in the atmosphere [54]. It is precisely the solar radiation scattered by the atmosphere that

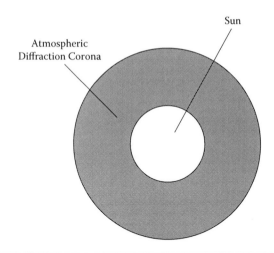

Figure 3.7   Solar disk and its atmospheric diffraction corona.
(*Source:* **Adapted from [67].)**

Table 3.1   Solar Power Received from the Solar Disk and Its Diffraction Corona for Varying Cloud Conditions (Data obtained from [67])

Angle to Sun	0°	0.5°	1°	1.5°	2°	2.5°	3°	3.5°	4°
Solar Power on Detector at 2:50 p.m.	6.5	−2.5	−12	−17	−20	−22	−24	−25	−25
Solar Power on Detector at 3:37 p.m.	−2.5	−10	−17	−20	−22	−24	−25	−26.5	−27

Note: The power values are given in dBm.

creates the *diffraction corona* of the sun (a corona of light around the image of the sun as seen from the receiver). This corona is larger than the angular extent of the solar disk, which is only 0.5°. An illustration of the sun and its atmospheric diffraction corona is presented is Figure 3.7. The intensity and the size of the corona depend on the environmental conditions of the location where the sun is being observed. Table 3.1 shows a number of values of received solar power at different angles. This table was obtained with a 780-nm receiver at different times; and, due to the variation of the atmospheric conditions with time (the different amounts of cloud at the moment of taking the measurements), the powers presented in the second and third rows are different [67].

Table 3.2   Ratio of Diffuse to Direct Solar Radiation as a Function of Wavelength

Wavelength (nm)	Ratio ($E_{diffuse}/E_{direct}$)
480	0.52
550	0.36
650	0.3
850	0.2
1700	0.13
2200	0.1

Note: Norway, solar Zenith angle = 45°, at ground level. Values selected from [52, 68].

The characteristics of the atmospheric scattering depend on geometrical considerations, on meteorological conditions (which determine the number of scatterers in the atmosphere), on the position of the sun in the sky, and on geographical location. The higher the amount of direct radiation present, the more diffuse radiation is created. Because the scattering of solar radiation is wavelength dependent, the amount of diffuse radiation is also dependent on wavelength. This explains why the ratio of diffuse to direct solar radiation, which is shown is Table 3.2 for different wavelengths, is much higher for visible wavelengths than for near-IR radiation. The values shown in Table 3.2 were measured by Bunnik [52, 68] at ground level with a solar zenith angle of 45°. Here it can be seen that the longer the wavelength, the smaller the ratio. Bunnik observed that the ratio of diffuse to direct radiation changes also with angle (the ratio increases at larger values of solar zenith angle). At a wavelength of around 850 nm, for example, the value of the ratio for a solar zenith angle of 45° is about 0.2, while for the same wavelength, the value at 68° is around 0.28. The larger variations occur at visible light wavelengths and at IR wavelengths close to the visible part of the spectrum.

It must be noted that, due to the change of position of the Earth with respect to the sun during the year, the solar radiation on the Earth also changes; and, in some cases, it is important to take this into account when designing an optical wireless link. The maximum irradiance is present in the months of December and January, while the minimum irradiance is observed in June and July. The variation from the maximum to the minimum irradiance throughout the year is approximately 4 percent. With regard to the geographic latitude, the higher the latitude, the higher the probability of conjunction. The probability of conjunction in Mexico City for a randomly oriented link, for example, is only around 0.32, while the probability of conjunction in Helsinki is more than 0.6 [67].

The solar threshold angle due to solar conjunction is also determined by the amount of solar energy that is blocked by objects surrounding the communication link. In many cases, buildings, hills, and trees surrounding the communication link may be in the path of the solar energy, thus reducing the contribution from sunlight. The amount of solar energy impinging on the detector is directly related to the percentage of blockage presented by these objects.

The attenuation of solar radiation as it enters the Earth's atmosphere changes the overall radiation spectrum curve according to the transmission windows of the atmosphere (presented in Figure 2.2). The solar radiation spectrum at ground level, taking into account the atmospheric absorption windows, is illustrated in Figure 2.1. This graph represents solar radiation at mid-latitude in a rural area during the summer time with a 5-km visibility. Here, a strong absorption in the near-IR part of the spectrum can be seen. This absorption derives from water vapor [52].

The radiation power P_b originating from direct sunlight at the photodetector's sensitive area can be calculated from the solar radiation flux at ground level at a given location. According to Sidorovich [69], this solar radiation power can be calculated as:

$$P_b = F_\lambda S_R \eta_R \Delta\lambda \tag{3.5}$$

where F_λ is the spectral density of the radiation flow at normal incidence, S_R is the effective area of the receiver, η_R is the transmittance, and $\Delta\lambda$ represents the bandpass of the filter. This expression assumes that the entire sun disk is within the detector's FOV. In situations where the FOV of the receiver (Ω_R) is not large enough to cover the entire sun disk (which has a mean solid angle $\Omega_S = 6.8 \times 10^{-5}$), this expression becomes:

$$P_b = \frac{F_\lambda S_R \eta_R \Omega_R \Delta\lambda}{\Omega_S} \tag{3.6}$$

The high power contribution from sunlight (which is much larger than the transmitted information signal) and the possibility of having the sun illuminating the photodetector at normal incidence have motivated some researchers to produce tables and graphs that predict the probability of disruption of outdoor wireless IR links, depending on the geographical location. Example of these graphs can be found in [67, 69].

Specular reflections (originating from objects surrounding the receiver of a wireless IR link) also introduce noise in the detector, and their effect must be considered when estimating the noise contribution from solar illumination. The background noise level originating from reflections from these objects and surfaces depends on their size, their reflectivity at the wavelength of interest, and their distance from the receiver. Despite the fact that the optical power caused by these reflections and reaching the sensitive area of the detector is only around one to two orders of magnitude smaller than the one generated by direct solar illumination, it may still be large enough to dominate the power of the transmitted signal [69]. It also has the effect of reducing the dynamic range of the system and making it more sensitive to other problems, such as link misalignment and path loss.

The time that the photodetector is exposed to reflected sunlight is equivalent to the time the photodetector is exposed to direct sunlight, with the difference that instead of depending on the orientation of the sun with respect to the photodetector, these reflections depend on the position of the reflecting surface with respect to the receiver. The background power caused by sunlight reflections can be calculated as [69]:

$$P_{bre} = \frac{\rho F_\lambda S_R \eta_R \eta_a R^2 \Delta\lambda}{\Omega_S} \tag{3.7}$$

where ρ is the reflectivity of the surface. From this expression, it is possible to calculate the minimum mean signal power required (at the APD of a wireless IR receiver) to achieve a specific BER (based on the levels of background illumination). According to Sidorovich [69], if a BER $\leq 10^{-12}$ is required, the mean signal power can be calculated as:

$$P_S = 2\sqrt{\left(\frac{\pi Fe\Delta f P_{bre}}{2S}\right)\ln\left(\frac{1}{BER}\right)} \tag{3.8}$$

Here, F is the noise factor of the receiver, e is the electron charge (1.6×10^{-19} C), Δf is the receiver bandwidth in hertz (Hz), P_{bre} is the power of the background illumination at the APD's active area in watts (W), and S is the APD's sensitivity (in A/W^{-1}) at the wavelength of interest.

It must also be mentioned that, because the photodetector of a wireless IR receiver is not only exposed to reflections from surrounding surfaces, but also to sunlight scattered by hydrometeors (such as clouds and fog), a factor can be incorporated into Equation (3.8) to take this into account. The density value q assigned to these aerosols can take values between 1 and 10 (with the maximum value corresponding to dense fog).

Scattered sunlight is particularly problematic when the path between the sun and the detector forms a narrow angle θ_p with the transmitter-detector path, as shown in Figure 3.8a. Fortunately, cloud cover also reduces the contribution from direct solar energy in the receiver. Depending on the position of the hydrometeor with respect to the link, the scattering can occur in two different ways. In one case,

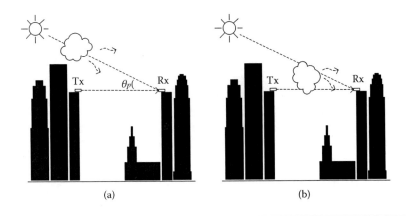

Figure 3.8 Atmospheric conditions on a wireless infrared link: (a) scattering in clouds and (b) scattering in fog.

the scattering aerosol may be above the link but not directly between the transmitter and the receiver — a cloud above the link as shown in Figure 3.8a (if this is the case, the brightness B of the scattered sunlight is practically the same for different aerosols with different optical densities and different values of q). However, the value of B varies, depending on the position of the sun with regard to the link and the value of the angle θ_p. The value of B (for wavelengths in the near-IR region corresponding to 780 to 830 nm), for example, is around 0.667 Wm^{-2}nm^{-1}sr^{-1} for angles θ_p equivalent to 15° and sun elevations of less than 45°. As the angle θ_p increases, the brightness of the scattered light decreases. If, on the other hand, the angle θ_p decreases, the value of the brightness increases, reaching values of up to of 2 Wm^{-2}nm^{-1}sr^{-1} for an angles $\theta_p \leq 3°$.

The value of the maximum background power at the photodetector, taking into account the brightness of the scattered sunlight, is given by [69]:

$$P_B = BS_R \eta_R \Omega_R \Delta\lambda \qquad (3.9)$$

In the second case, the aerosol is at link level and within the transmitter-receiver path (as illustrated in Figure 3.8b. This means that the hydrometeor does not only scatter the light from the sun, but also attenuates the information signal. If this is the case, the maximum background power at the photodetector can be calculated from Equation (3.9), but the signal power at the detector must be calculated taking into account the attenuation per kilometer at the relevant wavelength according to the specific meteorological conditions (the meteorological visibility length S_M). The aerosol attenuation Λ for an arbitrary communication path length R is given by:

$$\Lambda = e^{kR} \qquad (3.10)$$

where k is an aerosol attenuation factor that depends on wavelength. The limit to the path length achievable in the presence of fog (taking into account the background signal power and the visibility length factor) can be calculated as [69]:

$$R_L = \frac{1}{K_\lambda} \ln \frac{P_{S0}}{0.45\sqrt{P_B}} \qquad (3.11)$$

where K_λ is approximately $3/S_M$ (at 800 nm) and P_{S0} is the power of the transmitted information signal at the receiver in the absence of fog.

Rollins et al. [67] have produced link disruption statistics based on how meteorological conditions affect the amount of solar energy incident on the detector of a wireless IR link. They did this analysis using a parameter called *insolation* to account for solar atmospheric losses. Insolation in this case is defined as the amount of solar irradiance at the Earth's surface, and it is particularly relevant in cases where the sun presents low elevations, which make more probable the occurrence of solar conjunction.

Some objects around the receiver present, rather than specular, diffuse reflections. When this is the case (and if these objects are within the FOV of the receiver), the sunlight scattered by them is also detected by the receiver. Despite the fact that the power (at the receiver) from the diffuse reflections from these objects is lower than the power received from the information signal, their contribution cannot be neglected as it is contributes to reduce the SNR of the system. The radiation power at the receiver from diffuse reflections can be calculated as [69]:

$$P_B = \frac{\rho F_\lambda S_R \eta_R \Omega_R \Delta\lambda}{\pi} \qquad (3.12)$$

This expression assumes that the reflecting object's surface is normal to the path formed by the transmitter and the receiver, and that the sun is behind the receiver. It also assumes that the object fills almost entirely the FOV of the receiver and that the reflection from the object (which is assumed to have a very high reflectivity) presents Lambertian angular distribution (i.e., light is reflected with equal intensity in all directions).

3.1.2 Multipath Dispersion

Multipath dispersion is another important limitation to the data rate of an optical wireless system. This effect is especially significant in the diffuse configuration where the wide FOV of the receiver and the wide beam of the transmitter increase the number of possible paths that the transmitted energy can follow. The multiple reflections of the transmitted energy when traveling from the transmitter to the

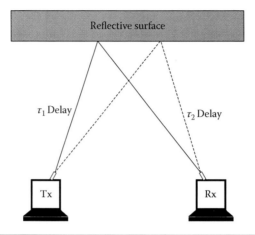

Figure 3.9 Illustration of two different paths with their corresponding time delays (τ_1 and τ_2) of a nondirected wireless IR link. (*Source:* Adapted from [66].)

receiver result in a broadening of the signal by the optical channel, which creates temporal dispersion and leads to intersymbol interference (ISI). The temporal dispersion process can be explained as follows. Given a system such as the one illustrated in Figure 3.9, where a reflective surface A (reflecting an IR signal) is radiating R watts per unit surface area, the optical power P_R incident on a photosensitive area A_p can be calculated as [66]:

$$P_R = A_p R \sin^2 \theta_a \tag{3.13}$$

where θ_a is the acceptance angle of the photosensitive area ($0° \leq \theta_a \leq 90°$). From this equation it can be observed that as long as the illuminating area is within the FOV of the photosensitive area and covers it completely, the received power does not depend on the angular orientation or the position of the photosensor with respect to the surface. Thus, the photodetector not only detects the direct signal from the transmitter, but also the energy reflected from surfaces within its FOV. This means that the energy reaching the detector can follow different paths and that different rays undergo different delays.

Two different paths are illustrated in Figure 3.9, each one having a delay τ_1 and τ_2. The delay related to a particular path can be calculated as [66]:

$$\tau = \frac{dp}{c_\lambda} \tag{3.14}$$

Here, dp is the path length and c_λ is the speed of the transmitted optical signal. The differential gain for each path can be calculated as $F(\tau)d\tau$, where $F(\tau)$ is given in [9, 66]:

$$F(\tau) = \frac{2\tau_0^{\,2}}{\tau^3 \left(\sin\theta_a\right)^2} \tag{3.15}$$

Equation (3.15) is valid for $\tau_0 \leq \tau \leq (\tau_0 / \cos\theta_a)$. If, on the other hand, $(\tau_0 / \cos\theta_a) < \tau < \tau_0$, then $F(\tau) = 0$. From Equation (3.15), the wave $g(t)$ received at the photosensitive area after having been reflected by a surface (where the incident wave is $f(t)$) can be calculated as [9, 66]:

$$g(t) = \int_{-\infty}^{+\infty} f\left(t - \tau\right) F(\tau) d\tau \tag{3.16}$$

This means that the transmitted symbol is spread in time, potentially creating ISI and limiting the data rate. As explained in Chapter 1, Gfeller et al. [10] have calculated a theoretical maximum speed —related to the size of a room. According to these authors, the maximum theoretical transmission speed is 260 Mb·m·s^{-1}, which means that the larger the room, the lower the transmission rate.

Kahn et al. [21, 70] have presented experimental results showing the multipath distortion in non-directed infrared channels. Their experiments, carried out in an empty conference room of 5.5 × 7.5 × 3.5 m (in this case, 3.5 m corresponds to the height of the ceiling) at different positions, show that short initial pulses dominate the impulse response of an LOS configuration, with the strongest reflections arriving between 15 and 20 ns after the initial pulse. The diffuse impulse response, on the other hand, presents a smaller height (10 percent) with a wider initial pulse of around 12 ns width due to the different path lengths of a number of reflections from the walls and the ceiling. Their experimental results also show that on shadowed channels, the diffuse configuration is less susceptible to suffer from shadowing than the LOS topology, which is explained by the different propagation paths available between the transmitter and the receiver. The effect of shadowing in a diffuse configuration is a broadening of the impulse response, while in the LOS configuration the effect is that the dominant impulse of the impulse response disappears.

According to Kahn et al., the channel root-mean-square (rms) delay spread D can be used to measure the severity of intersymbol interference (ISI) created by a multipath channel $h(t)$. This delay spread can be calculated from the impulse response using Equation (3.17) [21]:

$$D = \sqrt{\frac{\int_{-\infty}^{\infty} (t - \mu)^2 h^2(t) dt}{\int_{-\infty}^{\infty} h^2(t) dt}} \tag{3.17}$$

where μ is the mean delay, which can be calculated as [21]:

$$\mu = \frac{\displaystyle\int_{-\infty}^{\infty} t h^2(t)dt}{\displaystyle\int_{-\infty}^{\infty} h^2(t)dt} \qquad (3.18)$$

3.2 Eye Safety

In theory, increasing as much as possible the optical power emitted by a wireless IR transmitter could help overcome some of the data transmission limitations suffered by optical wireless technology. A high emitted optical power, for example, could compensate for the high attenuation suffered by the optical signal when transmitted through air, increasing the range of the system and improving its SNR.* A high power level at the transmitter could also allow the use of smaller and faster low-capacitance photodetectors, which present a smaller active area. Unfortunately, the optical power level at the transmitter is restricted due to eye safety considerations and to power consumption limitations [25].

Eye safety is one of the most important restrictions to the optical power level emitted by a wireless IR transmitter. Infrared, visible, and ultraviolet (UV) radiation can cause damage to the human eye if the energy emitted by optical sources at their respective wavelengths exceeds specific safety levels [39, 71]. Near-IR radiation, for example, can cause retinal burns, while medium- and far-infrared thermal radiation can cause corneal burns. The fact that the cornea is opaque to infrared radiation beyond 1.4 μm has raised the question of the possibility of using the 1.55 μm wavelength for wireless IR communication systems to avoid the restrictions imposed by eye safety considerations. Unfortunately, even if the retina is protected from near-IR radiation at this wavelength, it remains to be seen if the cornea would be exempt from any damage. In addition, as indicated by Kahn et al. [21], despite the availability of InGaAs and Ge photodetectors that are sensitive to EM (electromagnetic) energy at around 1.55 μm, these detectors are more expensive and present higher capacitances.

Damage to the eye from a near IR transmitter (operating at wavelengths between 760 and 1400 nm) can occur if excessive power from an optical source reaches the eye and the energy is focused on the retina. This focused energy creates a high-energy spot that increases the temperature of the tissue. The smaller this spot is, the higher the temperature of the tissue (and the worse the damage). Whether a lens can focus energy down to a point or to a spot of small diameter depends on its imperfections. When calculating the size of the spot created on the retina by

---

* This is due to the fact that in a system based on IM/DD, the SNR is proportional to the square of the received optical power [21].

the lens of the eye when focusing the energy from a collimated source, the eye is generally considered to be ideal and diffraction limited. In reality, the capability of the eye is not diffraction limited and the smallest spot that can be focused on the retina from a collimated source ranges between 50 and 100 μm (which is considerably larger than the 10-μm spot that could be created with a diffraction-limited eye) [72].

In addition to the power level emitted by an optical source, other factors that also need to be taken into account when evaluating the potential damage of the emitter to the eye include the wavelength of operation, the length of time that the eye is exposed to the source, and the flux density (the power per unit area).

3.3 Extended versus Collimated Sources

The most common optical sources used for the transmitter of a wireless IR link are light-emitting diodes (LEDs) and laser diodes (LDs). LDs are preferred in high-speed communication applications because they can be modulated at higher data rates than LEDs. They are also favored over LEDs when longer transmission distances are required, which can be achieved due to the fact that the energy from a laser source is collimated [73]. However, LDs must comply with stringent regulations in order to be used in a safe manner under all operating conditions. For this reason, eye safety standards have been created to define maximum exposure limits for different types of radiation and to classify optical sources according to the power and the characteristics of the beams they emit [21, 74].

The different sources of IR radiation used in optical wireless communication transmitters can be classified either as *extended* or *point sources*. If the light emitted by a specific device radiates in all directions (or within a very wide angle), the source is considered *extended*; whereas if the energy radiated by the device emits within a very narrow angle, the transmitter is considered a *point source* [63, 72]. Figure 3.10 illustrates the two types of sources. While an imaging system, such as the human eye, can focus a point source down to a point (a spot of diameter equal to nearly zero), an extended source can only focus the energy down to a finite spot. Lasers are a good example of a point source. Despite the fact that lasers are not perfect collimated light sources, they are very close to being so (due to the low divergence angle of their emitted beam, which can be sometimes as small as 0.2 mrad). For this reason, they are subject to more rigorous power limitations than other optical sources. LEDs, on the other hand, are generally considered safe to the human eye due to the fact that their emitted power is low, and because they form an extended image on the retina. However, this situation could change in the future due to the improvements in LED efficiency and power. Further details of LEDs and LDs are presented in Section 3.5.

As discussed above, the damage produced by an IR source depends on factors such as the exposure time, the wavelength, and the power of the signal. The

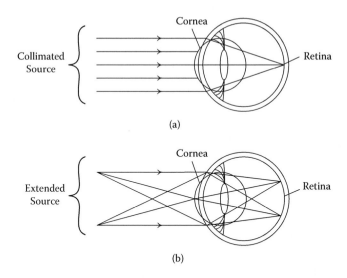

Figure 3.10 Eye transmission for different sources of illumination: (a) point source and (b) extended source. (*Source:* Adapted from [63, 72].)

limits to the output power levels of lasers are set by the International Electrotechnical Commission (IEC*), which describes allowable exposure limits (AELs). These AELs ensure that optical sources emitting IR and other types of radiation are safe and do not require warning labels. Any product that employs LEDs or LDs is subject to the IEC 60825-X standard in countries regulated by the European Committee for Electrotechnical Standardization (CENELEC). The limits imposed on the emitted power of optical transmitters are a function of the size of the optical source, the wavelength of the optical signal, and the viewing time. The maximum permissible exposure (MPE) levels, which define the levels of radiation to which a person can be exposed without risk to the eyes or the skin, are very low in general.

3.3.1 Class 1 Lasers

Class 1 products are defined as inherently safe, which means that they are safe even when viewed with an optical instrument. They are not supposed to present any hazard to the human eye independently of their wavelength of operation and the exposure time. Optical wireless communication systems are required to fall into this category.

---

* The IEC, founded in 1906, is a voluntary international standards organization whose objective is to promote international cooperation on all questions of standardization in the fields of electrical and electronic engineering [75].

Because laser diodes are point sources, they are required to emit a very low power to be rendered Class 1. LEDs, on the other hand, can be operated safely at larger emission power levels because they are extended sources [20, 72].

3.3.2 Class 2 Lasers

Class 2 applies to sources between 400 and 700 nm (visible light), and it states that lasers in this category are safe if the blink or aversion response of the eye operates (the blink or aversion response is the natural ability of the eye to protect itself by blinking. The blink response time is approximately 250 ms). The blink response of the eye does not operate with energy in the near-IR part of the electromagnetic spectrum; the reason for which this classification is only valid for visible light sources [72].

The new version of the standard also includes a 2M category for lasers that are safe if blink or aversion response operates and no optical instruments are used.

3.3.3 Class 3 Lasers

The power range of a Class 3 laser is between 1 mW and 0.5 W. The energy emitted by this type of source is dangerous not only if seeing a direct beam, but also when seeing specular reflections. Damage may occur in a period of time shorter than the blink response of the eye. For this reason it is important to wear the right protective eyewear when using this class of laser to avoid potential eye damage if accidentally crossing the beam path. It is also common practice when setting up Class 3 laser systems to put neutral density filters in the beam path to reduce the laser power to a Class 2 during laser alignment [72].

3.4 Holographic Diffusers

Holographic diffusers have been proposed as a way to overcome the eye safety problem when using laser sources [63, 76–78]. Transmitters that incorporate this type of diffuser achieve higher emitted optical powers, higher control over the pattern of the emitted beam, and safer distributions than the ones without holograms. A Class 3 laser source, for example, can be reclassified as a Class 1 emitter structure if it is combined with the appropriate hologram. This is possible because, as the collimated laser beam passes through the hologram, the optical element breaks up the emitted beam and diffuses the image of the laser spot on the retina. In addition, the use of holograms is advantageous because they give the system designer increased control over the shape of the emitted optical beams, something that is not possible when using diffusing screens.

Optical wireless LANs, for example, can benefit from the use of holographic diffusers because, in this type of application, a holographic diffuser can be used to

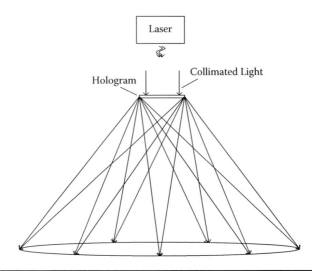

Figure 3.11 Side view of a room illuminated by a holographic diffuser. Here, the collimated light of a laser is split into a number of beams that cover the desired floor area.

illuminate a room of any shape in an optimum way, without creating intersymbol interference. Multipath distortion can be minimized by creating a footprint that prevents the illumination of walls. This has the additional advantage of optimizing the transmitted power. Figure 3.11 shows a holographic diffuser splitting a collimated laser beam into an array of different beams with uniform intensities that cover a desired room area.

Holography, in general terms, is the storage of the amplitude and phase information of a wavefront. It is commonly used as a way to produce three-dimensional (3-D) images of objects through interference between a coherent reference beam and a wavefront diffracted by the desired object. In the context of optical wireless communications, the hologram is a diffractive surface that modifies the phase of the energy passing through it. It alters the wavefront of the energy impinging on it thanks to a two-level structure that changes the optical path length of every ray traveling through it. This provides a different phase delay for each ray. Figure 3.12 shows the effect of a diffusing hologram on a set of rays from a collimating source.

Holograms can be generated from a physical object or from a mathematical description. When a mathematical description is used, any wavefront can be produced. If this mathematical description is implemented with the help of a computer, the hologram is called a computer-generated hologram (CGH), and the wave propagation can be computed using Fraunhofer or Fresnel approximations [77].

Holograms can be fabricated from quartz, using electron-beam lithography followed by reactive ion etching to create the two-level surface relief with the correct resolution. Unfortunately, this process and the required materials are expensive; and producing holograms via this technique in large quantities takes a long time. However, thanks

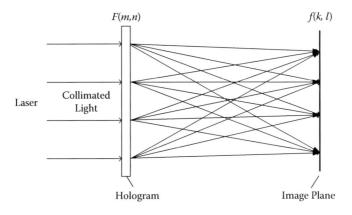

Figure 3.12   Effect of a diffusing hologram on a set of rays from a collimated source.

to the improvements in holograms for credit cards, the same technique (which implies creating a negative copy from a quartz hologram in nickel and using it to imprint sheets of transparent plastic) can be applied to create holographic diffusers [63].

There are two basic encoding schemes to generate holograms: (1) phase-only and (2) detour phase [76, 77, 79]. The first scheme controls the phase of the transmitted light only, and holograms generated under this scheme have a high efficiency because they allow all the energy to pass through them. Detour phase holograms, on the other hand, have a smaller efficiency due to fact that their optimization process uses both the phase and the magnitude in the hologram domain as variables, and the variation in magnitude contributes to the blockage of a percentage of the energy, which reduces the hologram transmittance to values below 20 percent.

Phase-only holograms can be generated using a small surface relief optical element called *kinoform*. This micro-optical element is also used as a filter for data processing because of its wide FOV and its use of incoherent light that enables parallel processing of two-dimensional data [80].

Yao et al. [77] have analyzed three iterative techniques to find the solution to the problem of satisfying the magnitude constraint in the image domain through phase variations in the hologram domain on a phase-only holographic diffuser. These techniques are (1) error reduction, (2) input-output, and (3) simulated annealing. Their results revealed that the simulated annealing technique produced better results than the other two. This was concluded from computer simulations, where a pattern reconstructed using simulated annealing produced brighter and better-defined spots than the reconstructions based on the error reduction and on the input-output methods, which indicates that the diffraction efficiency of holograms produced using simulated annealing is higher than the diffraction efficiency of the other two techniques. The disadvantage of this method is that, compared to the error reduction and input-output techniques, it also requires more computing time.

Eardley et al. [63] and Dames et al. [76] have also described the use of simulated annealing with the help of a computer, where the spot of light corresponding to the desired pattern is divided into pixels that change the phase of the energy by 0 or π. The optimization of the desired output is achieved by modifying the pixels in an iterative process until the light outside the required far-field shape is minimized and consistent light intensity within the pattern is achieved. The size of the pixels is approximately λ/sinθ, where θ is the half-angle of the beamwidth at the output of the source.

The methods proposed by Yao et al. [77] to design a hologram are based on the use of a set of variables and a cost function that corresponds to the characteristics of the system. The purpose of the algorithms is to find a global minimum in the cost function that corresponds to the optimized system.

There are a number of other ways to create an extended source from a point source. For example, a ground-glass diffuser can be placed at the output of a laser to change it from a point source to a large area source. Unfortunately, this does not give any control over the output intensity distribution. Lambertian diffusers have a cosine dependency with scatter angle; that is, the intensity of light scattered from a point on a reflecting surface follows the equation $I_o(\theta_s) = I_i \cos(\theta_s)$, where I_o *is the intensity of the scattered light,* θ_s is the angle of the scattered light, and I_i is the intensity of the incident light at the point. In addition, the spread angle of conventional diffusers is not only very difficult to control — but it is also small. Holograms, on the other hand, can be used to create a beam of any desired shape. An extended image also can be created using an array of LEDs.

3.5 Light-Emitting Diodes versus Laser Diodes

The low cost of IR sources in the short-wavelength part of the near-IR region of the electromagnetic spectrum and the high sensitivity of low-capacitance and low-cost silicon detectors at these wavelengths make the band of 780 to 950 nm the preferred option for a large number of optical wireless systems [21]. Unfortunately, as explained in Section 3.3, IR radiation is potentially dangerous at these wavelengths, which makes the selection of the right IR source and of the right optical elements (if, for instance, a collimator is used) of great importance.

The two most important emitters used for the transmitter of a wireless IR communication system are (1) the laser diode (LD) and (2) the light-emitting diode (LED). The selection of one over the other depends on their specific advantages and limitations and on their suitability for a particular system. In terms of eye safety, for example, LEDs are preferred over LDs because they are extended sources and they are generally rendered safe (cf. Section 3.2). This makes them the preferred option for indoor applications where users may inadvertently interrupt the transmitter beam. LDs, on the other hand, need to comply with eye safety constraints, which limit their emitted power and therefore also the range of the system. However, the spectral linewidth of LEDs is wide compared to the linewidth of LDs (25 to 100 nm for an LED compared to linewidths as narrow as 5 nm for LDs). Therefore, systems based on LDs can benefit from the use of narrow-band, thin-film optical filters and present a better background illumination noise rejection [21, 27].

When a single IR LED is used (for indoor applications), it is generally employed for directed-LOS or hybrid systems where the 10° to 30° divergence half-angles of the LED are enough to satisfy the angular requirements of such applications. For applications requiring a wider emission angle, LEDs are arranged in clusters where each one can be oriented in a different direction in such a way that the emission angle of each one is complementary to the others. LDs, by comparison, are generally only employed in directed point-to-point links due to the fact that the divergence in this type of source is very narrow. Alternatively, they may be pointed to a reflective surface in such a way that the reflected energy provides a diffuse configuration [21].

For most applications it is also important to compare the modulation bandwidth of each source, as this may also define a limit to the maximum bit rate achievable by the system. While the modulation bandwidth of LEDs ranges from tens of kilohertz to a few hundred megahertz, the modulation bandwidth of LDs extends from tens of kilohertz to tens of gigahertz. This means that for high data rate applications, LDs are the only transmitters capable of supporting high communication speeds.

LDs also present a higher electro-optical conversion efficiency than LEDs. While the electro-optical efficiency of the latter is just between 10 and 20 percent, the efficiency of LDs ranges between 30 and 70 percent [21]. The high data rates achievable by LDs make them the preferred option for outdoor systems where longer distances between transmitter and receiver are necessary (the slightly divergent beam of laser diodes concentrates the energy, allowing for longer transmission distances); and, due to the fact that transmitters and receivers are installed in high places inaccessible to unaware bystanders, higher class optical sources are sometimes used.

Other important factors that must be taken into account when choosing the source of a wireless IR transmitter include the cost of the component, its simplicity of implementation, and the potential complexity of its operation. In both cases, LEDs present an advantage when compared to LDs. While the cost of LEDs is generally low, the cost of LDs goes from moderate to high. In addition, LDs also require more complex drivers than LEDs. Further information about LEDs and LDs is presented in Section 6.3.

3.6 Special Considerations for Outdoor Systems

As explained at the beginning of this chapter, contrary to indoor optical wireless communication links that do not suffer from meteorological conditions, outdoor FSO systems need to consider conditions such as rain, fog, and clouds that affect the performance of the system in different degrees. In addition, the fact that outdoor systems generally operate over longer distances and that the sun's contribution in outdoor systems is much stronger than that present in indoor systems make the design of the FSO transmitter and receiver more challenging.

Outdoor FSO links generally employ collimated sources of light to concentrate the energy of the beam and in this way minimize losses and reach longer distances. Despite the fact that collimated sources make the alignment of point-to-point LOS links more difficult, they have the advantage of being able to use very directive receivers, which offer the benefit of having a very high gain (this topic is explored in more detail in Chapters 4 and 5). In addition, by defining a very narrow FOV at the receiver, it is possible to reject the ambient illumination that would otherwise reach the receiver at wider angles. This is particularly relevant in highly sensitive receivers because, in most cases, they not only receive and amplify the energy of the desired signal, but also that of background illumination. Unfortunately, restricting the FOV of the receiver is not always enough to have a proper rejection of solar energy. Sunlight may reach the detector not only directly, but also from reflections (due to the fact that the position of the sun relative to the receiver is not fixed).

It must be mentioned that outdoor IR receivers generally have FOVs wider than the divergence of the emitter source. This not only helps collect the incoming energy within the whole range of angles, but also to allow for building motion.

Even when using thin-film optical filters, outdoor wireless IR communication systems are at risk of being disrupted if the sun is within or near the receiver's FOV (because the power of the sun will always overwhelm the power of the transmitter at the entrance of the receiver). The time that the disruption lasts depends on the position and orientation of the link, as well as on the time of the year. The sun path, for example, may present an important variation, depending on the time of year, having very different effects on the performance of a specific link.

To improve the sensitivity of the receiver, FSO links generally use fast, highly sensitive photodetectors, such as avalanche photodetectors, that are about 10 dB more sensitive than PIN detectors [56], and that can be combined with transimpedance amplifiers to provide adequate bandwidth. This is necessary to compensate for the high attenuation suffered by the transmitted signal with distance. Unfortunately, APDs do not only amplify the desired signal, but also the background illumination (in this case contributed mainly by sunlight) that is within the receiver's FOV. Due to the high gain of APDs, even small amounts of background illumination can introduce noise in the receiver that is difficult to filter afterward. In addition, if the amount of solar illumination at the receiver is too high, the transimpedance amplifier is at risk of saturation. Rollins et al. [67] have presented values of receiver sensitivity degradation with increase of solar background illumination. In their analysis, they present measured values of background illumination in the range of –60 to –10 dBm, which reduce the receiver sensitivity to around 33 dB in a silicon APD detector receiver operating between 780 and 850 nm. The same variation in background illumination power introduces a receiver sensitivity loss of about 28 dB in an InGaAs APD receiver operating at 1.55 μm. From these results, Rolling et al. concluded that receivers operating at 780 to 850 nm are more susceptible to solar illumination than the 1.55μm versions. This is logical, taking into account that in general, photodetectors that are sensitive in the near-IR part of the spectrum are also highly sensitive to energy in the visible part of the spectrum, and that, as discussed above, the power contribution of sunlight is higher at visible wavelengths and decreases at longer wavelengths. In addition, in this specific case, InGaAs detectors are less sensitive than Si detectors.

An example of a line-of-sight FSO link for outdoor use can be found in [56]. There, Eardley et al. presented a 1-Gbps wireless IR system operating over 40 m. Their transmitter consisted of a collimated source of light (a GaAs/GaAlAs Fabry-Perot laser) operating at 820 nm with a beam divergence of only 0.5° that created a 30-cm spot at a distance of 40 m. Thanks to the use of a collimated source, they could also make use of a very directive receiver. The receiver consisted of an inverted telephoto system that concentrated the energy into a photodetector through two imaging lenses. The total FOV of the receiver was ±0.35° and its total focal lens was 75 mm, which is extremely large when compared to the non-imaging options used for indoor wireless IR communications. The receiver contained a silicon avalanche photodetector of 1 mm diameter (active area) that required a bootstrapped transimpedance amplifier or BTA (a positive feedback transimpedance amplifier) to minimize the thermal noise and to compensate for the high capacitance associated with the large area of this component — thus producing an adequate data bit rate while maintaining high sensitivity. Here, the APD was preferred over a PIN detector for the reasons explained above; and, due to its lower dark current, silicon was preferred over germanium. The receiver also integrated a narrow-band (70 nm) optical filter to minimize the noise contribution due to solar illumination.

Another example of an outdoor FSO link can be found in [83]. There, Mendieta et al. presented a wireless IR link for outdoor use that operated at distances greater than 2 km, but at a limited bit rate (2.048 Mbps). The transmitter consisted of a GaAlAs laser diode operating at 830 nm, followed by a collimating lens that restricted the divergence of the beam. The receiver consisted of a mirror, an optical filter, and a silicon PIN detector, followed by a high input impedance amplifier.

Other aspects worth considering when designing a directed-LOS link for outdoor use include the possibility of obstructions and the geographic location. The obstructions may be due to, for example, birds, trees, buildings, or objects that may be placed between the transmitter and the receiver obstructing the link's path. The formation of frost at the receiver or the transmitter window provides another example of a possible obstruction in the communication path.

When designing an outdoor link, it is also important to be aware of atmospheric turbulence, which consists of random irregularities in the properties of the air. These irregularities contribute to fluctuations in the received amplitude, focal point blurring, wander (random variations of position of the digital pulses), and broadening of the received pulses [56]. This effect becomes relevant at link distances greater than 500 m.

3.7 Summary and Conclusions

This chapter presented some of the major factors affecting the transmission of data in an optical wireless link. Background illumination, for example, reduces the SNR and the range of the system. In outdoor systems, the highest contribution to this background illumination is sunlight, sometimes called SBR (solar background radiation). This SBR can reach the receiver directly or from reflections by surrounding objects. Solar radiation reaching the detector directly can saturate the photodetector, interrupting the communication completely. Scattered solar illumination, on the other hand, can reduce the system's SNR , thus limiting its distance and reducing its transmission speed.

Some of the recommendations presented by Gebhart et al. to reduce the effect of solar illumination in outdoor FSO links include [54]:

- *Optical filtering.* By introducing optical filters that allow mainly energy at the wavelength of interest to impinge on the detector and reject energy at unwanted wavelengths, the effect of solar illumination can be significantly minimized. The fact that outdoor systems have very narrow emission angles and reduced FOVs at the receiver favors the use of narrow-band thin-film optical filters.
- *Positioning and location.* It is possible to further reduce sunlight noise by taking into account the location where the system is deployed and the position where the receiver is mounted. This is due to the fact that surrounding objects and structures can act as barriers between the sunlight and the link. The elevation of the receiver is also important.
- *Restricted FOV.* The fact that outdoor systems are, in general, highly directive also favors the use of a very narrow FOV at the receiver, which introduces a first stage of sunlight discrimination. Furthermore, as narrow emission angle transmitters concentrate the energy into a narrower beam, this helps to increase the SNR at the receiver for a given distance.
- *Modulation techniques.* The use of appropriate modulation schemes also contributes to increasing the SNR (cf. Chapter 8).

Eye safety is one of the most important considerations when designing a wireless IR link, especially for indoor applications. For this reason, the selection of the right source of illumination, the wavelength, the power emitted by the transmitter, and the optics of the system are of great importance.

Table 3.3   Safety Classification for Laser Sources

Class	880 nm	1310 nm	1550 nm
Class 1	<0.5 mW	<8.8 mW	<10 mW
Class 2	Applies only to visible wavelengths		
Class 3A	0.5–2.5 mW	8.8–4.5 mW	10–50 mW
Class 3B	2.5–500 mW	4.5–500 mW	50–500 mW

Source: From [74].

IR radiation is potentially dangerous to the retina because it can be focused by the eye without activating its blink response when exposed to a high optical power. That is why the International Electrotechnical Commission (IEC) establishes parameters under which different sources of illumination can be operated without risk of injury.

Light sources are classified according to whether the emitted energy forms an extended or a point source in the retina of the eye. If the eye can focus the emitted light, the source is classified as a point source and deemed potentially dangerous; if the eye cannot focus the emitted energy (i.e., if it creates an extended image in the retina), the source is classified as an extended source and is considered safer than a point source. Laser sources are considered more dangerous than LEDs due to the fact that the emitted energy of a laser is collimated into a slightly divergent beam. LEDs, on the other hand, radiate energy in a wider beam that produces a larger spot on the retina. The level of damage introduced by an IR source depends on the wavelength of the energy, its optical power, and the exposure time. Therefore, the optical power emitted by a wireless IR transmitter is restricted, which limits the range of the link.

The International Electrotechnical Commission (IEC) [75] establishes allowable exposure limits (AELs) under which an optical source is safe for different viewing times and at different wavelengths. These AELs are a function of the parameters that define the level of damage to the eye (wavelength, diameter, emission semi-angle of the source, exposure time). The IEC also classifies sources of illumination according to their emitted powers into four main categories: Class 1, Class 2, Class 3a, and Class 3b (Table 3.3 presents the safety classification of laser sources). For an optical wireless system that uses a laser source at the transmitter to be considered safe, it must fall within the Class 1 classification of eye safety, which establishes that the system is safe under all conditions of operation and its emitted power does not exceed 0.5 mW (at 880 nm). The AEL of point sources, for example, is about an order of magnitude lower than the one allowed for extended sources. For modulated sources, the IEC 825-I specifies that [20]:

a) In a pulse train, the exposure from any single pulse must not exceed the AEL for a single pulse.

b) For a pulse train of duration T, the average power must not exceed the AEL table values for a single pulse of duration T.

c) For any single pulse in a pulse train, the exposure value shall not exceed the AEL for a single pulse multiplied by the correction value C_5 as follows:

$$AEL_{train} = AEL_{single}C_5 \tag{3.33}$$

Here $C_5 = N^{-0.25}$, with N being the number of pulses in the pulse train during the applicable time base. Restrictions a), b), and c) above apply to wavelengths between 400 nm and 10^6 nm; the selection of the AEL is taken from the most restrictive condition of a) and b). In the case of wireless IR communications, the most restrictive condition is generally that of restriction b) [20].

Holographic diffusers have been proposed as a way to overcome the power restriction limitations imposed on lasers because they can make a point source operate as an extended source. They also offer the benefit of allowing the designer to better control the shape of the emitted beam and therefore to define the area covered by the transmitted beam.

Different sources of illumination for optical wireless communication are preferred, depending on the application. LEDs are generally inexpensive, safe, and cover semi-directive emission angles, which makes them the preferred option for indoor applications. Unfortunately, their electro-optic conversion efficiency is low, their spectral width is large (which makes the filtering of unwanted sources more difficult as the bandpass of the optical filters needs to be larger) and their modulation bandwidth is limited to a few hundred megahertz. LDs, on the other hand, are more expensive and require more complicated drivers. In addition, they are subject to stringent power transmission limitations due to eye safety but offer a number of benefits that compensate for these drawbacks and make them the preferred option for outdoor applications. The modulation bandwidth of LDs, for example, is larger than that of LEDs. In addition, the electro-optical conversion of LDs is higher, and the spectral width of these sources is narrow, which allows the use of narrowband thin-film optical filters to provide high reduction of background illumination noise. Furthermore, the eye safety of LDs can be improved by using diffusing elements such as holographic diffusers, which spread the radiation and create an extended image on the retina.

Chapter 4

Fundamentals of Optical Concentration

4.1 Overview

As explained in Chapters 2 and 3, one of the main drawbacks of infrared (IR) as a medium to transmit information wirelessly is the high attenuation suffered by the optical signal when transmitted through air. This has motivated research (from manufacturers and academic institutions) aimed at compensating this attenuation and maintaining the noise levels introduced by background illumination as low as possible — that is, maintaining a high signal-to-noise (SNR) at the receiver. A simple way to reduce the attenuation consists of decreasing the distance of the link (because attenuation increases with distance) or increasing the optical power emitted by the source. Unfortunately, while for some applications reducing the range of the link is not a problem, for others it may be a critical issue. Some systems need to provide maximum coverage and achieve the maximum possible range; and in these cases, reducing the range of the link to increase the optical power level at the receiver is not viable. Increasing the optical power emitted by the source, on the other hand, is not always possible due to eye safety considerations. As explained in Chapter 3, there is a limit to the amount of power that can be safely emitted by an optical transmitter, and the specific power limit for a given emitter depends on parameters such as the type and operation wavelength of the source and the exposure time.

Another way to compensate for the high attenuation suffered by the optical signal when transmitted through air is by using a large-area photodetector (because a larger collection area allows more energy to be detected by the receiver). Unfortunately, the capacitance of these devices is directly proportional to their area, and the use of large-area photodetectors reduces the bandwidth of the system. An alternative way to increase the collection area of the receiver without reducing its bandwidth is by using an optical concentrator. This optical element, which can be of the imaging or the non-imaging type, collects the energy that impinges on its entrance aperture and redirects it to the photodetector, which is placed at the smallest aperture of the concentrator (in the non-imaging case) or at the focal point of the lens (in the case of an imaging one). The gain obtained is proportional to the ratio of entrance-to-exit areas. This means that for a given system, the sensitivity of the receiver can be increased by optical means without having to replace the photodetector. This also means that the data rate of the systems can be increased (if the limiting factor is the speed of the photodetector) without sacrificing the sensitivity of the receiver. This can be done using a smaller photodetector combined with an optical concentrator that has an entrance aperture equivalent to the collection area of the original bare detector. This allows the SNR to be maintained (or even improved, depending on the gain of the concentrator) while increasing the bandwidth of the system.

This chapter, which is based on some of the principles of non-imaging concentrator design presented by Welford and Winston [84] in their work on *High Collection Nonimaging Optics*, introduces concepts that are fundamental to the design of the optical front end of a wireless IR receiver. It includes a description of geometrical optics and ray tracing, Fermat's principle (and its relevance to non-imaging optics design) and the Lagrange invariant, which is used to define the theoretical maximum limit of concentration. It also explains the edge ray principle and the concentration ratio (here, different versions of the concentration ratio and gain are introduced). These concepts and tools are used in the following chapters to explain the design of thin-film optical filters and of some of the most common optical concentrators currently employed in wireless IR communication systems.

4.2 Geometrical Optics and Ray Tracing

A variety of concepts and principles are used in the design of the optical front end of a wireless IR receiver. One of these concepts is *geometrical optics* (GO). Here, a ray is used to represent the propagation path of an electromagnetic wave. GO is widely used to analyze the refraction and reflection of light as it interacts with different media (reflecting surfaces, concentrators, and filters) during its trajectory from the transmitter to the receiver.

Another useful tool widely employed in the design of optical filters and concentrators is *ray tracing*. This technique — closely related to GO and based on Snell's laws of refraction and reflection — is used to model the path created by an

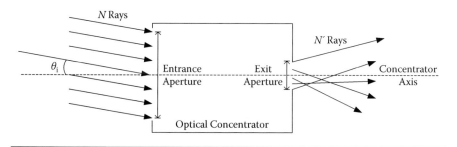

Figure 4.1 Total power and power density in an optical concentrator (represented through geometrical optics). (*Source:* From [84]).

electromagnetic (EM) wave (in the visible or near-IR part of the spectrum) as it interacts with different materials; and, thanks to the availability of computers to calculate ray-tracing algorithms, it can be used to trace a very large number of rays within an optical system and in this way simulate its behavior. The larger the number of rays traced, the more accurate the analysis of the paths within the optical system.

When designing a non-imaging optical concentrator, it is necessary to represent the total power and the power density of the energy impinging on it. This can be done in geometrical optics through the number and density of rays intersecting it [84]. Thus, the total power reaching the entrance of a concentrator can be represented by the number of ray intersections with it; and the power density can be represented by the density of rays. This is illustrated in Figure 4.1, where the incident energy is represented by N rays distributed uniformly over the large entrance aperture of the concentrator. These rays form an angle θ_i (called the angle of incidence) with the axis of symmetry of the optical element. The energy at the exit (the smallest aperture) of the concentrator is represented by N' rays, which emerge with a variety of angles (whose values do not exceed that of *the maximum output angle*). In this example, the power transmission at a given angle of incidence is given by N'/N; and the transmission losses in the concentrator are represented by the difference between N and N'.

More detailed information about geometrical optics, ray tracing, Snell's laws, and the definition of the index of refraction can be found in [85].

4.2.1 Snell's Laws

Snell's laws of refraction and reflection present a relationship between the angles of incidence, refraction, and reflection of an EM wave impinging on the boundary of two media with different indices of refraction. These indices indicate the speed of light in each particular medium with respect to the speed of light in vacuum [85]. The laws are based on the boundary condition, which defines a wave to be continuous across a boundary (which requires the phase of the wave to be constant on any given plane).

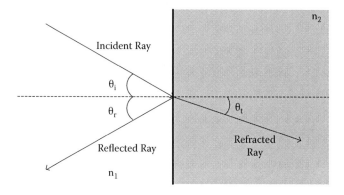

Figure 4.2   Illustration of Snell's laws of refraction and reflection.

If the angles of incidence, refraction, and reflection are represented by θ_i, θ_t, and θ_r, respectively, and the incident wave intersects a boundary between two media (with indices of refraction n_1 and n_2, respectively, as illustrated in Figure 4.2, then Snell's law of reflection can be expressed as [86]:

$$\theta_i = \theta_r \tag{4.1}$$

and Snell's law of refraction can be represented as [86, 87]:

$$n_1 \sin \theta_i = n_2 \sin \theta_t \tag{4.2}$$

These expressions are generally used in vector form (when employed for the design of the optical elements of a wireless IR receiver) due to the fact that it is easier to implement ray-tracing algorithms in vector-based design software (where each ray is represented by a vector) than in a scalar-based one.

To use Snell's law of reflection in vector form, it is first necessary to find the point of incidence of the incoming ray on the reflecting surface. Once this point has been obtained, a unit vector normal to the surface (here called the *normal vector*) at the point of interest, and a unit vector incident into that point (denominated *incident ray vector*) can be defined as shown in Figure 4.3a. Here, *i* is the unit vector along the incident ray, and *n* is the unit vector normal to the surface. The vector *r* (the unit vector along the reflected ray) can then be calculated using the vector formulation for the law of reflection, which is defined as [84]:

$$r = i - 2(n \bullet i)n \tag{4.3}$$

The same algorithm can be applied to all the other rays used in the ray-tracing computation that undergo a reflection. The calculation of the refracted rays

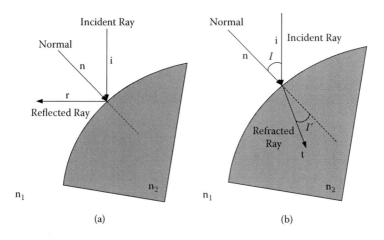

Figure 4.3   Vector formulation of Snell's laws: (a) law of reflection, and (b) law of refraction.

is performed in a similar way. First, for a given ray, the point of incidence at the boundary of the refracting media must be found, and a unit vector corresponding to the incident ray must be traced, as shown in Figure 4.3b. Here, n_1 represents the index of refraction of the input media, n_2 represents the index of refraction of the concentrator (the media of the refracted ray), i is the unit vector along the incident ray, and t is the unit vector along the refracted ray.

The vectors of the rays are analyzed in a numerical computation (both for the law of refraction and the law of reflection) through their vector components. The vector t is calculated according to [84]:

$$n_2 \, t \times n = n_1 \, i \times n \tag{4.4}$$

This expression assumes that the incident and the refracted rays in the figure are coplanar.

It can be seen that this expression is equivalent to the scalar form of Snell's law of refraction, where $t \times n = \sin\theta_t$ and $i \times n = \sin\theta_t$. A more useful expression for ray tracing can be obtained from Equation (4.4) by multiplying it (vectorially) by n [84]. This gives:

$$n_2 \, t = n_1 \, i + (n_2 \, t \bullet n - n_1 \, i \bullet n)n_1 \tag{4.5}$$

Here it can be seen that:

$$n_2 \, t \bullet n - n_1 \, i \bullet n = n_2 \cos\theta_t - n_1 \cos\theta_i \tag{4.6}$$

and θ_t can be calculated as:

$$\theta_t = \sin^{-1}\left(\frac{n_1 \sin \theta_i}{n_2}\right) \tag{4.7}$$

4.3 Optical Path Length and Fermat's Principle

Two other important concepts used in the design of the optical front-end elements of a wireless IR receiver are the *optical path length* and *Fermat's principle*. The former refers to the product of the geometrical length of the path followed by light when traveling through a system and the index of refraction of the medium through which light propagates. The latter refers (roughly) to the minimum time required by light to travel from one point to another [85].

It is well known that in a medium of refractive index n, the speed of light is c/n (where c is the speed of light in vacuum). Therefore, due to the fact that $v = d/t$ and in this case $t = nd/c$, the time it takes light to travel a distance d in a medium of refractive index n is proportional to nd, better known as the optical path length of d [84]. A more accurate definition of Fermat's principle can be stated as follows: from a number of possible paths between two points, the one followed by a beam of light is the one traversed through the one with the minimum optical path length [85]).

The *geometrical wavefront* is another important concept used in the design of non-imaging optical concentrators. This is illustrated in Figure 4.4, where an optical source is shown illuminating a sequence of optical elements, and different rays with the same optical path length (from the point where the source is situated — the origin o) along possible paths through the system are marked by the dotted lines. The path of all the rays can by calculated using Snell's laws of refraction and reflection, and the points with the same optical path length can be obtained by making the sum of the optical path length from the origin in each medium the same [84].

$$\Sigma \, nd = \text{constant} \tag{4.8}$$

The segmented lines shown in Figure 4.4 indicate places with the same optical path length from the origin (p_1, p_2, p_3). When these points are joined, they form a surface of constant phase called the *geometrical wavefront*, or simply *wavefront*.

4.4 The Étendue or Lagrange Invariant

One of the most important concepts in the design of optical concentrators is the *Lagrange invariant*, also known as *étendue* or *acceptance*. This quantity is a measure

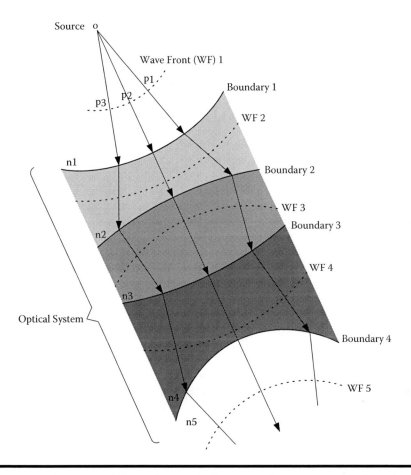

Figure 4.4 **Rays traced through an optical system where** o **corresponds to the source (origin), and points with the same optical path length are represented by** p**. The dotted lines illustrate optical wavefronts. (*Source:* Adapted from [84, 85].)**

of the power accepted by an optical system, and it refers to the product of the angular extent and the diameter of a beam of light. Through this, it is possible to define the concentration ratio C of an optical concentrator.

When applied to an optical source, the étendue is defined as the product of the source's area and the solid angle subtended by the optical system as seen from the source of light; and when applied to an optical concentrator, it is defined as the product of the angle subtended by an object at infinity and the diameter of a lens. This is illustrated in Figure 4.5, where a set of rays coming from an object of a specific size at a great distance is shown impinging on a thin converging lens (a lens whose thickness can be considered negligible). The lens brings each one of the rays from the object to a separate focal point forming an image of size $2f\theta$, where f is the focal length, 2θ is the angular extent and $2r$ is the diameter of the optical

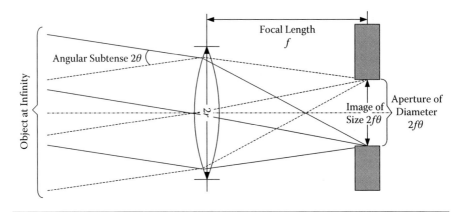

Figure 4.5 Illustration of an optical system with étendue = $\theta^2 r^2$.
(*Source:* Adapted from [84].)

system. This leads to the product shown in Equation (4.9) — the étendue in a two-dimensional (2D) system-, where the factor 4 has been removed [84].

$$étendue_{2D} = \theta \, r \tag{4.9}$$

If the losses inherent to the material of the optical system are ignored and the beam of light is not obstructed in its trajectory, this quantity does not change — hence its name. In a three-dimensional system the étendue is proportional to the square of the value of the two-dimensional version — that is, to $\theta^2 r^2$. If a flux of radiation B watts per square meter per steradian solid angle impinges on the entrance of a three-dimensional lens (such as the one shown in Figure 4.5), and an aperture of diameter $2f\theta$ is placed at its focus, the system then accepts a total flux $B\pi^2\theta^2 r^2$ W (in this case the system only accepts rays inside the diameter $2r$ and within the angular range $\pm\theta$), which confirms that the étendue is indicative of the power flow that can pass through the system (the power accepted by it). When the refractive index of the lens has a value different to unity, the three-dimensional (3D) version of the étendue is written as [84]:

$$étendue_{3D} = n^2 r^2 \theta^2 \tag{4.10}$$

As explained above, the étendue is also important because, through it, the concentration ratio limit of an optical system can be obtained (in its general form) for rays at finite angles with respect to the system's axis. This is illustrated by the next example, as described by Welford and Winston [84]. If an optical system of any shape and conformed by any number of elements has a homogeneous medium of refractive index n at the entrance, and another medium (also homogeneous) with refractive index n' at the output; a ray can be traced from a point A at the

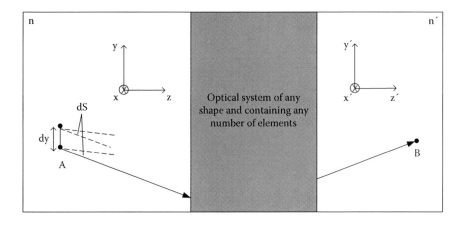

Figure 4.6 The generalized étendue for an optical system of any shape. (*Source:* Adapted from [84].)

entrance to a point B at the exit to consider the effect of small displacements of A and of small changes in direction of the ray segment through A on the emerging ray as shown in Figure 4.6. This is done by placing Cartesian coordinate systems at arbitrary positions at the input and the output media, which helps to define both the beam's angular extent and its cross-section. The segment corresponding to the input ray is defined by the direction cosines of the ray (R,S,T) and by the coordinates of point $A(x,y,z)$; and the segment corresponding to the output ray is specified by the direction cosines of the output ray (R',S',T') and by the coordinates of $B(x',y',z')$. This way, the effect of the emergent ray due to differential increments in the position of point A can be determined by expressing differential displacements of the point with respect to each of the axes (the displacement of each coordinate, given by dx for the x coordinate and by dy for the y coordinate). The small variations in the direction of the ray are represented in a similar way by the increments dR and dS corresponding to the direction cosines for the x and y axis. As a result of the increments dx, dy, dR, and dS, analogous increments to the output ray's direction and position (dx', dy', dR', dS') occur. Figure 4.6 shows the étendue in the y axis. The area of the beam generated by the variations at the system's entrance is given by $dxdy$, and the angular extent by $dRdS$. The invariant in this case is defined as $n^2dxdydRdS$, where [84]:

$$n^2dx'dy'dR'dS' = n^2dxdydRdS \qquad (4.11)$$

From this equation it can be observed that small variations in the beam's rays generate equivalent small variations at the other end of the optical system or structure. The generalized étendue can also be written in terms of the optical direction cosines (where $e = nR$ and $f = nS$) as follows [84]:

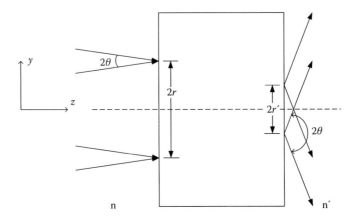

Figure 4.7 Optical system illustrating the theoretical maximum concentration in two dimensions. (*Source:* From [84].)

$$dx'dy'de'df' = dx\, dy\, de\, df \tag{4.12}$$

Figure 4.7 shows a two-dimensional concentrator with acceptance angle θ (half-angle or semi-angle) and output angle θ'. As the generalized étendue can be used to calculate the theoretical maximum concentration ratio, Figure 4.7 can be used to illustrate how the étendue can be used to calculate this ratio in a two-dimensional concentrator.

It is known from Equation (4.11) that, in a two-dimensional system, for any ray that crosses the system:

$$n\, dy\, dS = n'dy'dS' \tag{4.13}$$

which, integrating over *y* and *S,* gives [84]:

$$4\, n\, r\, \sin\theta = 4\, n'r'\sin\theta' \tag{4.14}$$

where θ is the input semi-angle. The concentration ratio (for a two-dimensional concentrator) can be obtained from this expression (Equation 4.14) as follows:

$$\frac{r}{r'} = \frac{n' \sin\theta'}{n \sin\theta} \tag{4.15}$$

Here, θ' is the widest angle possible for the rays emerging from the exit of the optical system and *r*' is an output aperture of such size as to allow any ray that reaches it to exit the structure. It can be seen from Figure 4.7 that the maximum output semi-angle cannot exceed 90°, which simplifies Equation (4.15) to:

$$C_{\max_{2D}} = \frac{n'}{n \sin \theta} \tag{4.16}$$

which is the two-dimensional version of the formula employed by a number of researchers to calculate the maximum concentration ratio of optical concentrators [20]. The theoretical maximum concentration expression for a three-dimensional axisymmetric structure (which can be obtained in a similar way to the two-dimensional one) is given by [20]:

$$C_{\max_{3D}} = \left(\frac{r}{r'}\right)^2 = \left(\frac{n'}{n \sin \theta}\right)^2 \tag{4.17}$$

There are different versions of concentration ratio (cf. Section 4.6.1) defined according to the dimensions, the geometry, and the losses of optical concentrators. The expressions in Equations (4.16) and (4.17) are known as the theoretical limits of optical concentration; and because they are only based on the index of refraction of the media and on the acceptance angle, they are only an indication of the maximum concentration possible. In reality, due to losses within the optical concentrator (incurred by unwanted reflections, absorption, etc.), these values may not be achieved. In some systems, the diameter of the exit aperture is such that some of the rays impinging on the entrance aperture within the concentrator's acceptance angle cannot cross it. This has motivated the creation of other definitions of concentration, which are discussed in more detail in Section 4.6.1.

4.5 The Edge Ray Principle

The edge ray principle is a concept widely used in the design of some types on nonimaging concentrators (such as the dielectric totally internally reflecting concentrator [DTIRC] and the compound parabolic concentrator [CPC]). This principle states that all the rays impinging on the entrance aperture of a concentrator at the extreme angle emerge from the exit border of its aperture. If this is the case, the exit rays are also extreme and form a maximum output angle with respect to the concentrator's axis. In addition, all the rays within the acceptance angle of the concentrator (that is, all the rays impinging on the entrance aperture at narrower angles than those of the extreme angle) exit the concentrator at narrower angles than those of the extreme rays at the exit. That is, concentrators based on the edge ray principle present the next property: rays that enter at the extreme angle θ_a leave from some point of the rim of the exit aperture at an extreme output angle θ_o. This is in clear contrast to imaging concentrators, which require that all the rays impinging on the entrance aperture at the extreme angle create clear images at the border of the exit aperture [84].

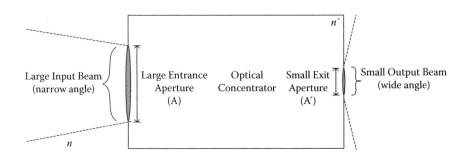

Figure 4.8 Side view of an optical concentrator represented as a box. Here, the entrance aperture corresponds to area *A*, and the exit aperture corresponds to area *A'*.

Optical concentrators whose designs are based on the edge ray principle present very high concentration ratios. For this reason (and because they can be designed for any FOV), they have been favored for wireless IR communication systems over other types of concentrators.

4.6 Concentration Ratio

The concentration ratio is one of the most important parameters defining an optical concentrator. It is generally defined as the ratio of the input to the output areas of the concentrator. For example, if a non-imaging concentrator is modeled as an optical element with flat input and output exit apertures of areas A_i and A_o, respectively, as illustrated in Figure 4.8, then the concentration ratio (usually denoted by C) can be defined as the ratio of input to output diameters (in the case of a two-dimensional concentrator) or as the ratio of the input area divided by the output area (in the case of a three-dimensional concentrator) [88]:

$$C = \frac{A_i}{A_o} \tag{4.18}$$

In this model, the size of the exit aperture is large enough to permit all transmitted rays to exit from it. The input and output surfaces may face in any direction [84]. In the three-dimensional version of the concentrator, the compression of the input beam is assumed to occur in two dimensions (corresponding to the plane created by the intersection of the beam by the flat entrance aperture of the concentrator), while in two-dimensional concentrators, the compression occurs in only one of the dimensions transverse to the direction of propagation of the energy.

If the radiation impinging on the entrance of the concentrator comes from a circular source at infinity subtending a semi-angle θ_i, and the value of the index

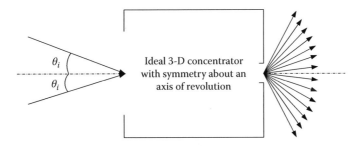

Figure 4.9 Schematic diagram of an optical concentrator with FOV = ± θᵢ° and exit aperture diameter equivalent to sin θᵢ. Here, the rays at the exit can reach angles of up to 90°. (*Source:* From [84].)

of refraction of the input and output media is one, then the theoretical maximum concentration in a three-dimensional (3D) concentrator is defined as [84]:

$$C_{m3D} = \frac{1}{\sin^2 \theta_i} \qquad (4.19)$$

Here, rays at the output aperture can exit with any angle from 0 to 90° with respect to the axis of the concentrator, as shown in Figure 4.9 (where a concentrator with symmetry about an axis of revolution and having an exit aperture diameter *sin* θᵢ times the diameter of the entrance aperture is illustrated). If the indices of refraction of the input and output media are different, Equation 4.19 can be defined in a more general form as:

$$C_{m3D} = \left(\frac{n'}{n \sin \theta_i} \right)^2 \qquad (4.20)$$

where *n* and *n'* are the indices of refraction of the input and output media, respectively. As can be seen, this corresponds to Equation (4.17). The maximum concentration value for a two-dimensional concentrator (when the indices of the input and output media are the same) is given by [84]:

$$C_{m2D} = \frac{1}{\sin \theta_i} \qquad (4.21)$$

and if the value of the indices of refraction are different, this equation becomes (cf. Equation (4.16)):

$$C_{m2D} = \frac{n'}{n \sin \theta_i} \qquad (4.22)$$

If a number of conditions are fulfilled (slight changes in the shape of a concentrator —which occur in the fabrication process — are avoided, all refracting surfaces are perfectly coated with an antireflection material and all reflecting surfaces satisfy the condition of total internal reflection (TIR) without any losses), it is possible to design concentrators that approach the theoretical maximum limit of concentration. In the case of two-dimensional concentrators, it is possible to design for the theoretical maximum concentration; and in the case of three-dimensional concentrators, it is possible to approach the theoretical maximum concentration value.

4.6.1 Different Versions of Concentration Ratio

As discussed in Sections 4.5 and 4.6, there are different versions of concentration, defined according to the geometry, type and refractive index of the concentrator. These and other definitions of concentration and gain can be found in a number of publications related to imaging and non-imaging optics. For this reason, it is important to recognize them and differentiate them to avoid confusion. One of the concentrations already presented in this text is the *theoretical maximum concentration ratio*, which is defined by Equations (4.20) and (4.22). This limit depends exclusively on the refractive index values of the input and output media, and on the acceptance angle of the optical element.

Another version of concentration is the one defined by the ratio of entrance to exit apertures. These apertures include the width in two-dimensional structures and the area in three-dimensional structures. The (d/d') and (A/A') ratios are generally called the *geometrical concentration ratio* and they depend on geometrical considerations only [88]. The value is independent of the theoretical maximum concentration ratio.

Another important definition of concentration is the one related to the proportion of incident rays (within the acceptance angle) that emerge from the exit aperture of an optical concentrator. Tracing rays through the system helps to determine this. The name assigned to this type of concentration is *optical concentration ratio*; and if not just the amount of rays that exit the concentrator (quantified as a percentage) is taken into account, but also the losses (due to impurities, manufacturing errors, attenuation due scattering, absorption and unwanted reflections), the concentration is called the *optical concentration ratio with allowance for losses* [84].

The *experimental optical (or opto-electronic) gain* is another version of concentration. It is obtained by dividing the electrical signal generated by a photodetector with the concentrator attached to its active area (or placed in front of it) by that of the photodetector without the concentrator (that is, a photodetector whose active area is directly exposed to the incoming radiation) [89]. The value of the opto-electronic gain (similar to the case of the optical concentration ratio) is always lower than the value of the theoretical maximum concentration ratio.

4.7 Summary and Conclusions

This chapter introduced some of the most important definitions and formulas used in the design of the optical front end of a wireless IR communication receiver (and sometimes of the transmitter too) and in their evaluation. Geometrical optics, Snell's laws, ray tracing, and Fermat's principle are fundamental concepts for the design and understanding of optical filters and concentrators (such as the hemispherical, the CPC, and the DTIRC).

GO is a helpful tool to represent a wave as a ray that denotes its direction of propagation. It forms the base of ray tracing, where the path created by light as it interacts with the different surfaces (of different materials, shapes, and indices of refraction) of an optical system can be modeled by a set of rays. These rays represent the power of the radiation being transmitted, and their interaction with different boundaries obeys Snell's laws of refraction and reflection. Ray tracing is a particularly useful technique when designing imaging and non-imaging concentrators that are combined with thin-film optical filters, whose response characteristics depend on the angle of incidence of the incoming energy (cf. Chapters 5 and 6). It allows the designer to verify the angles of the rays at the exit of the concentrator; and, in this way, calculate the wavelength shift suffered by the filter. In addition, ray tracing is a useful tool to calculate the *gain with allowance for losses* of an optical concentrator.

Snell's laws of refraction and reflection are also fundamental in the creation of optical systems as they allow the designer to calculate the angles of the refracted and reflected rays when the transmitted energy reaches the boundaries of media with different indices of refraction and shapes. Whenever light impinges on an interface, it can be reflected from it or transmitted (refracted) into the next medium. Snell's law of reflection states that the angle formed by the reflected ray with a line drawn from the point of incidence on the interface and normal to it (the normal) is equal to the angle formed by the incident ray with the normal.

If the medium containing the incident ray has an index of refraction n_1, and the medium containing the transmitted ray has an index of refraction n_2, the sine of the angle of the refracted ray with respect to the normal can be calculated as the ratio of the product of the index of refraction n_1 and the sine of the angle of the incident ray (with the normal) divided by the index of refraction of the transmitted ray. It must be noted that in geometrical optics, the angles of the incident, reflected, and refracted rays are always measured with respect to the normal to the boundary.

Two other fundamental concepts widely used in the design of optical concentrators are Fermat's principle and the optical path length. The Fermat principle states that the time required by light to travel from one point to another (in an optical system) is the minimum when compared to the time required from neighboring paths (a more accurate definition is given in Section 4.3). This is particularly relevant in GO analysis and when producing ray-tracing algorithms. The optical path length refers to the geometrical length of the path followed by light when traveling through an optical system.

The étendue or Lagrange invariant presented in Section 4.4 is a measure of the power accepted by an optical system, and it is a useful tool to calculate the theoretical maximum limit of concentration achievable in an imaging or a non-imaging collecting system. The étendue in a two-dimensional system is calculated as the product of the system aperture and the acceptance half-angle. In a three-dimensional system, the étendue is proportional to the product of the square of the acceptance half-angle and the square of the entrance aperture diameter of the system.

The edge ray principle is another useful tool for the design of CPCs and DTIRCs. It states that all the rays impinging on the entrance aperture of a non-imaging concentrator at the extreme angle emerge from the exit's border at an extreme output angle. All the rays impinging on the concentrator with angle values smaller than that of the critical angle reach the exit aperture and exit the concentrator within the exit critical angle. In an imaging system, on the other hand, it is also necessary that the incident rays at the extreme angles form a clear image at the border of the exit aperture to ensure that all the rays incident on its entrance aperture and within its acceptance angle reach the exit aperture and exit from it.

Section 4.6 presented the different versions of concentration and gain. They include the theoretical maximum concentration ratio (which depends exclusively on the acceptance angle and on the indices of refraction of the input and the output media), the geometrical concentration ratio (based on the ratio of entrance to exit apertures), the optical concentration ratio (related to the proportion of the incident rays that emerge from the exit aperture), the optical concentration with allowance for losses (based on the ratio of entrance to exit apertures taking into account attenuation and manufacturing errors), and the opto-electronic gain (which is the ratio of the electrical signals generated by a photodetector with and without a concentrator).

Chapter 5

Optical Concentrators

5.1 Overview of Optical Concentrators

Despite the fact that the majority of the techniques and technologies employed by optical fiber communication systems can be used for wireless infrared (IR) communications, there are parts of the optical wireless system that present different requirements and challenges. Two of the main differences between wired and wireless optical communication systems are (1) the characteristics of the emitted beam and (2) the main source of shot noise in the detector. In an optical wireless link, for example, the beam emitted by the transmitter may be spread over a broad cone to cover a wide area. In these situations, the optical power density of the information signal at the detector is low, and the receiver requires a large effective collection area to capture as much energy as possible. Optical fiber systems, on the other hand, can make use of small-area photodetectors due to the fact that the emitted beam propagates through an optical fiber, which presents minimum losses. In addition, while in optical fiber systems the main source of shot noise is generated by the dark current of the photodetector or by the signal itself, in optical wireless systems the dominant source of noise at the receiver is created by background illumination.

One of the solutions employed in wireless IR receivers to compensate for the high attenuation suffered by the IR signal when transmitted through air involves the use of imaging or non-imaging optical concentrators. By carefully designing and using an optical concentrator in conjunction with a photodetector, it is possible to increase the effective collection area of the receiver without necessarily using a large-area photodetector. This optimizes the system power budget, improves eye safety (because it makes possible the use of a lower emitted power at the transmitter for a given distance), or allows the distance between the transmitter and the receiver to be increased. Furthermore, the use of an optical concentrator decreases the receiver's capacitance by allowing the use of smaller photodetectors, which increases the receiver's bandwidth).

Another element that can be combined with an optical concentrator and a photodetector to further improve the performance of a wireless IR receiver is an optical filter. Thin-film bandpass or longpass filters improve the signal-to-noise ratio (SNR) at the receiver by rejecting the out-of-band radiation introduced in the detector by incandescent and fluorescent lamps or by solar illumination. As discussed in Chapter 3, these sources of background illumination are present in the majority of the environments where optical wireless systems operate. In addition, antireflection (AR) coatings and index-matching gels can be used at the concentrator surface, the filter, and the photodetector (and at boundaries of media with different indices of refraction) to improve even further the overall performance of the receiver's optical front end by minimizing or eliminating unwanted reflections.

This chapter examines the optical front-end requirement of wireless IR receivers for indoor and outdoor applications. Different options of optical filters are described (including bandpass thin-film filters and longpass filter-Si detector structures), and a variety of imaging and non-imaging concentrators are presented. The main characteristics of Fresnel lenses are introduced and the principal features of hemispherical lenses, CPCs, and DTIRCs are explained and compared. In addition, the way in which different concentrators can be used in combination with a variety of filters is explained. Finally, a summary and conclusions are presented in Section 5.9.

5.2 Wireless IR Receiver Requirements

Chapters 1 and 3 introduced some of the drawbacks and limitations of wireless IR communications systems. In some configurations, IR links can suffer from blocking (from persons or objects), which may result in the attenuation of the signal at the receiver or in the disruption of the communication link. In addition, some configurations suffer from a high path loss because the transmitted energy is spread over a large area. Moreover, optical wireless receivers operating indoors and outdoors are affected by the noise introduced by different sources of background illumination.

A great number of applications use intensity modulation/direct detection (IM/DD) as the transmission reception technique. In a DD receiver, the electrical SNR is proportional to the square of the received optical power [21]. Therefore, it is important that the detector receives as much radiation as possible. Unfortunately, as explained in Chapter 4, the transmitter power cannot be increased beyond specific levels due to eye safety and power consumption considerations. One way to overcome this problem consists of using receivers with large effective collection areas and narrow optical passbands. This reduces the problems of path loss and background illumination noise. Unfortunately, large-area photodetectors have a large capacitance related to them, which leads to a reduction in receiver bandwidth and affects the receiver thermal noise indirectly. An alternative to using large-area photodetectors is to use optical concentrators, which improve the collection efficiency of the receiver by transforming a set of rays (corresponding to the incoming near-IR radiation) that impinge over a large area into a concentrated set of rays that emerge from a smaller area (the exit aperture of the concentrator in the non-imaging case) [20]. This smaller area is generally in contact with the active area of a photodetector and in many cases of the same size, which implies that smaller photodetectors can be used — thus decreasing the capacitance and the cost of the receiver and improving its sensitivity.

The noise introduced by the background illumination can be reduced by incorporating a thin-film optical filter to the optical front end. This technique is particularly useful when the transmitter presents a narrow optical linewidth. It has been demonstrated in the past that the use of omni-directional and directed concentrators, used in combination with optical filters, effectively rejects unwanted ambient radiation and increases the effective area of the photodetector [47, 90]. Examples of concentrators used in the optical front end of the receiver of commercial systems include hemispherical concentrators and Fresnel lenses.

5.3 Optical Filters

High-speed wireless IR communication systems require transmitters whose emitted energy presents a narrow linewidth. This allows the system's receiver to incorporate high-performance, narrowband optical filters to reject out-of-band radiation, which as discussed above contributes to the shot noise in the detector.

Wireless IR receivers can use either bandpass or longpass filters to allow the passage of light at the wavelength used by the transmitter only. Thin-film optical filters, such as the one illustrated in Figure 5.1, are usually constructed from multiple layers of thin dielectric film, therefore relying on the phenomenon of optical interference [47, 85]. The Fabry-Pérot resonator is the basic element of the dielectric

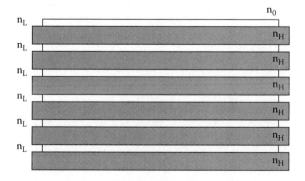

Figure 5.1   Schematic diagram of a thin-film optical filter.
(**Source:** Adapted from [20, 47, 85].)

thin-film optical filter. It behaves like a comb filter that, when combined with other Fabry-Pérot resonator structures, creates a very narrow optical passband, allowing the passage of light within narrow sets of wavelengths. Many wireless IR systems require passbands of under 100 nm, for which typically two or three of these Fabry-Pérot resonators must be coupled together to produce appropriate noise rejection [20]. Fortunately, thin-film optical filters can achieve passbands as narrow as 1 nm [21].

The response of thin-film optical filters reveals a strong dependence on the angle of incidence of the incoming energy, which means that this must be taken into account during the design of the optical front end of a wireless IR receiver when combining them with optical concentrators and photodetectors. It has been shown that the transmission characteristics of a bandpass thin-film optical filter can be approximated to an mth-order Butterworth response of the form [47]:

$$T_f(\lambda_0, \theta_i) \approx \cfrac{T_{f0}}{1 + \left(\cfrac{\lambda_0 - \lambda_c(\theta_i)}{\Delta\lambda/2}\right)^{2m}} \tag{5.1}$$

where θ_i is the angle of incidence, λ_0 is the center wavelength at normal incidence, T_{f0} is the peak transmission, and $\Delta\lambda$ is the optical bandwidth of the filter. This approximation fails to take into account the degradation in the filter profile and the passband ripple due to polarization effects, but shows clearly the strong dependence of the center wavelength on the angle of incidence. The variation of the center wavelength of the filter (λ_c) with the angle of incidence (θ_i) can be approximated by [91]:

$$\lambda c(\theta_i) = \lambda_o \sqrt{1 - \left(\frac{n_1}{n_s}\right)^2 \sin^2\theta_i} \tag{5.2}$$

Here, λ_o is the center wavelength at $\theta_i = 0°$, n_s is the effective refractive index of the spacer, and n_1 is the refractive index of the input layer. Figure 5.2 shows a typical spectral transmission of an optical bandpass filter using a Butterworth approximation, and Figure 5.3 shows the angular dependence of the center wavelength on the angle of incidence. Thin-film bandpass filters achieve a high rejection of ambient light because, as explained above, they can have very narrow bandwidths. It can be observed in Figures 5.2 and 5.3 that, as the angular value of the incident rays increases, the central wavelength of the filter shifts to smaller values. For this reason, this kind of filter is generally used for reduced FOV detectors. To maximize the SNR of a wireless IR receiver incorporating a thin-film optical filter, the optical spectrum of the transmitter must lie within the filter bandwidth [21]. Therefore, applications requiring maximum noise rejection and incorporating narrow passband optical filters rely on the use of laser diodes (LDs) at the transmitter rather than on light-emitting diodes (LEDs).

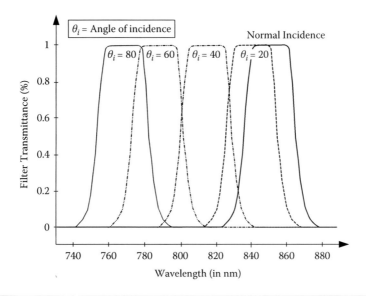

Figure 5.2 **Angular transmission characteristics of a bandpass thin-film filter using a Butterworth approximation (obtained from Equation 5.1).**

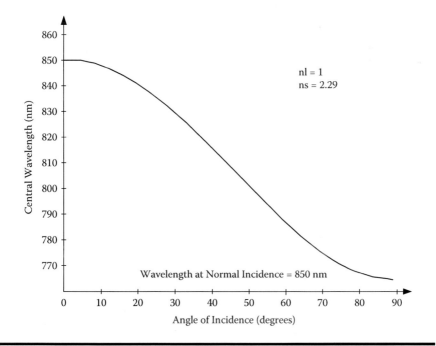

Figure 5.3 **Variation of central wavelength with angle of incidence of a thin-film optical filter (obtained from Equation (5.2)).**

An alternative way to reject background illumination noise — in situations where the use of thin-film optical filters is not favored due to economic reasons (the better the characteristics of the thin-film optical filter, the more dielectric layers are required, which increases the cost) or because the dependence of the filter characteristics on the angle of incidence needs to be avoided — consists of combining the natural responsivity of a silicon (Si) photodetector with the transmittance of materials that present a longpass response. This combination behaves as a bandpass filter [9]. Longpass filters allow the passage of light beyond a specific wavelength (which acts as one of the edges of the filter); and when used in combination with a Si photodiode (from which the roll-off of their responsivity can be used as the other edge of the filter), the combination behaves as an optical bandpass filter. This is illustrated in Figure 3.3, where the responsivity curve of a typical Si photodiode is shown superimposed on the response of a material exhibiting longpass characteristics. Here it can be observed that the photodiode responds only to wavelengths below 1100 nm, while the filter passes light at wavelengths beyond (about) 875 nm. This creates an overall optical bandpass of approximately 200 nm. The diagram of a filter-detector structure incorporating AR coatings is shown in Figure 5.4.

Longpass filters, which are generally fabricated from colored plastic or glass (or from a GaAs substrate), are widely used in commercial systems. One of the most attractive features of these filters is the independence of their transmission characteristics from the angle of incidence. This angular independence plus their simplicity and low cost have made them one of the favorite filter options for a number of applications.

Compared to colored plastic or glass where the edge of the filter is difficult to control, GaAs longpass filters present the additional advantage of giving the designer some control over the position of the edge of the filter, which can be done by varying the levels of doping in the material. This makes it possible to bring the passband of the filter-detector combination to values on the order of 100 nm [20]. An alternative material that can be used with the same purpose and in a similar way is indium phosphide (InP). This material allows a further reduction of the passband of the filter-detector combination to about 80 nm. The fact that the characteristics of materials exhibiting a longpass response are primarily independent of the angle of incidence has the additional advantage of providing a greater flexibility when combined with optical concentrators and photodetectors.

One way to reduce the losses due to unwanted reflections at the entrance of optical filters or optical concentrators consists of using antireflection (AR) coatings, which contribute to a better overall transmission of the optical signal. This can be observed from the values presented in Table 5.1, where the transmittance of AR-coated and uncoated filters is shown for a number of angles of incidence (at a wavelength of 980 nm). As observed, the characteristics and efficiency of the AR-coated material are practically insensitive to small variations in the angle of incidence, becoming noticeable only for angles of incidence beyond 60°. This, plus the fact that AR coatings are easy to fabricate and can be produced at a relatively low cost, makes them an important addition to the optical front end of the receiver.

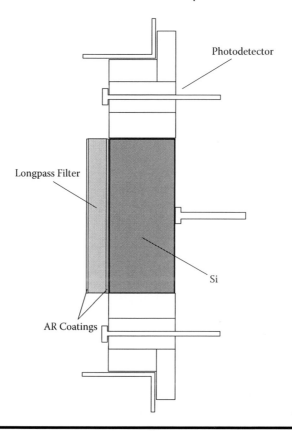

Figure 5.4   Detector-filter structure consisting of an AR coating, a material with long-pass filtering characteristics, an index matching gel, and a Si photodetector.

When designing the optical elements of a wireless IR receiver, it is also important to take into account the degradation of the filter profile due to polarization effects. If the transverse magnetic (TM) and the transverse electric (TE) polarization states of the incident radiation have equal power, the total fraction of power transmitted through the filter can be calculated as[47]:

$$T(\theta_1) = 1 - \frac{|\rho_{TE}|^2 + |\rho_{TM}|^2}{2} \tag{5.3}$$

As can be seen, this equation is a function of the angle of incidence of the incoming energy; where, according to Barry et al. [47, 91], the reflection coefficients ρ_{TE} and ρ_{TM} can be defined as:

$$\rho = \frac{N_1 - \eta_2}{N_1 + \eta_2} \tag{5.4}$$

Table 5.1 Dependence of Transmission Values on Angle of Incidence for GaAs and Antireflection-Coated GaAs (Data obtained from [20])

Angle of Incidence	GaAs (%)	AR-Coated GaAs (%)
Normal incidence	47	99
10°	47	99
20°	47	99
30°	47	99
40°	47	97.5
50°	49	95
60°	52.5	87
70°	54.5	69
80°	47	34
90°	0	0

Here, N_k can be calculated as $n_k / \cos\theta_k$ for the TE state, and as $n_k(\cos\theta_k)$ for the TM state (for $k \in \{2,...,k\}$). n_k can be calculated as [91]:

$$\eta_k = N_k \frac{\eta_{k+1} \cos\beta_k + jN_k \sin\beta_k}{N_k \cos\beta_k + j\eta_{k+1} \sin\beta_k} \qquad k \in \{2,...,k\} \tag{5.5}$$

$$\theta_k = \sin^{-1}\left(\frac{n_{k-1}}{n_k} \sin\theta_{k-1}\right) \tag{5.6}$$

Here, η_k is the effective complex-valued index seen by the IR wave as it enters the medium k, θ_k is the angle formed by light rays when passing from a medium k to a medium $k+1$, and β_k is given by [47]:

$$\beta_k = \frac{2\pi \cos\theta_k n_k d_k}{\lambda} \tag{5.7}$$

The variation of spectral shape with angle of incidence observed in a thin-film optical filter (especially at wide angles) derives from these polarization effects. This

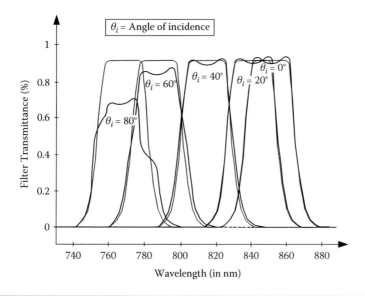

Figure 5.5   Profile degradation in a thin-film filter due to polarization effects. (*Source:* From [47].)

is illustrated in Figure 5.5, where it is shown how, as the angle of incidence of light becomes wider with respect to the normal filter's surface, the degradation gets worse. The rate of change of the spectral response as a function of θ_i is different for the TE and TM polarization modes.

5.4 Optical Concentrators

As discussed in Chapter 4, the objective of using an optical concentrator is to transform a set of rays incident on a large area into a set of rays emerging from a smaller area to improve the collection efficiency of the receiver. This is important due to the fact that the optical power detected by an IR receiver is proportional to the effective collection area of the detector. In theory, large-area photodetectors can be used as a way to increase the effective collection area of the receiver. Unfortunately, as noted above, the large capacitance and noise related to large-area detectors make them impractical for high-speed applications because they reduce the bandwidth of the system. In addition, large-area photodetectors are also expensive in general. For these reasons, wireless IR communication systems favor the use of optical concentrators at the receiver, which improves the collection efficiency by optical means, allowing the use of small and inexpensive low-capacitance photodetectors.

Optical concentrators can be of two different types: (1) imaging and (2) non-imaging. The selection of one over the other depends on the topology on which a specific system is based. Imaging optics, for example, are generally employed in directed-LOS

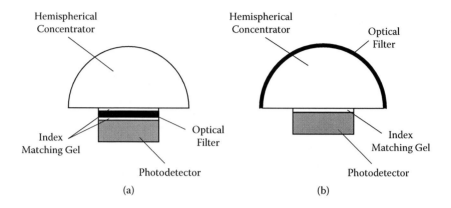

Figure 5.6   Schematic diagram of two hemispherical concentrator structures: (a) incorporating a flat longpass filter, and (b) incorporating a thin-film optical filter coated onto the concentrator's surface. (*Source:* Adapted from [20, 21].)

(outdoor; LOS = line-of-sight) links that provide point-to-point communication. In these situations, the migration of the focal point of the imaging lens is limited because the transmitter and the receiver are generally fixed.* Non-imaging concentrators, on the other hand, are preferred for indoor applications where a larger FOV (field of view) is required to provide higher mobility or flexibility of alignment.

One of the most popular non-imaging concentrators is the hemispherical lens (Figure 5.6). Its simplicity, low cost, and wide FOV make it the preferred option for a number of commercial systems. Hemispherical lenses, for example, are specified as the optical concentrating element of the IrDA (Infrared Data Association) transceivers, which means that they are found in the IR port of a number of mobile phones, laptop computers, and PDAs (personal digital assistants). Hemispherical concentrators with a filter coated onto the surface, like the one shown in Figure 5.6b, achieve a narrow passband and a wide FOV. For this reason, different versions of semi-hemispherical concentrators have been proposed for the diffuse configuration [92, 93]. Unfortunately, they present a reduced concentration, achieving a gain close to n^2 (where n is the index of refraction of the material at the wavelength of interest) only if the hemisphere is large enough compared to the photodetector [21, 47].

Another non-imaging concentrator proposed for wireless IR communications is the compound parabolic concentrator (CPC). This optical element presents a number of advantages when compared to the hemispherical concentrator. CPCs can incorporate flat thin-film optical filters, which are much easier to fabricate than their hemispherical versions. In addition, the central wavelength can be tuned by tilting the filter, which allows a certain degree of error in the manufacturing process. Moreover, receiver structures using CPCs can also be designed to provide

---
* In reality, there may be small variations to the position of the focal point due to the natural oscillation of high buildings and wind.

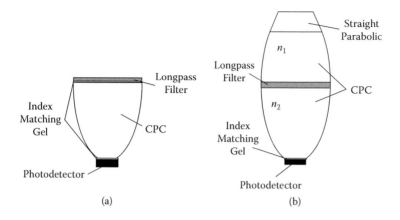

Figure 5.7 **Combined CPC structures: (a) CPC with longpass filtering material, and (b) dielectric and hollow CPCs with straight section at the end.** (*Source:* **Adapted from [21, 90].)**

any FOV between 0 and 90°, which allows them to achieve optical gains greater than n^2 when the FOV is less than 90° [90]. This means that smaller and cheaper photodetectors can be used, which improves the speed of the system and reduces the receiver cost. Two different receiver structures, combining flat, thin-film filters with CPC-based concentrators, are shown in Figure 5.7. Figure 5.7a shows a single CPC, allowing a small acceptance angle; and Figure 5.7b illustrates a combination of dielectric and hollow CPCs having a straight section at the end (acceptance angle less than 90°). The principal drawback of CPC structures is their excessive length, which makes them unsuitable for practical applications.

Another way of achieving a high gain without incurring an excessive length is by using a rotationally symmetric, dielectric totally internally reflecting concentrator (DTIRC). This optical structure achieves concentrations close to the theoretical maximum limit presented in Chapter 4. The operation of DTIRCs is based on a combination of front surface refraction and total internal reflection from the concentrator sidewall [89].

Compared to the hemispherical concentrators, DTIRCs offer higher concentrations and the possibility of using flat optical filters; and compared to CPCs, DTIRCs present two advantages: (1) a smaller size (due to the use of a curved entrance aperture, which decreases the size of the lens according to the value of the front surface arc angle), and (2) a higher concentration (which increases with the value of the index of refraction of the dielectric material). For large front surface arc angles, the size of DTIRCs is significantly smaller than that of CPCs. Furthermore, DTIRCs present the additional advantage of being able to incorporate thin-film optical filters at their exit (between the concentrator and the active area of the photodetector), which reduces the amount of material required for the filters and facilitates the deposition process.

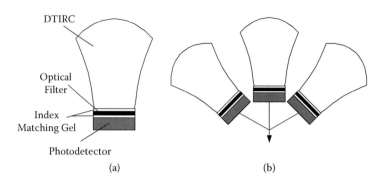

Figure 5.8   DTIRC structures with filters: (a) DTIRC with planar filter, and (b) array of DTIRC-filter structures.

Figure 5.8 shows the cross-sectional view of a variety of structures that combine DTIRCs with thin-film optical filters. Figure 5.8a shows an optical concentrator structure incorporating a flat bandpass filter between the concentrator and the detector; and index matching gel at both sides of the filter. The DTIRC required for this type of structure needs to limit the maximum output angle at its exit to minimize the wavelength shift in the thin-film optical filter. Figure 5.8b shows an array of DTIRC-filter-detector structures that provide, at the same time, a wide FOV and a high gain. This is possible due to the fact that each DTIRC can have a narrow FOV (and therefore a higher gain). The overall wide FOV is provided by the combination of the FOV of the individual concentrators.

5.4.1 *Ideal Concentrators*

Optical concentrators permit a higher receiver bandwidth because smaller low-capacitance detectors can be used, improving the sensitivity of the receiver and decreasing its cost. The effect of reducing the capacitance also reduces the f^2 noise component of the photodetector. According to Kahn et al. [21], the effective collection area of a photodetector (ignoring reflection losses) without filter or concentrator for an angle of incidence θ_i within the range: $0 \leq \theta_i \leq \pi/2$ is:

$$A_{\mathit{eff}}^{\mathit{nc}}(\theta_i) = A\cos\theta_i \qquad (5.8)$$

where A is the active area of the detector. If $\theta_i \geq \pi/2$, the effective collection area of the photodetector is equal to zero. When the concentrator and the filter are included, the effective collection area for an angle of incidence θ_i within the range $0 \leq \theta_i \leq \Phi_c$ (where Φ_c is the concentrator FOV half-angle) becomes [21]:

$$A_{\mathit{eff}}^{c}(\theta_i) = Ag(\theta_i)Ts(\theta_i)\cos\theta_i \qquad (5.9)$$

where $g(\theta_i)$ is the concentrator gain and $Ts(\theta_i)$ is the average transmission of the filter over the wavelength range for the specific angle of incidence. If $\theta_i \geq \Phi_c$, then $A_{\mathit{eff}}^c(\theta_i) = 0$.

As explained in Chapter 4, the geometrical gain of a non-imaging concentrator can be defined as the ratio of the areas of its entrance and exit apertures. This ratio is inversely proportional to the FOV of the concentrator, with narrow FOVs providing large entrance apertures (relative to the exit aperture) and therefore high gains; wide FOVs correspond to low gains.

According to Street et al. [20], the maximum concentration ratio of a non-imaging concentrator placed on top of a large-area Si photodetector is given by:

$$C_{\max} = 0.82 \left(\frac{n}{\sin \theta_a} \right)^2 \tag{5.10}$$

where θ_a is the acceptance angle (half-angle) of the concentrator and n is the index of refraction of the concentrator's material at the wavelength of operation. This expression is valid for $\theta_a \leq 65°$.

5.4.2 Imaging Concentrators and Fresnel Lenses

Directed LOS links such as the point-to-point wireless IR communication systems connecting buildings generally employ imaging or Fresnel lenses as the collimating/concentrating elements at their optical front ends. These lenses were created to minimize the amount of material employed in the fabrication of imaging lenses. This is possible due to the fact that the focal properties of a lens depend on the contour of its refracting surface rather than on the bulk of material that constitutes it. In Fresnel lenses, a set of coaxial concentric rings is used to simulate the shape of the imaging lens. This is illustrated in Figure 5.9, where the aspherical shape of a convex imaging lens has been replaced with a set of concentric rings that produce an equivalent focal point.

Fresnel lenses are named after the French physicist and mathematician Augustin Jean Fresnel, who in 1822 proposed an aspherical lens in which, to minimize aberrations, the center of curvature of each ring receded along the axis of the set of rings according to its distance from the center of curvature. These lenses became particularly popular in lighthouses due their large aperture, the low volume of material required (compared to conventional lenses), and low absorption losses.

Presently, Fresnel lenses are used in a variety of applications. In wireless IR communication systems, for example, they are used either as optical concentrators or as collimators. As concentrators, they can be used to bring IR radiation to a focal point (or area), which corresponds to the position of the active area of a photodetector. The configuration of the Fresnel lens for this sort of application is with the grooved side toward the infinite conjugate. This means that as the angle

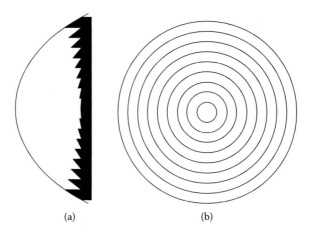

(a)   (b)

Figure 5.9   Construction of a Fresnel lens from an aspherical shape: (a) side view, and (b) front view. (*Source:* Adapted from [94].)

of incidence of the incoming rays with respect to the axis of the concentric rings increases, some losses are introduced.

The surfaces used in each cylindrical ring to approximate the shape of the imaging lens are called *grooves*. As it can be seen in Figure 5.9, each groove has an inclination corresponding to the curvature of the aspherical shape, but translated to the side of the lens.

The inclination of these grooves also presents a slight modification that compensates for their spatial translation. The grooves closer to the axis of the set of rings present a lower inclination than the grooves corresponding to the external rings (which is particularly the case in lenses with a low focal distance). This is due to the fact that the inclination of each groove corresponds to the curvature of the imaging lens at that point. Figure 5.9a shows a cross-sectional view of the profile of a Fresnel lens; and Figure 5.9b illustrates the front view of the concentric rings.

Modern manufacturing methods (such as compression, injection molding, and computer-controlled machining) and low-cost plastics of high quality allow the fabrication of high-quality Fresnel lenses for a wide variety of applications. This has prompted their adoption for point-to-point wireless IR communication systems.

A variety of materials can be used to fabricate Fresnel lenses (and other types of optical concentrators). They include polycarbonate, acrylic, and rigid vinyl, which are suitable for near-IR transmission. Polycarbonate, for example, presents a higher index of refraction than acrylic and rigid vinyl. It is preferred in applications that require a high impact and a high temperature resistance. It also offers a relatively high transmittance (close to 90 percent) at visible and near-IR wavelengths up to around 1.6 μm (for sample thickness = 3.2 mm) [94]. Acrylic is used in a wide variety of applications due to its high transmittance (around 92 percent for sample thickness = 3.2 mm [94]) and to the fact that this transmittance is practically constant for a range of wavelengths from the visible to the near-IR (up to around 1.2 μm). Rigid vinyl is an inexpensive fire-retardant material that presents a relatively high index of refraction (around 1.54, compared to the 1.49 of acrylic [94]).

5.4.3 Hemispherical Concentrators

The hemispherical lens is one of the most popular non-imaging concentrators currently employed in indoor and mobile wireless IR communication equipment. This is due to the fact that, as explained above, hemispherical concentrators can be manufactured easily and are simple to use, which explains their inclusion in the IrDA protocol (cf. Chapter 9). Hemispherical lenses offer a wide FOV, which makes them ideally suited for mobile applications [92, 93]. Unfortunately, their wide FOV corresponds to a moderate gain and accepts more background illumination noise than concentrators with limited FOVs. Figure 5.6 illustrates two possible concentrator-detector arrays incorporating hemispherical concentrators.

The FOV of a hemispherical concentrator is approximately 90°, with an optical gain given by [21]:

$$g(\theta_i) = n^2, \quad \text{for } 0 \leq \theta_i \leq 90° \tag{5.11}$$

where n is the index of refraction of the concentrator's material and θ_i is the angle of incidence of the incoming energy on the concentrator. This gain value is equal to n^2 over the entire FOV of the hemispherical concentrator.* This relationship is true only when the radius of the hemisphere is large enough compared to the radius of the active area of the photodetector. According to Barry [47] and Kahn [21], to satisfy this relationship, the radius of the hemisphere must be larger than the square of the index of refraction of the concentrator's material multiplied by the radius of the photodetector's active area (that is, $R > n^2r$).

The combination of different optical filters with hemispherical concentrators has been proposed in the past [47, 91, 95]. Planar longpass filters (which, as discussed in Section 5.3, have characteristics that do not change with angle of incidence), for example, can be used at the exit (the flat surface) of the lens and at the entrance of the photodetector. Index matching gel can be used in this type of structure (between the lens and the filter and between the filter and the photodetector) to decrease unwanted reflections, as illustrated in Figure 5.6a. Planar thin-film optical filters, on the other hand, cannot be placed between the concentrator and the detector because light rays reach the detector with a wide range of angles. A possible solution to the angular shift of thin-film optical filters, in this case, consists

---

* Despite the fact that hemispherical concentrators are considered to have omnidirectional gain, this is not exactly the case. Their effective area varies with angle of incidence according to the formula $A_{eff}(\theta_i) = An^2 \cos \theta_i$ [21].

of using nonplanar filters coated to the hemispherical surface. Here, the rays that reach the detector have impinged on the concentrator (and the filter) within a narrow range of angles, which minimizes the bandpass shift in the filter.

The main disadvantage of hemispherical concentrators is that hemispherical thin-film optical filters are difficult to manufacture, and coating these filters to a hemispherical concentrator may prove complicated. In addition, the area of the filter is much larger when coated on the concentrator's hemispherical entrance than on the planar exit, which increases the cost. Furthermore, the low gain of hemispherical concentrators and the fact that their FOV is wide (and fixed) means that they are not necessarily the best solution for all types of applications.

5.4.4 Compound Parabolic Concentrators

Compound parabolic concentrators (CPCs) present an alternative to hemispherical concentrators for wireless IR communication applications. One of their advantages, when compared to hemispherical lenses, is their higher gain, which is obtained when their FOV is narrow. If, on the other hand, a smaller gain can be tolerated, the FOV of the CPC can be increased to cover a wide area. In addition, they offer the possibility of incorporating planar thin-film optical filters and can be designed for a variety of FOVs [90]. Various compound parabolic concentrators are illustrated in Figure 5.7.

The fact that the entrance aperture of the CPC is planar allows the use of planar filters; and, in directive links where a narrow FOV and a high gain are required, the filter can be a narrow-band, thin-film optical filter (the narrow FOV of the structure means that the wavelength shift in the filter will be minimal). Moreover, two CPCs can be combined as illustrated in Figure 5.7b to reduce the angular shift of the thin-film optical filter placed between them while providing some gain and a not too narrow FOV. Unfortunately, CPCs are not without their drawbacks. While this type of non-imaging concentrator provides higher gains than hemispherical concentrators (especially for reduced FOVs) and offers the possibility of using planar thin-film optical filters that are easier to manufacture than their hemispherical counterparts, their excessive length makes them impractical for many applications (particularly the ones related to portable devices).

A typical CPC consists of three parts, as illustrated in Figure 5.10: (1) a flat entrance aperture, (2) a totally internally reflecting profile, and (3) an exit aperture. Its design (in two dimensions) is based on the edge ray principle, which, as explained in Chapter 4 (see Section 4.5), states that it is necessary that all rays entering the concentrator at the acceptance angle $\theta_a = \theta_i$, — where θ_i is the extreme angle — exit from the corner x (the edge) of the exit aperture as shown in Figure 5.10a [84]. This condition is easily fulfilled because, for a simple parabolic shape (like the one shown in Figure 5.10b) with the incident rays parallel to the parabola axis and its focus at the edge of the exit aperture, the rays are directed to the point x. Once

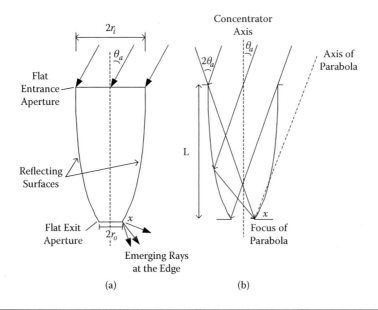

(a)

(b)

Figure 5.10   (a) The edge-ray principle in a CPC, and (b) profile construction of a CPC. (*Source:* Adapted from [84].)

the two-dimensional design has been created, it can be rotated 360° around its axis of symmetry to form the three-dimensional concentrator. As explained by Welford and Winston [84], the focal length f of the parabola and its entrance aperture d_1 can be calculated as:

$$f = r_o \left(1 + \sin \theta_a\right)$$ (5.12)

$$d_1 = \frac{2r_o}{\sin \theta_a}$$ (5.13)

and the total length is given by:

$$L = \frac{r_o \cos \theta_a}{\sin^2 \theta_a} \left(1 + \sin \theta_a\right)$$ (5.14)

From the definition of C for a two-dimensional concentrator with flat entrance and exit apertures (cf. Section 4.6), the geometrical concentration of a CPC can be calculated as [84]:

$$C = \frac{r_i}{r_o} = \frac{1}{\sin \theta_a}$$ (5.15)

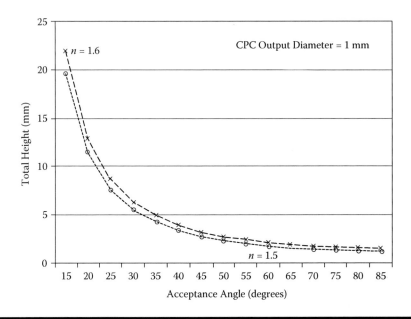

Figure 5.11 Total length variation with acceptance angle and index of refraction in a CPC.

and from Equations (5.13) and (5.14), the total length can be obtained as:

$$L = \left(r_i + r_o \right) \cot \theta_a \qquad (5.16)$$

This means that the acceptance angle and the diameter of the exit aperture determine the total length of the concentrator. As illustrated in Figure 5.11, the narrower the acceptance angle, the larger the concentrator becomes.

It must be noted that the maximum gain offered by an optical concentrator is obtained only if all the rays within the acceptance angle of the concentrator emerge from its output aperture. This is not always possible in the case of the three-dimensional version of a CPC due to the fact that some of the rays that cross it may undergo multiple internal reflections (due to skew rays) and may be reflected back (even if they entered the concentrator within the acceptance angle).

5.4.5 Dielectric Totally Internally Reflecting Concentrators (DTIRCs)

A method for designing DTIRCs was first proposed more than a decade ago by Ning et al. [89]. They presented a new family of non-imaging concentrator elements that were capable of achieving concentrations close to the thermodynamic

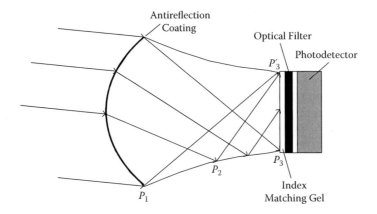

Figure 5.12   Cross-sectional view of a DTIRC where the refracted extreme rays reaching section P_1- P_2 converge into P'_3 and rays reaching section P_2- P_3 exit in parallel. (*Source:* Adapted from [89])

limit by making use of the refractive and reflective properties of a dielectric material. These concentrators were initially proposed for solar energy and fiber-optic applications; and later on, the use of DTIRCs combined with optical filters was proposed for optical wireless communication receivers [96, 97].

DTIRCs derive their name from the use of total internal reflection in the boundary of a dielectric material (similar to what occurs in an optical fiber), which in this case is the profile of the concentrator. They consist of three basic parts: (1) a curved front surface, which constitutes the entrance aperture; (2) a totally internally reflecting profile created through two different curves; and (3) an exit aperture. When a set of rays comes in contact with the front curved surface at the critical angle as shown in Figure 5.12, these rays are refracted and directed to the sidewall (profile), where they are totally internally reflected to the exit aperture according to the design method. The only rays that reach the exit aperture are those within the designed acceptance angle of the concentrator. If the rays incident on the front surface are beyond the acceptance angle, they exit from the side profile, missing the detector [89].

The design method of a DTIRC is very similar to that of the CPC described above. In a way, the DTIRC may be considered a variation of the CPC, where, by curving the flat entrance aperture of the CPC (in the case of the traditional DTIRC this curve corresponds to a semi-hemisphere), it is possible to obtain designs that achieve higher concentrations and are more compact than their CPC counterparts.

The best way to design a three-dimensional concentrator is to solve the two-dimensional case because, once the two-dimensional solution has been obtained, a three-dimensional version of the concentrator can be created by rotating the two-dimensional profile about its axis of symmetry. In a typical DTIRC, the front surface is a portion of a sphere, which simplifies the design and the manufacturing process. However, other shapes such as parabolas and ellipses have also been proposed [98].

To produce the sidewall, the slope of the side profile is set to values that allow the condition of total internal reflection (TIR) for the rays that are being directed to the photodetector. In addition, it may be necessary to take into account other requirements. It may be necessary, for example, to minimize the maximum output angle of rays at the exit of the concentrator if a thin-film optical filter will be used between the concentrator and the photodetector. Another possible restriction may relate to the space available (the casing) and therefore to the size of the concentrator.

DTIRCs can incorporate planar thin-film optical filters to reduce background noise. This means that, if the filter is placed between the collector and the photodetector, the concentrator's profile must be designed taking into account that the reflected rays must not exceed a maximum value. DTIRCs can also incorporate colored, GaAs, or InP longpass filters, which can be placed at the entrance or at the exit of the concentrator due to their insensitivity to angle of incidence.

The design of the DTIRC based on the phase conserving method of Ning et al. [89] requires that the exiting rays at the maximum angle (which corresponds to the critical angle) form a new wavefront (in a two-dimensional design) after having been reflected by the sidewall. This allows one to design the concentrator in such a way as to allow the reflected rays to exit within a predetermined maximum angle value and to be combined with a thin-film optical filter.

The general DTIRC design method is based on the geometrical wavefront definition (cf. Chapter 4), which requires that the total optical path length be a constant for every ray connecting the initial and the final wavefronts. The side profile of a DTIRC can be divided into two parts: P1-P2 and P2-P3 as shown in Figure 5.12. Rays impinging on the entrance aperture at the critical angle are refracted and directed to section P1-P2 of the profile, where they are reflected to the corner P3′. Rays reflected at P2 also exit from P3′, but rays reflected from the profile beyond that point (from section P2-P3) exit the concentrator at different points of the exit aperture, forming a new wavefront [89]. Note that this method assumes a single reflection for every ray within the concentrator (in the two-dimensional design).

The design of a DTIRC for a specific application is produced by taking into account a number of input variables. These include the FOV of the concentrator, its front surface arc angle (the curvature of the entrance aperture), the refractive index of the material used to fabricate the lens, a trial entrance aperture diameter, and the diameter of the exit aperture (which corresponds to the active area of the photodetector). After deciding on the input parameters, a computer program can calculate the design based on the preferred algorithm and on a trial height. This aids in calculating the concentrator's profile coordinates from a set of rays entering the concentrator at the critical angle, where the number of rays depends on the precision required. Having done this, the program then compares the trial aperture with the entrance aperture obtained. The new aperture and height are calculated from the difference between the two entrance apertures. The program iterates until the difference between both apertures is within a predetermined value [89]. Once the final entrance aperture is defined, the coordinates of the two-dimensional version of the DTIRC can be obtained.

The theoretical design can be verified by fabricating a template that simulates a two-dimensional cross section of the concentrator. This template can then be placed over a printed version of the rays traced for the specific design at the critical angle, and a visible light laser can be fired at the template to compare the path of the real rays with those of the ray-tracing simulation. The refractions and reflections at different angles of incidence can be verified by rotating the concentrator template at the desired angle.

5.5 DTIRC Characteristics

There are a number of restrictions that affect the design of an optical wireless receiver: the economic budget, the power budget, the size, the weight, the availability of components, etc. In the case of the optical front end, some of the most important restrictions are the size of the photodetector (directly related to its cost and its capacitance), the size of the concentrator, and the index of refraction of the concentrator's material. The latter is chosen, depending on the receiver front-end requirements: maximum concentration, small size, or reduced maximum output angle at the exit of the concentrator. In the specific case of the DTIRC, two of its most important parameters (in addition to the index of refraction) are the front surface arc angle and the acceptance angle.

The selection of the DTIRC input parameters is not always straightforward, as choosing an input parameter value that is beneficial to one of the characteristics of the concentrator may not necessarily be advantageous to another of the output parameters. Therefore, it is important to know the implications of the selection of specific values for any of the input parameters in order to make the right compromises between the remaining design variables.

This section presents the effect that variations in the different input parameters have on the output parameters (such as size, maximum output angle, and gain) of a DTIRC. It also presents opto-electronic gain measurements performed (at different angles of incidence) on a variety of DTIRCs, as well as a comparison of their ideal and their real gain responses. In addition, comparisons of gain, size, and angular response of traditional DTIRCs with the output parameters of concentrators based on hemispherical lenses and compound parabolic concentrators are presented.

5.5.1 Geometrical Gain

As discussed in Chapter 4, one of the most important properties of an optical concentrator is its geometrical gain, which defines the factor by which the size of the detector (and thus its bandwidth) can be improved. Figure 5.13 shows the effect of

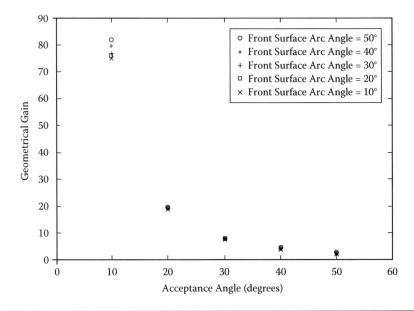

Figure 5.13 Variation of the geometrical gain of a DTIRC with acceptance angle for a variety of front surface arc angles (for an index of refraction = 1.5)

the variation of the acceptance angle on the geometrical concentration of a DTIRC (with a semi-hemispherical entrance aperture) for a number of front surface arc angles. Here, it can be observed that the geometrical concentration is inversely proportional to the acceptance angle and that variations in the front surface arc angle have a relatively small effect on the geometrical gain. The index of refraction of the material used for this set of data was 1.5.

Figure 5.14 presents a more complete view of the geometrical gain variation with different input parameters, where geometrical gain values are shown for a variety of acceptance angles, front surface arc angles and indices of refraction. As evidenced in this graph, the gain is almost the same for all the front surface arc angles; and, in the same way as in Figure 5.13, the concentration values are small for large acceptance angles and large for narrow FOVs. For this reason, when designing the receiver of a wireless IR link, a compromise must be made between concentration and acceptance angle, depending on the topology of the system. If the receiver of a diffusely illuminated room incorporates a DTIRC, the FOV of this concentrator should be wide in order to provide maximum overage, but at the expense of presenting a reduced gain. Directed-LOS links, on the other hand, benefit from the use of narrow FOV DTIRCs that offer large concentrations.

It is also evident from Figure 5.14 that DTIRCs fabricated with high index of refraction materials will present larger geometrical gains than those fabricated with lower indices of refraction materials. This is illustrated in Figure 5.15 also, where the effect of the index of refraction on the gain is shown for different acceptance

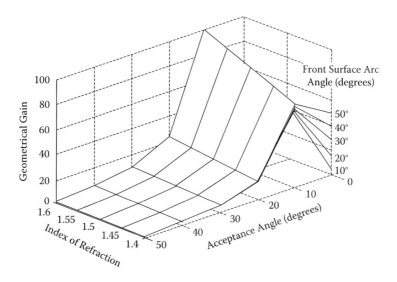

Figure 5.14   The DTIRC's geometrical gain variation with acceptance angle and index of refraction for various front surface arc angles.

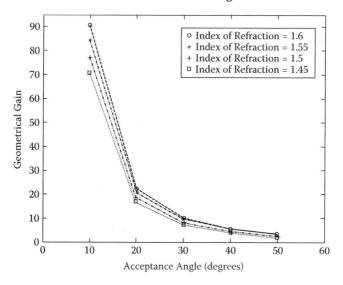

Figure 5.15   Geometrical gain variation of a DTIRC with acceptance angle for a variety of indices of refraction.

angles (this graph was obtained for a DTIRC with a front surface arc angle = 45°). This means that if the most important parameter of the receiver front-end design, regardless of the maximum output angle and the size, is the geometrical gain, the concentrator must be designed for the maximum index of refraction available. The

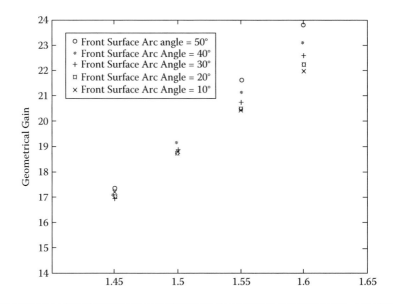

Figure 5.16 DTIRC geometrical gain variation with index of refraction for various front surface arc angles.

front surface arc angle can be selected depending on its effect on the rest of the output parameters of the concentrator.

Figure 5.16 shows the variation in the geometrical gain with the index of refraction for various front surface arc angles. Here, it can be observed that the variation of the geometrical gain with refractive index is almost linear. For this graph the value of the acceptance angle for all the plots was 10°.

One way to achieve a wide FOV while at the same time increasing the sensitivity of the receiver consists of using an array of DTIRCs such as the one shown in Figure 5.8b. In this case, each individual DTIRC can be designed for a reduced acceptance angle (with a large geometrical concentration); and by accommodating the individual concentrator's FOV in a complementary way, the overall acceptance angle of the array can be widened. The drawback to this approach is that the wider the FOV and the higher the concentration required, the larger the number of DTIRCs and detectors that need to be incorporated, which increases the cost of the receiver.

5.5.2 Maximum Output Angle

The strong dependence of a thin-film optical filter on the angle of incidence means that the maximum output angle of a concentrator, which defines the maximum angle of incidence on the filter, is of paramount importance in the design of the receiver optical front end. Figure 5.17 shows the maximum output angle of the rays

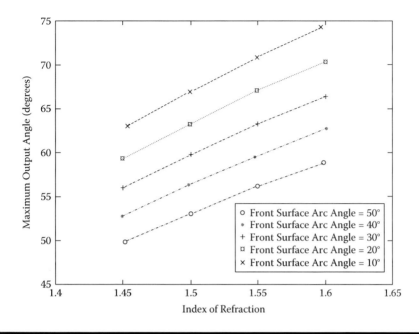

Figure 5.17   Maximum output angle variation of a DTIRC for a variety of indices of refraction and front surface arc angles.

exiting a DTIRC for different indices of refraction and different front surface arc angles. The acceptance angle used to generate this graph was 10°. One of the first things that can be observed here is that the maximum output angle is wide for large values of index of refraction. This is explained by the fact that high gain concentrators present a large entrance-to-exit aperture ratio (a large gain) for which the rays at the output form a wide angle with respect to the axis of the lens (as illustrated in Figure 4.7). As explained in the previous section, a narrow FOV corresponds to a large geometrical gain, and this is reflected in the large maximum output angles presented in Figure 5.17 for a DTIRC.

An alternative graph, a three-dimensional version of Figure 5.17, is presented in Figure 5.18, where the variation of the maximum output angle of a DTIRC with index of refraction and acceptance angle is presented for various front surface arc angles. It is observed that high index of refraction values (related to high gains) contribute to large maximum output angles. Therefore, one way to reduce the maximum output angle of rays at the exit of a DTIRC is by choosing a material with a low index of refraction for its fabrication (at the expense of reducing the gain).

As explained above, the selection of the input parameters of a concentrator is not always straightforward. The decision on the value of the index of refraction, for example, depends not only on the value of the desired gain, but also on the value of the maximum output angle and on the total height required. Thus, a compromise

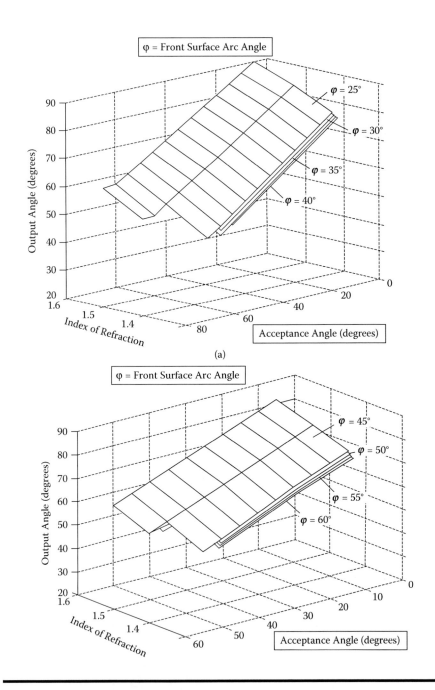

Figure 5.18 **DTIRC's maximum output angle versus acceptance angle and index of refraction for various arc angles.**

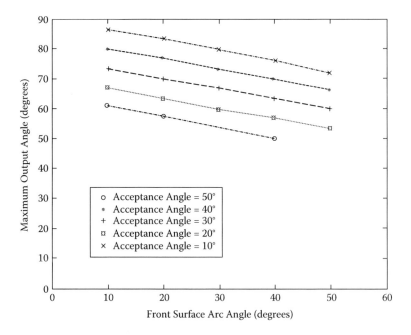

Figure 5.19   DTIRC's maximum output angle variation with front surface arc angle for various acceptance angles.

between the input parameters according to the characteristics of the concentrator that are more important for a specific application must generally be made.

Figure 5.19 shows the variation in the maximum output angle with different values of front surface arc angle. The index of refraction of the concentrator used to produce this graphs was 1.5. Here it can be observed that the maximum output angle decreases as the front surface arc angle increases. Thus, if the front surface arc angle influences the gain as little as presented in the previous section, a DTIRC benefits, at least with regard to the maximum output angle, from the selection of a wide front surface arc angle. As discussed in the next section, a large front surface arc angle is also beneficial to the size of the lens.

5.5.3 Total Height

Another very important parameter of a DTIRC is its size. This must be taken into account when designing the front end of an optical wireless receiver as the total height of a concentrator may be critical for portable units that need to be small and light (which is the reason why the excessive length of a CPC makes it inadequate for a number of applications). In addition, when the concentrator is attached to a photodetector (or a system), the space available in the detector's casing (a TO-CAN, for example) may be limited, needing to be taken into consideration.

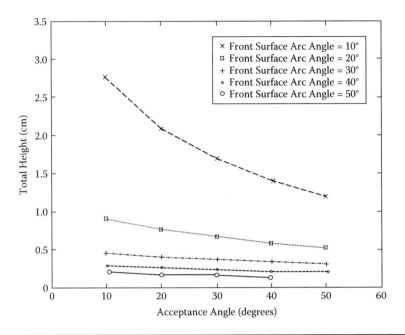

Figure 5.20 **DTIRC's total height variation with acceptance angle for various front surface arc angles (data obtained for a DTIRC with exit aperture diameter equal to 1 mm).**

Figure 5.20 shows the total height (the profile height plus the front surface height) variation of a DTIRC for different acceptance angles and for various front surface arc angles. The index of refraction of the concentrator's material in this case was chosen as 1.5. It can be seen that, just as in the CPC case, a high gain corresponds to a large concentrator. The main difference between the CPC and the DTIRC is that, in general, the latter is significantly smaller than the former. For small acceptance angles where the geometrical gain is high, the height of the concentrator also is large.

The effect of the front surface arc angle on the total height of a DTIRC is also of interest. Figure 5.21 shows the size variation of this type of concentrator with front surface arc angle for various indices of refraction. The acceptance angle value used to produce this graph was 10°, and the exit aperture of the concentrator was 1 mm. It can be seen from this figure that, as the front surface arc angle becomes larger, the total height of the concentrator decreases (which can also be observed in the three-dimensional graph presented in Figure 5.22).

Taking into account that the CPC can be considered a DTIRC with front surface arc angle equal to zero, it is understandable that the largest front surface arc angle values achievable are the ones that produce the most significant size difference between them. It can also be concluded from this figure that large front surface arc angles are advantageous in general to all the output parameters of a

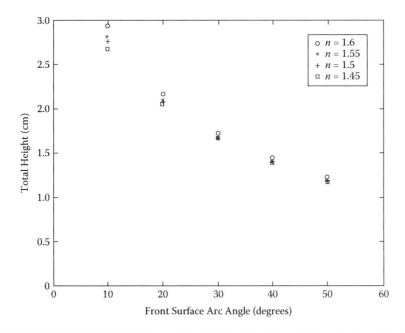

Figure 5.21   DTIRC's total height variation with front surface arc angle for various indices of refraction (data obtained for a DTIRC with exit aperture diameter equal to 1 mm).

DTIRC. Large front surface curvatures contribute not only to increase the gain and to reduce the maximum output angles, but also to reduce the size, the weight, and the amount of material required for the fabrication of the optical front end of the receiver. Figure 5.22 shows the height variation of a DTIRC with acceptance angle and front surface arc angle (for two indices of refraction values). The exit aperture diameter used as a reference in Section 5.5.3 was 1 cm.

The effect of the value of the index of refraction on the height of the concentrator is presented in Figure 5.23 where it can be seen that, as the index of refraction increases, the total height of the DTIRC also increases slightly. This is expected due to the fact that a high gain means a larger entrance diameter and therefore a larger structure. The front surface arc angle of the concentrator used to produce this graph was 40°.

5.6 Comparison of Concentrators

As explained in Section 5.4, DTIRCs present several advantages when compared to CPCs and hemispherical concentrators. Compared to hemispherical lenses, DTIRCs offer a higher concentration, the possibility of defining the FOV, and the

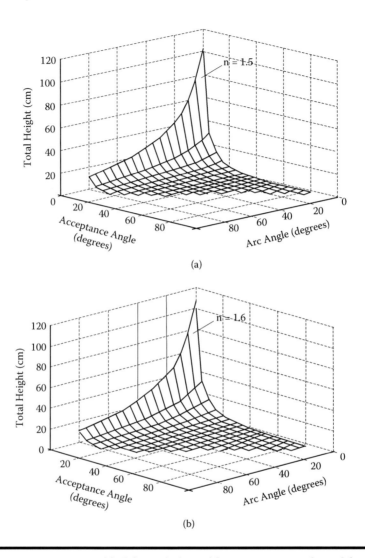

Figure 5.22 DTIRC's total height variation with acceptance angle and front surface arc angle for two indices of refraction: (a) *n* = 1.5, and (b) *n* = 1.6.

possibility of using flat thin-film optical filters that are easier to fabricate (and therefore allow the manufacturing process to be relaxed). Compared to CPCs, DTIRCs offer two advantages: (1) smaller size and (2) higher concentrations.

Figure 5.24 shows a size comparison between a DTIRC (with different front surface arc angles) and a CPC. It is observed that, for large front surface arc angles, the size of the DTIRC is significantly smaller than that of a CPC. This difference is particularly pronounced at small acceptance angles.

Figure 5.25 shows a superposition of a variety of DTIRCs over a CPC and a hemispherical concentrator. Here, as the front surface arc angle of the DTIRC

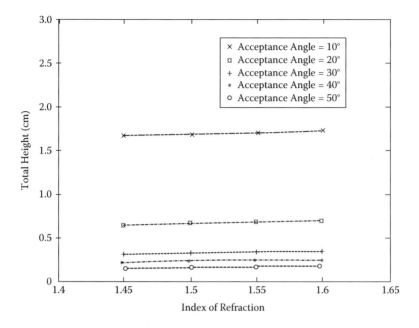

Figure 5.23   DTIRC total height variation with index of refraction for various acceptance angles (front surface arc angle = 40°).

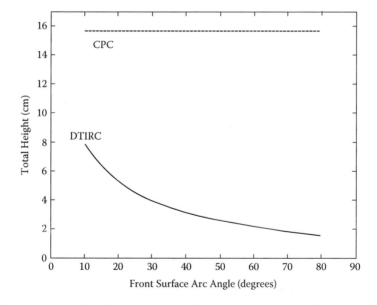

Figure 5.24   Total height comparison between a CPC and various DTIRCs with different front surface arc angles (for an exit aperture = 1 mm, *n* = 1.5, and an acceptance angle = 5°).

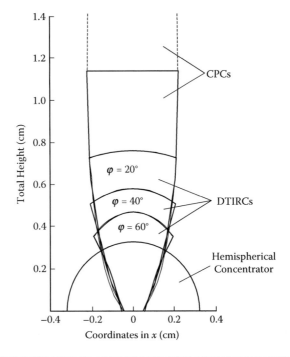

Figure 5.25   Size comparison of a CPC, a hemispherical concentrator, and three DTIRCs with different front surface arc angles. (*Source:* Adapted from [89].)

increases (from 0°, which corresponds to a CPC to 60°), the size of the concentrator decreases. It is also observed in this figure that, even if the height of a hemispherical concentrator is not too big, the amount of material used can exceed the one employed for a DTIRC with a large front surface arc angle.

A comparison of the opto-electronic gain provided by a DTIRC and the concentration offered by a hemispherical concentrator is presented in Figure 5.26. As explained in Subsection 4.6.1, the opto-electronic gain of an optical concentrator can be calculated by dividing the value of the electrical signal obtained from the concentrator-photodetector structure by the value of the signal generated by the bare detector. The angular response of the concentrator can be obtained by measuring its opto-electronic gain at different angles of incidence as shown in Figure 5.27. The values presented in Figure 5.26 were calculated in such a way. First, the concentrator was placed directly on top of the detector, and the optical gain was measured by dividing the electrical signal obtained from the concentrator-detector structure by that of the bare detector. This measurement was taken every 1° (starting at normal incidence = 0°) until it reached the critical angle (the acceptance angle). The index of refraction of the photodetector resin was 1.5, as this is the typical value of the resin on other detectors. The DTIRC used to obtain this graph had a front surface arc angle of 60°, an exit aperture diameter of 1 cm, a half acceptance angle of 25°,

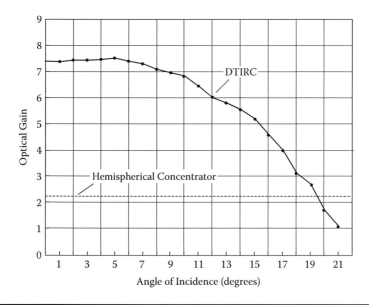

Figure 5.26 Opto-electronic gain measurements for a DTIRC and gain of a hemispherical concentrator.

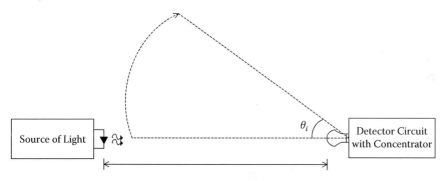

Figure 5.27 Top (aerial) view of the experimental setup employed to measure the angular response (opto-electronic gain variation with angle of incidence) of the DTIRC used to generate Figure 5.26.

and a geometrical gain of 12.65. The index of refraction of the material was 1.5. The cross-sectional view of the DTIRC used for this experiment can be observed in Figure 5.28a, and the three-dimensional version is shown in Figure 5.28b.

As discussed in Chapter 4, there are a number of factors that affect the opto-electronic gain of a concentrator. In this case, for example, the transmittance of the material used was not 100 percent, the front surface arc angle was not anti-reflection coated, an index matching gel was not used between the exit aperture of the concentrator and the detector, and some manufacturing errors were present in the sample.

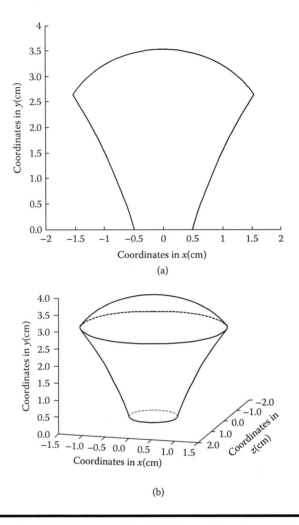

Figure 5.28   Schematic diagram of the DTIRC used for the experimental measurements presented in Figure 5.26: (a) cross-sectional view, and (b) three-dimensional version.

5.7 Practical Issues

The first step in the design and fabrication of an optical concentrator is the selection of a high-quality optical material with an appropriate index of refraction. These materials are chosen according to their optical, electrical, and mechanical properties. From these, one of the most important properties is the transmittance of EM radiation with wavelength, which defines the fraction of energy that reaches the exit of the material with respect to the amount of energy at the entrance. Figure 5.29 provides an example of a graph showing the transmittance of a material with wavelength.

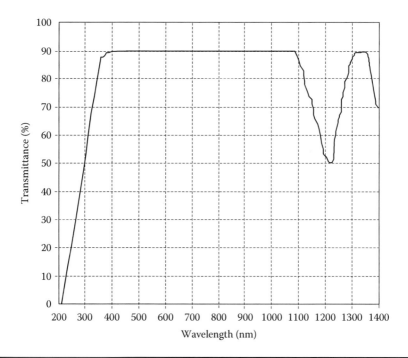

Figure 5.29 Transmittance variation with wavelength for the Zeonex 480R. (*Source:* **From [99].**)

Another very important parameter is the index of refraction of the material. Graphs showing these values are usually given for different wavelengths and different temperatures. An example of such a graph is presented in Figure 5.30, which shows the spectral variation of the index of refraction (for a number of temperatures) of Zeonex 480R from Zeon Chemicals [99].

One way to verify that an optical concentrator behaves according to the ray-tracing pattern predicted by the design software is by fabricating a "two-dimensional" template. Such templates can be used in combination with narrow divergence visible light lasers to verify that the rays (generated by the beam from the laser at different angles) are being refracted and reflected according to theory. A print of the ray-tracing pattern can be placed underneath the template for that purpose.

To improve the performance of an optical concentrator (by reducing the losses due to unwanted reflections), an index matching gel can be used between the exit aperture of the concentrator and the detector (or at the boundaries of different media). The index of refraction value of the index matching gel is calculated by taking into account the value of the index of refraction of the concentrator and the index of refraction of the active area of the detector (which in the case of a silicon detector is around 3.6 for a wavelength of 950 nm). Figure 5.31 shows a comparison of the opto-electronic gain generated by a concentrator-detector structure that

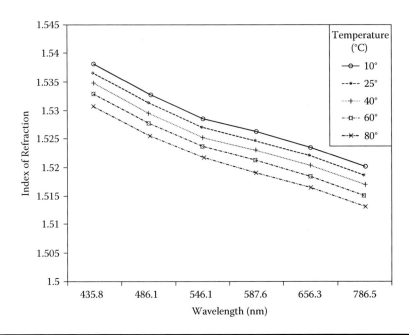

Figure 5.30 Index of refraction variation with wavelength for the Zeonex 480R at different temperatures. (*Source:* From [99].)

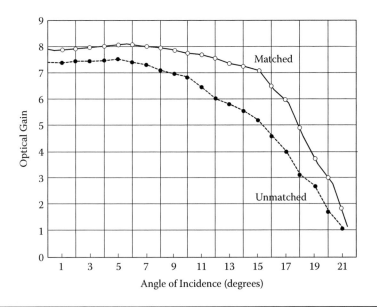

Figure 5.31 Opto-electronic gain comparison between an index matched DTIRC-detector structure and a structure without gel.

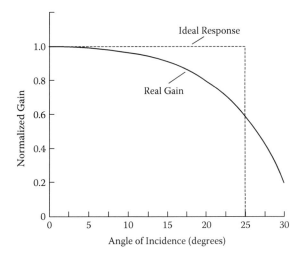

Figure 5.32   Normalized opto-electronic gain of a DTIRC. Comparison between the ideal and the real curves (FOV = ±25°).

incorporates an index matching gel to that of a structure without any such gel. This figure shows that there is an obvious improvement in the opto-electronic gain due to the elimination of the air gap between the concentrator and the detector.

It is interesting to notice that the electro-optical gain of a DTIRC does not have a flat response up to the critical angle. One of the reasons for this is that the incident rays impinging on the concentrator at wide angles create more skew rays, and some of these rays are reflected back to the entrance in accordance with the theory of non-imaging concentration presented by Welford and Winston [84]. This is illustrated in Figure 5.32, which compares the ideal and the real opto-electronic gains of a DTIRC.

Note that, in general, optical concentrators can be scaled to any size according to the requirements of the photodetector. The main limitation (especially in small designs) is imposed by the manufacturing process.

5.8 Other Shapes of DTIRCs

The effect that increasing the front surface arc angle has on the gain, the total height, and the maximum output angle of the traditional DTIRC has motivated the exploration of alternative shapes for the entrance aperture of the DTIRC. Some of the curves investigated thus far include parabolas and ellipses [98]. Each of them exhibits a different behavior (compared to the traditional DTIRC), depending on their specific design parameters. The parabolic DTIRC (PDTIRC) is presented and compared to the semi-hemispherical one in Subsection 5.8.1 and the elliptical DTIRC (EDTIRC) is analyzed in Subsection 5.8.2.

5.8.1 *Parabolic DTIRC*

The design of the PDTIRC is based on the same principles used for its semi-hemispherical version. The difference in this case is that because of the impossibility of defining a front surface arc angle in the same way as for the semi-hemispherical DTIRC, the parabola height and diameter are chosen as the input parameters for the entrance aperture, which makes the design slightly more complex. It is also important to take into account that aspherical entrance apertures make the manufacturing process more complicated. In the design algorithm used to generate a parabolic DTIRC, the base of the parabolic shape is placed on the x axis and its center at the origin of a coordinate plane. The p parameter of the parabola can be obtained from the equation of a parabola with the aperture facing down and the vertex in the coordinate (0, parabola height) as follows:

$$\left(x - x_0\right)^2 = -2p\left(y - y_0\right) \tag{5.17}$$

where x_0 and y_0 are the coordinates of the vertex. Thus,

$$p = \frac{\left(\dfrac{d_1}{2}\right)^2}{2b_{pa}} \tag{5.18}$$

where d_1 is the trial entrance diameter of the concentrator and b_{pa} is the parabola height. Once the p parameter is found, it is possible to obtain the corresponding y coordinates of the parabola in that position as follows:

$$y_{initial} = -\frac{x^2 - 2pb_{pa}}{2p} \tag{5.19}$$

where the number of $x_{initial}$ coordinates are assigned according to the desired numerical precision. This is done by gradually increasing the value of the $x_{initial}$ position between $-d_1/2$ and $d_1/2$, which creates an x vector that can be used to calculate the vector for the y position.

After defining the parabolic surface, an "equivalent" front surface arc angle can be obtained. A line joining the first two points of the parabola with respect to the x axis forms this angle. This is equivalent to the maximum front surface arc angle θ used to define the incident ray positions in the traditional DTIRC.

The importance of the equivalent front surface arc angle is that it helps define the maximum refracted angle with respect to the vertical axis, and thus the profile height. It also helps as a point of reference to compare different DTIRCs. In an equivalent way to the hemispherical DTIRC, the maximum refracted angle can be calculated as [89]:

$$\theta'_{max} = \sin^{-1}\left(\frac{\sin(\theta_a - \varphi)}{n}\right) + \varphi \tag{5.20}$$

where φ is the equivalent front surface arc angle, θ_a is the acceptance angle of the concentrator, and n is its index of refraction of the material used to fabricate the concentrator. The profile height can be subsequently calculated as [89]:

$$H = \frac{d_1 + d_0}{2}\cot\left(\theta'_{max}\right) \tag{5.21}$$

Once the height has been obtained, the real front surface coordinates can be calculated as:

$$x = x_{initial} \tag{5.22}$$

$$y = y_{initial} + H \tag{5.23}$$

As observed, the complexity of this method derives from the fact that the concentrator front surface is not defined in as straightforward a manner as it is in the semi-hemispherical one; but once this parabolic surface is defined, the coordinates of the concentrator profile can be calculated in the same way as in a traditional DTIRC (with a semi-hemispherical entrance aperture). When the refracted incident rays reach section P_1-P_2 of the profile, these rays are reflected into P'_3 (and $l_4 = 0$). This is illustrated in Figure 5.33. If, on the other hand, the refracted incident rays reach section P_2-P_3 of the profile, the extreme rays exit the concentrator with an angle θ_0 defined by the angle at which rays barely satisfy the condition at P_2.

The angles of refraction of every ray that comes into contact with the front surface of the concentrator can be calculated at every point with respect to the vertical axis. This allows the evaluation of the angles of refraction in every point of the front surface as follows:

$$\theta = -\tan^{-1}\frac{|y|}{|x|} \tag{5.24}$$

and

$$\theta' = \sin^{-1}\left(\frac{\sin\left(\theta_a + \theta\right)}{n}\right) - \theta \tag{5.25}$$

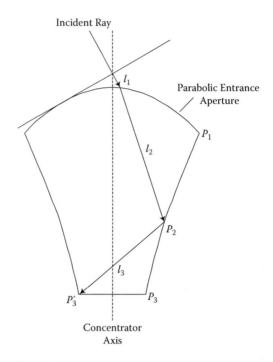

Figure 5.33 Optical path length sections of the PDTIRC design.

Here, θ is the angle of the normal of each point with respect to the vertical axis and θ' is the angle of the refracted rays with respect to the vertical axis.

The variation of the output parameters of a PDTIRC is presented in Figures 5.34, 5.35, and 5.36, where it can be seen that they exhibit similar behavior to the one presented by the traditional DTIRC. A concentrator designed with a narrow acceptance angle, for example, presents a high gain (Figure 5.34), but also a large maximum output angle (Figure 5.36) and a larger size than the ones designed for a wide acceptance angle. The latter, on the other hand, have the advantage of being compact and of offering wide coverage and greater mobility, but at the expense of a reduced gain.

In a PDTIRC, defining a large parabola height is equivalent to defining a wide front surface arc angle in the semi-hemispherical version. In fact, as the height increases, the equivalent front surface arc angle also becomes wider. In the graphs shown in Figures 5.34, 5.35, and 5.36, the parabola height is given in centimeters, and the exit diameter chosen for the concentrator is 1 cm. This facilitates the comparison with the graphs presented in Section 5.5.

The effect of a wide front surface arc angle on the maximum output angle is similar for the hemispherical and parabolic DTIRC cases. As this angle increases, the maximum output angle also increases.

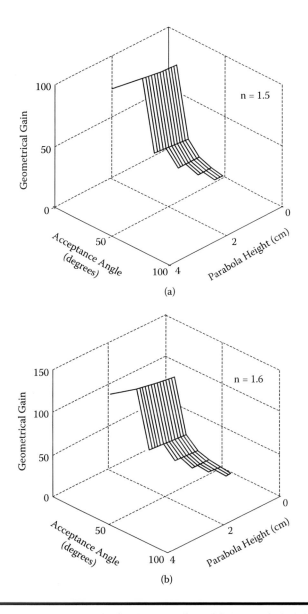

Figure 5.34 PDTIRC geometrical gain variation with parabola height and acceptance angle.

The effect of the variation in the input parameters on the total height of the concentrator can be observed in Figure 5.35, where it is shown that as the acceptance angle decreases (that is, the gain increases), the size of the concentrator increases. In addition, if the equivalent front surface arc angle decreases (that is, if the *a*

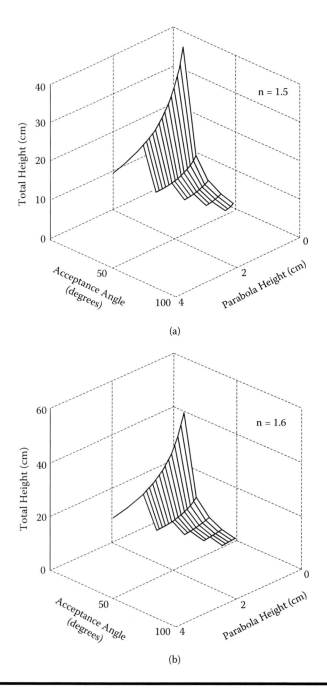

Figure 5.35 **PDTIRC total height variation with arc angle and parabola height.**

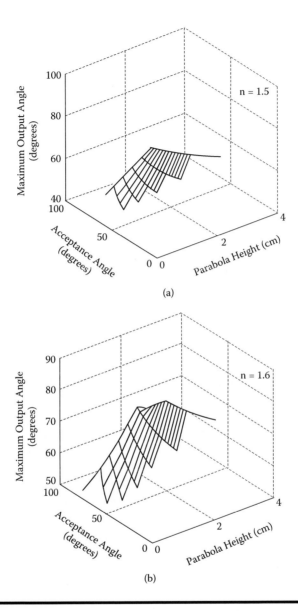

Figure 5.36 Maximum output angle variation with acceptance angle and parabola height.

parameter of the parabola decreases), the total height further increases. A high index of refraction also increases the size of the PDTIRC. The comparison between the PDTIRC and other DTIRCs is presented in Section 5.8.3.

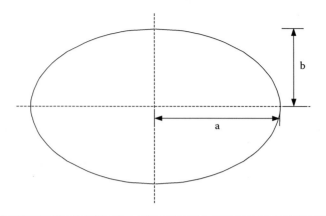

Figure 5.37   Ellipse defining parameters.

5.8.2 Elliptical DTIRC

An alternative shape to the parabola and the semi-hemisphere explored for the front surface of the DTIRC is the ellipse [98]. EDTIRCs have been proposed as a way to vary the size and gain of a DTIRC. In this case, just as for the PDTIRC, the profile design of the concentrator is based on the edge ray principle and on the phase conserving method.

The first step in the design of the two-dimensional version of the elliptical DTIRC is to define the front surface. Unlike its parabolic and semi-hemispherical counterparts, the elliptical DTIRC requires that two parameters be defined: (1) the ellipse height b and (2) the ellipse half base a, as shown in Figure 5.37.

The other parameters that should be taken into account to define the concentrator include the index of refraction of the chosen material, the exit aperture (the diameter of the detector), the acceptance angle, the numerical precision, and a trial entrance diameter.

One way of generating the EDTIRC is by, prior to calculating the concentrator profile coordinates, generating the points (on the positive side of the y axis) of the ellipse according to:

$$\frac{x^2}{a^2} + \frac{y^2}{b^2} - 1 = 0 \tag{5.26}$$

The points on the y axis can be calculated at increments in x defined by the numerical precision required. Once the original ellipse has been generated, it is possible to define the desired section of the ellipse to be used as the entrance aperture of the concentrator (the part contained within the defined entrance diameter). It is impossible to use the ellipse as originally defined because this would be equivalent to defining a hemispherical DTIRC with a front surface arc angle equal to 90°, for which designs are physically impossible to obtain.

The equivalent front surface arc angle can be calculated from the first two points of the elliptical section. In addition, the maximum refracted angle can be calculated through the equivalent front surface arc angle obtained and through the acceptance angle. Once this angle has been obtained, the profile height can be calculated, and the front surface coordinates can be redefined as follows:

$$x = x_{initial} \tag{5.27}$$

$$y = y_{initial} - y_{initial}(1) + H \tag{5.28}$$

where $y_{initial}(1)$ is the y coordinate of the first point of the ellipse. Here, as in the PDTIRC, the concentrator coordinates are more difficult to generate than in the traditional DTIRC with a semi-hemispherical aperture. This is due to the fact that the elliptical front surface cannot be defined in as straightforward a manner as in the semi-hemispherical case.

Figures 5.38, 5.39, and 5.40 show the most important output parameters (gain, total height, and maximum output angle) of an EDTIRC for different combinations of input parameters (that is, index of refraction, acceptance angle, and front surface arc angle). Here again, the diameter of the exit aperture was set to one to allow a direct comparison with the DTIRCs presented in Sections 5.5 and 5.8.1.

5.8.3 Comparison of DTIRCs

This subsection presents a comparison of the output parameters of the different DTIRCs. This is possible through the use of an equivalent front surface arc angle in the design of the elliptical and parabolic versions, which guarantees that the maximum angle of the rays refracted from the front surface is the same for all the designs. This also implies that the maximum output angle is the same for the different DTIRCs. The size and the geometrical gain comparisons of the different DTIRCs are presented for the same maximum output angle.

Figure 5.41 compares the size of a traditional DTIRC with that of the elliptical and parabolic DTIRCs. These graphs were produced for indices of refractions of 1.5 (Figure 5.41a) and 1.6 (Figure 5.41b), and for a variety of front surface arc angles and acceptance angles. Here it can be observed that, in general, the variation in size for the different DTIRC designs is small. It can also be observed that the DTIRC design that provides the smallest size is the elliptical one, while the parabolic DTIRC is the largest concentrator in the majority of the graphs. In addition, it can be seen that whereas the PDTIRC becomes larger than the SHDTIRC as the front surface arc angle increases, the EDTIRC becomes smaller for small arc angles

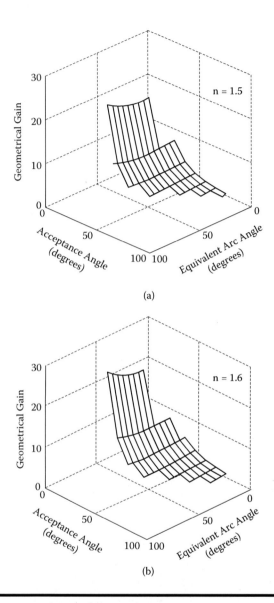

Figure 5.38 EDTIRC geometrical gain versus acceptance angle and equivalent front surface arc angle for two indices of refraction.

and larger for big ones. It is also interesting to notice that the difference in size between the different DTIRCs is more significant for large front surface arc angles. This is explained by the fact that the smaller the arc angle, the more all the designs resemble a compound parabolic concentrator. Another interesting aspect is that for small acceptance angles, the elliptical DTIRC presents a small size, whereas for wider ones, it becomes larger than the SHDTIRC.

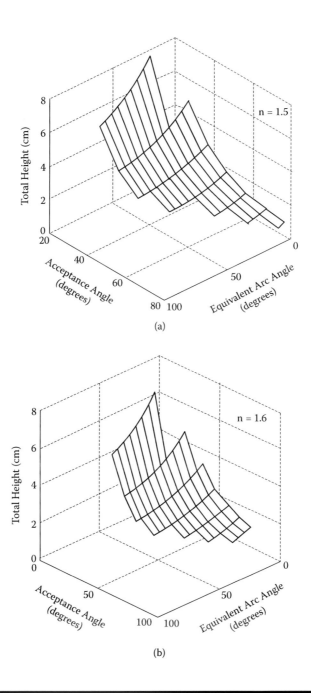

(a)

(b)

Figure 5.39   EDTIRC total height versus acceptance angle and equivalent front surface arc angle for two indices of refraction.

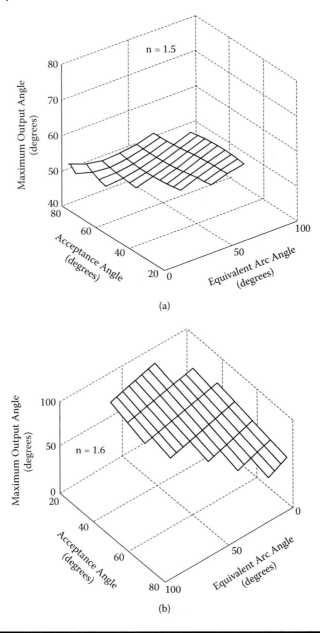

Figure 5.40 EDTIRC maximum output angle versus acceptance angle and equivalent front surface arc angle for two indices of refraction.

With respect to the refractive index, the variation between the different designs is not particularly significant. This can be noticed when comparing Figures 5.41a and b, where, for the same front surface arc angle, the variation between the different designs for the same acceptance angle is minimal.

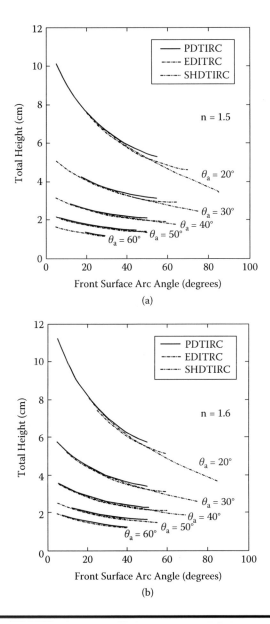

Figure 5.41 DTIRC comparison of the total height variation with arc angle for various acceptance angles.

The geometrical gain comparison for the different DTIRCs is presented in Figure 5.42. Here it is observed that, as expected, the larger size of the parabolic DTIRC is compensated by a higher gain. It can also be observed how, for the front surface arc angle values that make the PDTIRC and the EDTIRC larger than the

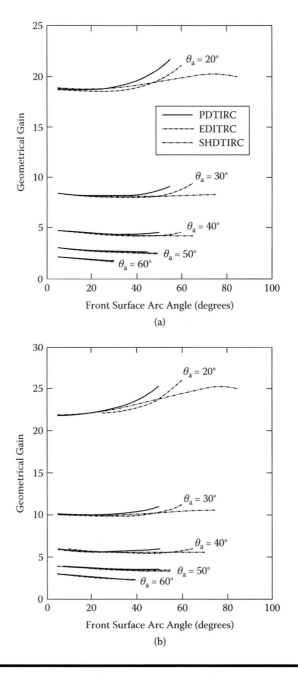

Figure 5.42 DTIRC comparison. Gain variation with arc angle for various acceptance angles and indices of refraction.

SHDTIRC, the gain values also increase. For values of the EDTIRC arc angle where the total height is smaller than that of the SHDTIRC, the geometrical gain is also lower. Also note in Figure 5.42 that the most significant differences between the three types of DTIRCs occur at large front surface arc angles.

It should be noted that the DTIRC that provides the largest number of designs for the different combinations of input parameters is the SHDTIRC, whereas the concentrator that allows fewer design possibilities is the parabolic one. This is due to the intrinsic characteristics of the design method used, which make impossible the design of a concentrator for some combinations of input parameters. This is observed particularly for designs with large front surface arc angles, which explains why the end of the lines presented in Figures 5.41 and 5.42 (at large front surface arc angles) for the different concentrators are missing.

5.9 Summary and Conclusions

This chapter explained that, despite the similarities between wireless and wired optical systems, the former presents particular challenges with regard to the high attenuation suffered by the IR signal when transmitted through air and to the background illumination noise to which optical wireless systems are generally exposed. Some of the particular requirements of wireless IR systems include the use of small-area detectors (to increase the bandwidth of the system and to reduce the cost), the use of low-power sources of light (to comply with eye safety regulations and to minimize the power consumption of the system), and the use of optical filters (to reject out of band radiation and to reduce the noise in the receiver).

With regard to the background illumination noise introduced in the receiver, two of the most common optical filtering techniques include (1) the use of a long-pass filter material combined with the responsivity of the photodetector, and (2) the use of a thin-film optical filter. Both of them help improve the SNR at the receiver; and one may be preferred over the other, depending on the level of filtering required and on the economic budget available. The combination of a silicon detector and a material exhibiting longpass characteristics, for example, is relatively economic and simple to implement. This combination also presents the advantage of being insensitive to the angular variations of the energy reaching the receiver. The main drawbacks of this technique are that, in general, there is little control over the edges of the filter; and that the passband of the filter is relatively wide.

Thin-film optical filters, on the other hand, can be designed to achieve very narrow passbands at the desired wavelength. Unfortunately, the cost of this type of filter is higher than that of their long-pass counterparts, which can be made from colored plastic or glass. Moreover, the characteristics of thin-film optical filters change with the angle of incidence, which must be taken into account when designing the remainder of the receiver's optical front end.

One of the techniques that allows for the use of small low-capacitance photodetectors consists of using an optical concentrator in such a way that the energy collected by its entrance aperture is redirected to the small active area of a photodetector. Optical concentrators, which can be of the imaging or non-imaging type, also help improve the power budget, the range, and the eye safety of the system.

Imaging optical concentrators are usually employed in directed-LOS links due to the fact that the focal point generated by these optical elements does not change position with angle of incidence. Because the acceptance angle of imaging concentrators is very narrow, this prevents them from being used in mobile applications. A good example of an imaging concentrator is the Fresnel lens.

Non-directed and non-LOS configurations favor the use of non-imaging concentrators that can provide a wide FOV. Examples of non-imaging concentrators include hemispherical lenses, CPCs, and DTIRCs. Hemispherical concentrators, for example, are easy to fabricate and offer a wide FOV. Unfortunately, a large FOV is not always desirable (especially in applications subject to intense background illumination) and their FOV cannot be controlled. A further disadvantage of hemispherical concentrators is that they cannot use flat thin-film optical filters; and, if a thin-film filter is required, the filter must be coated to the entrance aperture, thereby increasing the complexity of the manufacturing process.

Another popular non-imaging concentrator is the CPC, which can incorporate a flat thin-film optical filter and can be designed for any FOV. However, one of its disadvantages is its excessive length, which makes it impractical for a number of applications.

An alternative to hemispherical concentrators and CPCs is the DTIRC, which not only can incorporate flat thin-film optical filters and be designed for any FOV, but whose size is significantly smaller than that of the CPC, which makes it lighter, more portable, and less expensive (because less material is required for its fabrication). Compared with hemispherical concentrators, DTIRCs offer not just higher gains, but also greater flexibility in design (because they can be designed for a variety of FOVs) and the possibility of incorporating flat thin-film optical filters.

Because subsidiary conditions can be imposed, DTIRCs can be designed to meet different requirements. In the case of directed-LOS link topologies, they can be designed to provide very high concentrations and reduced exit apertures (required for small detectors). When intended for mobile applications, they can be designed with wide FOVs, and to limit the maximum angle of the rays emerging from their exit aperture (which is necessary in structures consisting of an optical filter positioned between the concentrator and the photodetector).

DTIRCs can incorporate either bandpass or longpass filters. If a flat thin-film filter is used, it can be placed either at the entrance of the concentrator (as a window, for receivers with narrow FOVs) or at its exit when the FOV of the receiver is wide.

From the characteristics of DTIRCs presented in Section 5.5, it can be concluded that concentrators designed for reduced FOVs present large geometrical and opto-electronic gains (which is in agreement with the theory of Welford et al. [84] presented in Chapter 4), which means that directed-LOS links benefit particularly from the use of optical concentrators. Designing these concentrators with wide front surface arc angles and fabricating them with materials having large indices of refraction are two more ways of increasing their gain.

Other important parameters of a DTIRC, such as the total height and the maximum output angle, are also affected by the selection of the acceptance angle, the front surface arc angle, and the material used to fabricate the concentrator. The maximum output angle, for example, can be reduced by selecting a material with a low index of refraction, by increasing the front surface arc angle of the concentrator, or by increasing its FOV. The total height can be minimized in a similar way.

As observed, different concentrator designs can be used according to the requirements of a specific application. In general, the most important trade-off that must be made is that between the FOV and the gain. While a directed link benefits from a concentrator with a small acceptance angle and a front surface arc angle as large as possible, for diffuse links, a receiver with a large FOV may be preferred, even if the gain is not high. Applications requiring both high gain and large FOVs can use an array of concentrator-detector structures, where each individual concentrator provides a high gain (reduced FOV) and where the large FOV is created by the combination of the FOV of each individual concentrator. Systems requiring maximum FOV and nearly constant gain may benefit from the use hemispherical concentrators, at the cost of sacrificing gain and accepting more background illumination noise. For applications requiring high gain, FOV design control, better background illumination noise rejection, and compact size, the DTIRC offers a more attractive alternative.

While different versions of DTIRCs with different entrance aperture shapes have been presented, with slight advantages in terms of either size, gain, or maximum output angle, the fact that the difference between their output parameters is not considerable may still make the traditional DTIRCs the preferred option. This is due to the fact that the manufacture of a semi-hemispherical entrance aperture is generally simpler.

Among the important considerations to account for when fabricating the optical front end of the receiver, two of the most important are (1) the wavelength and (2) the transmittance of the material. The first defines the index of refraction, and the second has a direct impact on the gain achieved by the concentrator. Thus, it is important to take into account the charts of index of diffraction and transmittance of the preferred material during the design of the concentrator. It is also relevant to take into account that, to avoid unwanted reflections at the exit of the concentrator, an index matching gel must be used between the concentrator and the photodetector. Reflections at the entrance aperture can be minimized using antireflection (AR) coatings.

Chapter 6

Optical Wireless Transmitter Design

6.1 Introduction to Optical Wireless Transceiver Design

In the generalized free-space optical (FSO) link illustrated in Figure 6.1, the information, prior to transmission from the source to the receiver, exists in electrical form (prior to modulation).

The transmitter, which consists of two parts; an interface circuit and a source drive circuit, converts the input signal to an optical signal suitable for transmission. The drive circuit of the transmitter transforms the electrical signal to an optical signal by varying the current flow through the light source. This optical light source can be of two types: (1) a light-emitting diode (LED) or (2) a laser diode (LD) (cf. Section 3.5). The information signal modulates the field generated by the optical source. The modulated optical field then propagates through a free-space path before arriving at the receiver. Here, a photodetector converts the optical signal back into an electrical form.

The receiver consists of two parts: (1) the optical detector and (2) the signal-conditioning circuit. The optical detector receives the optical signal, and the signal-conditioning circuit regulates the detector output so that the receiver output matches the original input at the transmitter. A good receiver amplifies and processes the optical signal without introducing noise or signal distortion. Noise effects and limitations of the signal-conditioning circuit introduce distortion in the receiver's electrical output signal. The optical detector used at the receiver can

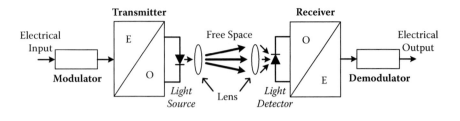

Figure 6.1   Block diagram of an optical wireless link showing the front end of an optical transmitter and receiver.

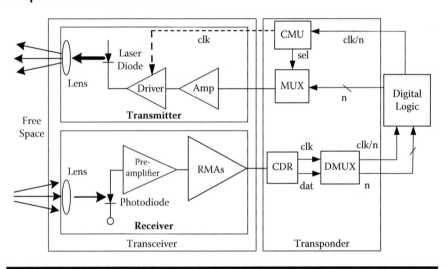

Figure 6.2   Schematic diagram of a typical optical transceiver followed by a transponder for data recovery.

be either a semiconductor positive-intrinsic-negative (PIN) diode or an avalanche photodiode (APD) [267].

In practical optical wireless links, both the transmitter and the receiver blocks are developed in a single circuit called an optical transceiver. The function of this transceiver is to provide full-duplex communication as illustrated in Figure 6.2. Some optical wireless transceivers are developed in single integrated circuits (ICs) using integrated opto-electronic IC (IOEC) technology. To do this, on the transmitter block, the parallel data from the digital logic block is merged into a single high-speed data stream using a multiplexer (MUX). To control the select lines of the MUX, a bit-rate clock must be synthesized from the slower word clock, performed by a clock multiplication unit (CMU). A laser driver or modulator driver drives the corresponding opto-electronic device. The laser driver modulates the current of the LD, whereas the modulator driver modulates the voltage across a modulator, which in turn modulates the light intensity of a continuous-wave laser.

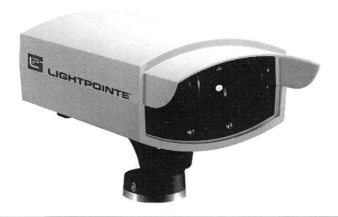

Figure 6.3   A commercial optical free-space transceiver module (FlightExpress 100) manufactured by LightPointe Communications, Inc. (*Source:* From [100].)

Some laser/modulator drivers require a bit-rate clock from the CMU to perform data retiming.

On the receiver block, the same process occurs in reverse order. The detected optical signal is received from the free space by a photodetector that produces a small output current proportional to the optical signal. This current is amplified and converted to voltage by the preamplifier. The preamplifier is followed by the receiver main amplifiers (RMAs) for further amplification of the voltage signal. The RMAs or wideband amplifiers can be limiting amplifiers (LAs) or automatic gain control (AGC) amplifiers. The amplified signal is fed into a clock and data recovery (CDR) circuit, which extracts the clock signal and retimes the data signal. In high-speed receivers, a demultiplexer (DMUX) converts the fast serial data streams into n parallel lower-speed data streams that can be processed conveniently by the digital logic block. Some CDR designs perform the DMUX task as part of their functionality. The digital logic block descrambles or decodes the bits, performs error checks, extracts the payload data from the framing information, synchronizes to another clock domain, etc. Figure 6.3 shows a commercial optical free-space transceiver module called FlightLite® 100 manufactured by LightPointe Communications, Inc. This optical transceiver is capable of transmitting uncompressed HDTV signals wirelessly at a full-duplex throughput of 100 Mbps for recommended distances up to 500 meters outdoors [100]. It can be an alternative to T1/E1 leased lines or 802.11 LAN solutions.

6.2 Transmitter Design Considerations

The development of optical fiber transmitter systems has spawned semiconductor lasers with broad bandwidths and high launch powers, features that should be equally attractive to optical wireless applications. Unfortunately, as explained in

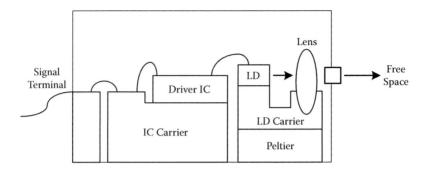

Figure 6.4 Structure of an optical transmitter module.

Chapter 3, one of the most important restrictions to optical wireless transmitters is precisely the optical power level emitted by the source, which, when exceeding specific levels, is potentially dangerous to the human eye. This situation must be taken into account, particularly for indoor free-space optic applications where lasers pose a particular safety hazard to unaware bystanders, which may walk through the path of a wireless IR link. The creation of eye safety standards is justified by the large number of people who may be exposed to the IR radiation of an optical wireless system not just during system operation, but also during system development, installation, and maintenance. Further information on the issue of eye safety can be found in Chapter 3. The reader is also encouraged to check the appropriate ANSI, OSHA, and FDA specifications.

The optical transmitter front end consists of a driver circuit along with a light source. The general structure of the optical transmitter may consist of a lens, an LD or an LED, a driver IC, a Peltier element for cooling, and a modulator block. All these components can be assembled into a mini-sized package, as illustrated in Figure 6.4. For outdoor use, this module package generally presents a special casing to protect it from the rain and from direct exposure to sunlight.

6.3 Optical Source Characteristics

To transmit light in an optical wireless communication link, a suitable light source is needed at the end of the transmitter circuit. The appropriate light source, which as discussed above can be a light emitting diode (LED) or a laser diode (LD), is chosen depending on the specific application of the system. These optical sources are often considered the active component in an optical communication system. Their basic principle of operation is discussed in detail in a number of reference books [101–104].

The output properties and the characteristics of the optical source used for the transmitter are important parameters to consider when designing and evaluating an

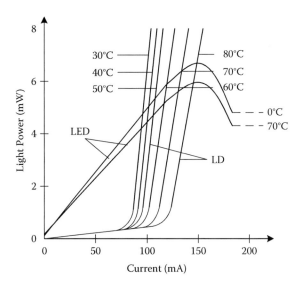

Figure 6.5 **Light power versus current (LI) characteristics of an LD and an LED showing the temperature dependency of the laser.**

optical wireless communication system. It is important, for example, that the light source launches its energy at angles that optimize the transmitted beam. It is also important that the frequency response of the light source exceeds the frequency of the input signal. Furthermore, the light source should have a long lifetime, present a sufficiently high intensity, and be reasonably monochromatic.

Both LEDs and LDs provide good brightness, small size, low drive voltage, and are able to emit a signal at a desired wavelength or range of wavelengths. The selection of one over another depends on the characteristic of the particular application in which they are to be used. When deciding whether to choose an LED or an LD as the light source in a particular transmitter system, one of the main features to consider is their optical power versus current characteristics. This is particularly important because the characteristics of these devices differ considerably (as illustrated in Figure 6.5). It can be seen in the figure that, near the origin, the LED response is linear, although it becomes nonlinear for larger power values. The laser response, on the other hand, is linear above the threshold. Sometimes, mode-hopping* creates a slightly nonlinear response above the threshold in a multimode laser. Single-mode lasers exhibit a linear response above the threshold. The linearity of the source is particularly important for analog systems. The power supplied by both devices is similar (about 10 to 20 mW) [102], but LDs are much more sensitive to temperature variations than LEDs. This is illustrated in Figure 6.5, where it can be observed that, as the temperature increases, the laser diode's gain decreases (for example, a laser that

---

* At certain current values, the maximum emission suddenly hops to an adjacent spectral line.

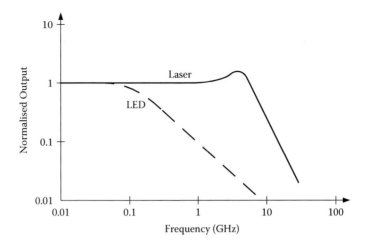

Figure 6.6   Small-signal frequency response of an LED and an LD with negligible parasitic effects.

at 30 °C requires 70 mA to output 2 mW of optical power may require in excess of 130 mA at 80 °C). This implies that more current is required before oscillation.

Another important feature that must be taken into account when deciding whether to use an LD or an LED for a specific application is the speed of the device. LDs, for example, are much faster than LEDs due to the fact that the rise time of an LED is determined by the natural spontaneous-emission lifetime of the material, whereas the rise time of the laser diode depends on the stimulated emission lifetime. Because an LED emits spontaneous radiation, the speed of modulation is limited by the spontaneous recombination time of the carriers. LEDs have a large capacitance, which means that their modulation bandwidths are not very large (a few hundred megahertz). Biasing the diode with a forward current can reduce the capacitance, resulting in an increase of the modulation speed. In the case of a laser above the threshold, the electrons remain in the conduction band for a very short time due to the stimulated recombination; therefore, very fast modulation is possible (up to 10 GHz). Figure 6.6 shows this characteristic.

The spectral emission of an LD remains more stable with temperature than that of an LED. Figure 6.7 shows the spectral shift due to temperature variation in a typical LD. Changes in the output power of the LD with temperature can be prevented by stabilizing the heat sink temperature with a Peltier element and a control circuit. This generally requires more complicated electronic circuits than the ones used for LEDs.

Laser diodes are semiconductor junction devices that contain etched or cleaved substrates, to act as reflecting facets for field reinforcements over the junctions. They therefore combine the properties of an LED and a cavity reflector, producing an external light radiation that is higher in power (10 to 50 mW) and has a better

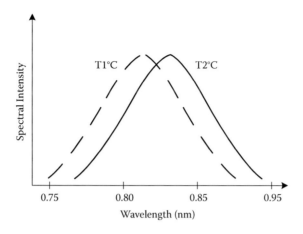

Figure 6.7   Typical LD spectral dependency with temperature.

focused beam than that of a simple LED. LED radiation, on the other hand, is projected outward in all directions, depending on its aperture. The ways in which light is emitted by the source can influence its apparent brightness.

First-generation optical communication sources were designed to operate between 800 and 900 nm. This is because, originally, the properties of the semiconductor materials used lend themselves to emission at these wavelengths. An LED is formed from semiconductor junctions that interact when subjected to an external current, which results in radiated light energy. The choice of junction material determines the emitted wavelength. These materials must emit light at a suitable wavelength if they are to be utilized in conjunction with commonly available detectors, whose spectral response is in the range 0.8 to 1.7μm. Ideally, to achieve emission at a desired specific wavelength, they must allow bandgap variation, which can be achieved through appropriate doping and fabrication. Semiconductor optical sources are typically formed from compounds of gallium arsenide (GaAs) and produce light as presented in Table 6.1 [102].

Most optical transmission technology is designed to operate at a wavelength of 850 nm. However, the latest technology includes 1.55-μm devices [24, 105], which are attractive due to the fact that, up to certain power levels, they do not harm the human eye as the cornea filters incoming light and allows only wavelengths ranging from 0.4 to 1.4 μm into the retina. Thus, transmissions at 1.55 μm do not pass through the corneal filter, and cannot harm the sensitive retina. This means that, at these wavelengths, the emitted power is allowed to reach values up to 10 mW [106] when the source is used as the transmitter of a wireless IR link.

As discussed in Chapter 3, IR sources pose a potential safety hazard if operated incorrectly. For this reason, safety standards have been established to classify optical sources according to their total emitted power [107]. LEDs, for example, do not produce a concentrated light beam. They are large-area devices that cannot be

Table 6.1 Material Combinations Used in the Fabrication of Optical Sources

Material Systems Active Layer/Confining Layers	Useful Wavelength Range (μm)	Substrate
GaAs/AlxGa1-xAs	0.8–0.9	GaAl
GaAs/InxGa1-xP	0.9	GaAs
AlyGa1-yAs /AlxGa1-xAs	0.65–0.9	GaAs
InyGa1-yAs /InxGa1-xP	0.85–1.1	GaAl
GaAs1-xSbx /Ga1-yAlyAs1-xSbx	0.9–1.1	GaAs
Ga1-yAlyAs1-xSbx/GaSb	1.0–1.7	GaSb
In1-xGaxAsyP1-y/InP	0.92–1.7	InP

Source: From [102].

focused by the retina. LDs, on the other hand, are collimated sources whose energy can be focused by the retina. This means that a much lower launch power can be used in order to be considered Class 1 (eye safe). This favors the use of LEDs for indoor applications. The penalty, however, is bandwidth. Whereas the speed of LDs extends to gigabits per second (Gbps), the speed of LEDs is limited typically to 10 Mbps, perhaps extending to 50 Mbps for some specialty devices [106].

Unlike indoor optical wireless systems, the design of an outdoor wireless link or line-of-sight terrestrial system must deal with propagation effects due to atmospheric phenomena (which attenuate the transmitted signal) and with high levels of atmospheric turbulence across the path (cf. Chapter 2). In outdoor environments, the properties of LDs — such as narrow spectra, high power launch capability, and higher access speed — make these devices the favorite optical source for long-distance and outdoor directed-LOS links. However, recent developments in vertical cavity surface emitting lasers (VCSEL), which offer a safer peak wavelength at 1.55 μm [108], is changing this situation. VCSELs are becoming an increasingly attractive option for outdoor and even indoor applications due to their well-controlled, narrow beam properties, high modulation bandwidth, high-speed operation, excellent reliability, low power consumption, and the possibility of having array arrangements. The most commonly developed VCSELs today are the selective oxide-confined ones. In these devices, the operation current is increased to obtain an optical output power that results in a multiple cone-shaped far-field pattern (FFP). A high-power VCSEL can be obtained by increasing the size of the current confinement aperture; however, this method raises concerns regarding device characteristics such as low-frequency performance and a large dip at the center of the FFP [109]. Simultaneously driven VCSEL arrays have been developed by [110] to overcome these issues. Figure 6.8

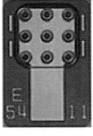 

Simultaneously Driven
3 × 3 VCSEL Array

Simultaneously Driven
4 × 4 VCSEL Array

Figure 6.8   Microscopic top views of the simultaneously driven 3×3 and 4×4 VCSEL arrays. (*Source:* From [110].)

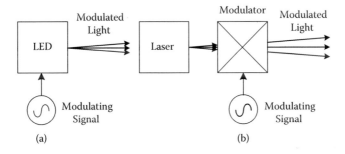

Figure 6.9   Optical modulators: (a) internal modulator and (b) external modulator.

illustrates the microscopic top views of two VCSEL arrays (3×3 and 4×4). These devices consist of several high-speed VCSELs placed 50 μm apart and integrated on a single chip. The epitaxial layers of these VCSELs consist of $Al_xGa_{1-x}As$ semiconductors grown by metal-organic chemical vapor deposition. The development and use of effective VCSEL arrays in optical wireless applications is a topic of increasing interest and ongoing research [111, 112].

6.4 Types of Optical Modulation

Optical modulation can be achieved in two ways: (1) by directly modulating the light source (also known as internal modulation) or (2) by external modulation. These two methods are illustrated in Figure 6.9. In each case, analog or digital modulation schemes can be employed. The modulation of the source information can be in the form of frequency modulation (FM), amplitude modulation (AM), or phase modulation (PM), each of which can be theoretically implemented at any carrier frequency (cf. Chapter 3).

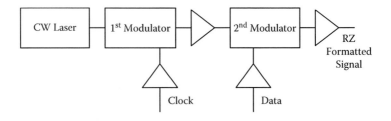

Figure 6.10   RZ signal reformation for an optical transmitter.

In the case of a digital modulation scheme, the first step involves the conversion of an analog signal into digital data (ON/OFF — bit "1" or bit "0" — pulses based on pulse code modulation, PCM). The ON/OFF pulses can then be line coded into different formats. In a practical optical system, line coding formats such as non-return-to-zero (NRZ) are generally used. To produce a very high-speed return-to-zero (RZ) modulated signal, a cascade of two optical modulators can be used. In this arrangement, illustrated in Figure 6.10, the first modulator modulates the light from the laser with an NRZ signal; and the second modulator converts the NRZ signal to an RZ signal in the optical domain. It has been shown [113] that the return-to-zero (RZ) format leads to higher receiver sensitivity, which in turn allows for larger system margins. Sensitivity can be further increased using more sophisticated formats, such as differential phase-shift keying (DPSK) [113]. This technique, recently improved and presented by [114] for next-generation optical transmission, operates at speeds of 40 Gbps or 43 Gbps utilizing a forward error correction (FEC) format. In the approach taken by [114], one device, the electrically synchronized self-pulsating PhaseCOMB laser, replaced the laser. The first modulator includes the electrical amplifier and the first optical amplifier. The self-pulsating laser, also used for all-optical clock recovery [115], is a compact (length <1 mm), three-section laser consisting of two detuned distributed feedback (DFB) gratings and an integrated phase tuning section.

In intensity modulation, the source itself is directly modified by the information signal (analog or digital) to produce a modulated optical field. The laser output intensity is proportional to changes in the injected current, which gives this technique its name. In the case of analog signals, where minimum signal distortion is required, the laser bias current must be higher than the threshold value. Frequency or phase modulation can be inserted on a laser tube by varying its cavity length. Pulse modulation is easily applied to a diode by driving it above and below threshold. Such modulations, in general, are limited to the linear range of the power characteristic. The concept of intensity modulation can be explained with the use of a graph illustrating the behaviors of a laser diode, for a digital or an analog signal, as shown in Figures 6.11a and 6.11b, respectively.

For digital modulation, the diode is modulated by a current source, which simply turns the LED on and off. As sketched in Figure 6.11a, a binary "1" is generated

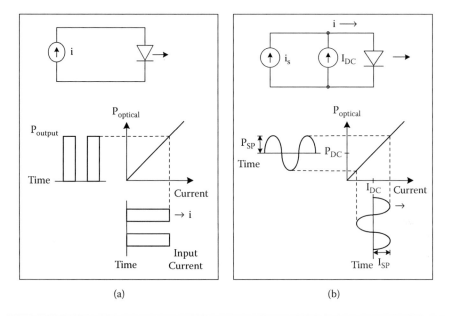

Figure 6.11 Optical modulation of an LED: (a) digital and (b) analog.

when the signal current contains a positive pulse. When biased near threshold, the diode turns on more quickly, and the signal current can be smaller than without the bias. For analog modulation, the DC bias is moved beyond threshold, so that operation is along the linear portion of the power-current characteristic curve. The linearity of the laser diode must be checked carefully if the analog signal is to be reproduced with low harmonic distortion. Analog modulation requires a DC bias to keep the total current in the forward direction at all times. Without the DC current, a negative swing in the signal current would reverse-bias the diode, shutting it off. The total diode current is given by:

$$i = I_{dc} + I_{SP} \sin \omega t \qquad (6.1)$$

and the corresponding optic output power is:

$$P = P_{dc} + P_{SP} \sin \omega t \qquad (6.2)$$

where, I_{dc} is the DC bias current, i_s is the signal current, P_{dc} is the average power, and P_{sp} is the peak amplitude of the modulated portion of the output power (AC power). The shape of the input-current variation is replicated by the optic power waveform due to the linear relationship between power and current. Deviations from linearity distort the signal. When very low distortion is required, the linearity of the proposed source must be evaluated.

In practice, internal modulation may lead to changes in the carrier density and cause a frequency shift in the carrier (optical chirp), which in turn leads to dispersion penalties. This problem can be overcome using an external modulator. In this case, the light generation and the modulation processes are isolated, so that the chirp and turn-on transient effects are avoided [101]. In an external modulation scheme, the light of the source focuses through an external device, whose propagation characteristics are altered by the modulating signal. Such systems have the advantage of utilizing the full power capability of the source. Modulation is achieved via the electro-optic or acousto-optic effect of the material, in which external currents can modify the transmission properties (index of refraction, polarization, direction of flow, etc.) of the inserted light. These effects produce delay variations (phase modulation) or polarization changes (intensity modulation) on the excited beam. Blocking or reflecting the light path can achieve pulsed outputs. Unfortunately, external modulators insert a limit to the modulation range, and generally require a relatively higher modulation drive power.

In general, the internal modulation has the advantages of simplicity, compactness, and cost effectiveness, whereas external modulation can produce optical pulses of higher quality and higher bit rates. The fact that economic budget is one of the main factors when developing an indoor optical wireless LAN makes internal modulation an attractive option for this type of application.

6.4.1 External Optical Modulators

The electro-absorption modulator (EAM) and the Mach-Zehnder modulator (MZM) are the two types of external optical modulators commonly used in optical transmitter systems, regardless of whether they are fiber based or free space. The EAM is small and can be integrated with the laser on the same substrate, whereas the MZM is much larger but exhibits superior chirp and extinction ratio (ER) characteristics. An EAM combined with a continuous-wave (CW) laser source is known as an electro-absorption modulated laser (EML). The EML consists of a CW laser followed by an EAM, which can be integrated monolithically on the same InP substrate, as shown in Figure 6.12a, leading to compact designs and low coupling losses between the two devices. The EAM consists of an active semiconductor region sandwiched between P and N doped layers, forming a PN junction. The EAM operation is based on a principle known as the Franz-Keldysh effect, according to which the effective bandgap of a semiconductor decreases with increments in the electrical field. Without bias voltage across the PN junction, the bandgap of the active region is just wide enough to be transparent at the wavelength of the laser light. However, when sufficient large reverse bias is applied across the PN junction, the effective bandgap is reduced to the point where the active region begins to absorb the laser light and thus becomes opaque [116].

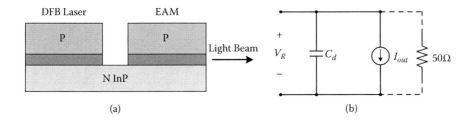

Figure 6.12 Schematic diagram of an integrated laser and EAM and its equivalent circuit.

The relationship between the optical output power P_{out} and the applied reverse voltage V_R of an EAM is described by the so-called switching curve (refer to Figure 6.14a) that shows the achievable extinction ratio (ER) for a given switching voltage V_S. The typical V_S is in the range of 1.5 to 4 V, and the dynamic ER is usually in the range of 11 to 13 dB for switching the modulator from the ON to the OFF state [117]. Due to the fact that the electric field in the active region not only modulates the absorption characteristics, but also the refractive index, the EAM produces some chirp (lower than 1), but usually smaller than that of a directly modulated laser. From the electrical point of view, the EAM is just a reverse-biased diode. Thus, when the CW laser is OFF, the EAM impedance is mostly capacitive. For a 10-Gbps modulator, the diode equivalent capacitor, C_d is about 0.1 to 0.15 pF. When the CW laser is ON, the photons absorbed in the EAM generate a photocurrent similar to that within a PIN photodetector. This current I_{out} is a function of the modulation voltage V_R. Thus, the photocurrent can be described by an equivalent voltage control current source (VCCS) as shown in Figure 6.12b. Hence, the capacitive load appears shunted by a nonlinear resistance, which has a low value during transition and a significantly high value when totally ON or OFF. The package of an EAM often contains a 50-Ω parallel resistor to match the modulator impedance to that of the transmission line that connects the EAM to its driver.

Another type of external optical modulator is the MZM known as an interferometer modulator. In this case, the incoming optical signal is split equally and sent down two different optical paths, as shown in Figure 6.13a. After a few centimeters, the two paths recombine, causing the optical waves to interfere with each other. If the phase shift between the two waves is 0°, the interference is constructive and the light intensity is high (1 state). If, on the other hand, the phase shift is 180°, the interference is destructive and the light intensity is zero (0 state). It is possible to control the phase shift difference and thus the light intensity by changing the delay through one or both optical paths by means of an electro-optical effect. This effect occurs in some materials such as lithium niobate ($LiNbO_3$) and other semiconductors or polymers whose refractive index changes in the presence of an electric field. Two RF waveguides such as titanium diffused into the substrate in the form of

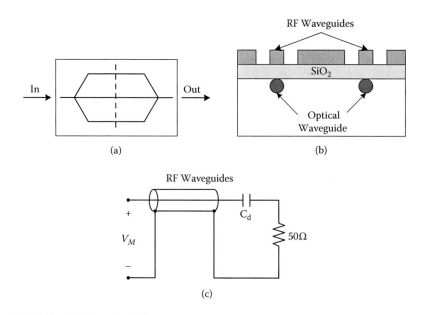

Figure 6.13 Schematic diagram of the dual drive MZM: (a) top and (b) cross-sectional view, and (c) electrical equivalent circuit.

coplanar transmission lines produce electrical fields, which penetrate into the two optical waveguides.

In the dual-drive MZM shown in Figure 6.13b, both optical paths are controlled by two separate RF (radiofrequency) waveguides. Dual-drive MZMs, which have two input ports, can be driven in a push-pull fashion with essentially zero optical chirp. Whereas, for a single-drive MZM, both light paths are controlled by a single RF waveguide, which produces a higher optical chirp. From an electrical point of view, the MZM is just a transmission line, as shown in Figure 6.13c. The transmission line impedance is usually 50 Ω, permitting the use of standard connectors and cables. The end of the transmission line is generally AC terminated within the package. The dynamic ER of an MZM is in the range of 15 to 17 dB and the chirp in this case can be made smaller than 0.1 [116]. The capability of a dual-drive MZM to modulate the intensity and the phase of light makes it a versatile device. It can generate the so-called *optical duobinary signal* (light ON with no phase shift; light OFF; and light ON with 180° phase shift). This property provides a spectral linewidth half the size of a two-level NRZ signal, increasing the receiver sensitivity. MZM modulators can also produce alternative modulation scheme formats such as chirped return-to-zero (CRZ), carrier suppressed return-to-zero (CSRZ), and return-to-zero differential phase shift keying (RZ-DPSK) [118].

The modulation and bias voltage range of the EAM and the MZM modulator drivers are illustrated in Figure 6.14. The modulation voltage V_M is the difference

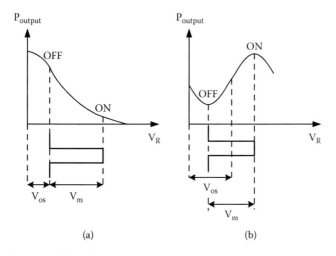

Figure 6.14   Modulation and bias voltage of (a) an EAM and (b) a MZM.

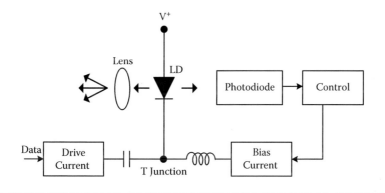

Figure 6.15   Block diagram of directly modulated laser.

between the ON and OFF state voltages supplied by the modulator driver. The DC offset voltage V_{OS} is the voltage supplied by the driver during the ON state in the case of an EAM driver. In the MZM driver, V_{OS} is the DC component supplied by driver.

6.4.2 Direct Digital Modulator

A block diagram showing an example of the direct/internal digital modulator of an optical wireless transmitter is presented in Figure 6.15. Here, the high-speed drive current is shown summed with the DC bias at the T-junction. In this arrangement, the back mirror monitors the light emitted from the laser, whereby the signal is

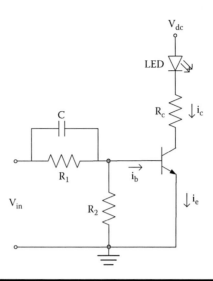

Figure 6.16   Transistor-switched LED digital modulator.

emitted to the atmosphere. The resulting photocurrent is used with a regulator circuit to produce the bias required to maintain high ER and constant output power. When designing a transmitter employing the external modulator presented earlier, it is also important to take into account that it must: supply high-speed current pulses and DC bias; provide logic-level compatibility; and include laser protection and alarms for abnormal operation. In practical digital modulation circuits, a transistor often provides the switching mechanism. Figure 6.16 shows the transistorized series-switched modulator of an LED.

High-speed current pulses at gigabits per second (Gbps) are usually generated by a differential current switch employing GaAs MESFETs (metal-semiconductor field effect transistors), which are faster than silicon BJTs (bipolar junction transistors). Figure 6.17 illustrates a high-speed FET modulator employing a differential current switch. The current switch can easily be designed to supply either constant or controllable peak currents. It can be driven at gigahertz (GHz) rates using controlled rise time, single-ended or differential input signals, and it is readily made compatible with standard high-speed logic families. Transistors F_1 and F_2 along with diodes D_1 and D_2 shift the logic-level input more negative to prevent F_4 from saturating, even if the voltage requirement of the laser rises (for example, due to high end-of-life current, increased series resistance, etc.). Single-ended shunt or series drivers using GaAs MESFETs for maximum speed are also used, but control of the bias or driver current is less straightforward than with the differential switch. Logic gates used at the input (shown in Figure 6.17) are often internally compensated to ensure logic-level tracking and higher noise immunity over the full operating temperature range and thus provide a simple way of obtaining compatibility

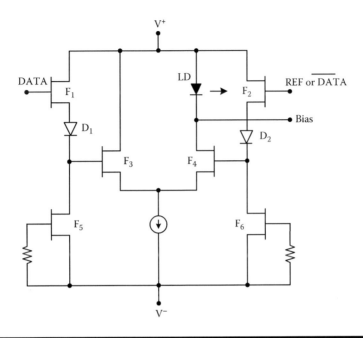

Figure 6.17 High-speed FETs modulator employing a differential current switch.

with associated circuitry. By using a D flip-flop instead, input data can be retimed to correct for pulse-shape degradation ahead of the transmitter.

Current-source drivers have also been found advantageous at high bit rates because they minimize rise time degradation due to lead inductance. For the typical circuit shown in Figure 6.17, operation between the ground rail and a negative voltage is convenient, as many lasers have their p contact connected to the package (that is, grounded). Current-source drivers using a common-base/common-emitter configuration can also be found [119]. In Figure 6.18a, the base current of Q1 sets the laser bias, while the high-speed drive is introduced via Q2. A variation of bias control is shown in Figure 6.18b. A PNP common-emitter stage determines the bias while the NPN common-emitter switch shunts a portion of current supplied by Q1 around the laser [120].

Transient effects such as chirp can be minimized at circuit level by shaping the drive pulse. A dual-pulsing scheme has been analyzed, wherein a low-amplitude pulse of width equal to the period of the relaxation oscillation precedes the main drive pulse [121].

It has been calculated that with proper timing and amplitude, the transient overshoot can be minimized or eliminated, thus reducing chirp. An alternative to reduce chirp has been presented by [122]. Here, a network with resonant characteristics is placed between the driver and the laser [122], reducing the chirp in a 1.5-μm DFB laser by a factor of 3.

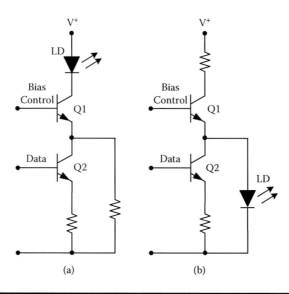

Figure 6.18 **Alternate current driving circuits used to control the laser bias.**

6.4.3 Direct Analog Modulator

A variety of LED analog modulation circuits exist that are relatively simple to develop. One of them is illustrated in Figure 6.19. In this modulator, the LED driving current is given by the transistor's collector current i_c. The supply voltage, together with resistor R_a and R_b, provides the DC base current i_b, which forward-biases the emitter base junction, turning the transistor ON. The resulting collector current $i_c = \beta\, i_b$, where β is the transistor's current amplification factor. The signal voltage V_{in} produces a time-varying base current that adds to i_b. The resistor R_e stabilizes the operating point. The Q point is chosen so that the total base current neither shuts off the transistor during a negative swing nor drives the transistor to saturation during a positive swing.

For linear operation, certain control circuitry design can be used to stabilize the bias in the lasing region of the LI curve (refer to Figure 6.5), usually at a point around which the laser exhibits maximum linearity. The time constant must be selected to be less than the lowest frequency component in the modulating signal. There are two means for minimizing the effects of laser or circuit nonlinearities: negative feedback and predistortion. An example of the negative feedback linearization technique is shown in Figure 6.20, where the input electrical signal is compared with the photocurrent derived from sampling the laser's output using a photodetector [120].

Predistortion or pre-emphasis can be added using a nonlinear circuit between the input signal and the laser. Predistortion that results in a linearized output is

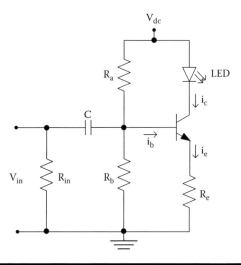

Figure 6.19 **Simple analog modulator/drive for LED.**

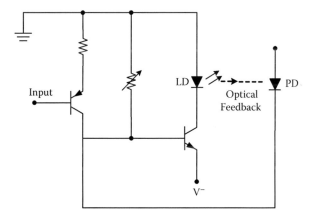

Figure 6.20 **Negative feedback linearization of an analog laser.**
(*Source:* **Adapted from [120].)**

adjusted during transmitter assembly while monitoring distortion products. For nonlinearities that change with time (for example, that are induced by aging or temperature), predistortion has limited effectiveness because it ordinarily does not track these changes. In view of the difficulty of linearization schemes for lasers, particularly for high-frequency applications, it is better to start with either a laser designed for adequate linearity (for example, DFB design) and carefully suppress the reflections or to use a subcarrier approach as described previously. Most recent linearization techniques are discussed and compared in Subsection 6.3.7.

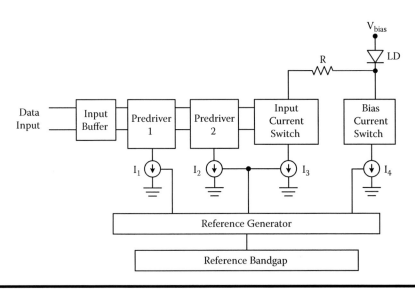

Figure 6.21   **Laser diode driver architecture. (*Source:* Adapted from [123].)**

6.5 Driver Circuit Design Concepts

Following the transmitter design consideration of the previous section, this section highlights laser/modulator driver specifications and presents a revision of general optical driver circuit design. Major design issues include (1) adequate switching speed to ensure minimum intersymbol interference, (2) high output current to launch optical power corresponding to a logical "1", and (3) tolerance of the voltage swing across the laser. In general, two circuit drivers cooperate with the optical transmitter: (1) the laser driver and (2) the modulator driver. The former is a current switch that turns the laser diode ON and OFF according to the logical value of the data. It consists of two main parts: (1) a high-speed modulation driver providing modulation current and (2) a bias network generating DC current, as shown in Figure 6.21 (detailed circuit schematics and operation can be found in [123]). The core circuit is operated under a 3.3-V supply, while the last stage is operated at less than 7 V for sufficient voltage headroom across the laser. Both biased and modulation currents are derived from a bandgap reference to maintain a more stable ER over temperature variations. The modulation driver consists of an input buffer, a two-stage predriver, and an output current switch. The input buffer provides input matching and waveform shaping. The predriver then enlarges the input signal for driving the large current switch stage.

6.5.1 Driver Specifications

The purpose of the transmitter driver is to convert an incoming electrical signal (either analog or digital) into the required current to drive a low-impedance light source. The choice of converter depends on the current requirement of the light source. If the signal is digital, the electrical signal is a high-speed electrical pulse that turns the light on and off. If the light source is analog, the driver must be able to supply an appropriate current to the light source to represent the variations of the desired signal. There are a number of system issues that must be taken into account when designing an optical transmitter driver. These issues are discussed in many references, but a brief description is given as follows:

- The range of modulation and bias current of a laser driver must be large enough to operate the laser under worst-case conditions. Uncooled lasers, for example, require a large current range.
- The range of modulation and bias voltage of modulator drivers must be large enough to allow the desired modulator to operate under worst-case conditions. High-speed MZMs, for example, require a large modulation voltage compared to EAMs.
- The power dissipation of a laser or modulator driver is quite large compared to other transceiver blocks, such as the preamplifier or the main amplifier. Low power dissipation is desirable because it reduces the heat generated in the driver IC and in the system. Excessive heating in the IC may require an expensive package, and excessive heating in the system may degrade laser performance or require a large power-consuming thermoelectric cooler to remove the heat.
- The rise and fall times must be shorter than the bit period — however, not as short as to limit the generation of optical chirp. The rise time is an important variable when selecting light sources and modulation techniques. LEDs have a rise time ranging from a few nanoseconds to a few hundred nanoseconds. The majority of laser diodes have much faster rise times than LEDs, while VCSELs are faster than edge-emitting laser diodes. In practice, edge-emitters can be directly modulated up to 2.5 Gbps and VCSELs can reach 10 Gbps.
- Jitter in the electrical output signals is caused by noise and ISI from the driver circuit. It also is caused by carrier mobility variations due to instantaneous temperature fluctuations. A low jitter generation is desirable because it improves the horizontal eye opening and makes the clock-recovery process at the receiver more robust. Furthermore, in some types of regenerators, the clock signal recovered from the received optical signal is used to retransmit the data. When cascading several such regenerators, jitter increases because of the jitter generated in each regenerator. The laser or modulator driver's jitter generation must be lower because the driver is one of several components contributing to the total jitter generation.

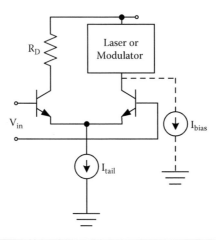

Figure 6.22   Current steering circuit used for driving a laser or a modulator (represented by the boxed load).

6.6 Current Steering Output Circuit

The steering circuits currently available are used mainly in the output stage of a laser driver and modulator. The base circuit of a current steering circuit is shown in Figure 6.22. The same arrangement can also be used with FETs. The differential input voltage swing V_{in} must be sufficient to obtain full or near-full switching in the circuit. The function of the dummy load R_D is to improve the symmetry of the output stage. An asymmetric load configuration results in an input offset voltage, which can cause pulsewidth distortion (PWD). It also results in an undesirable modulation of the voltage across the tail-current source.

Various lasers and modulators can be driven with the current steering circuits illustrated in Figure 6.23. This figure shows (a) a DC-coupled laser, (b) a DC-coupled EAM, (c) an AC-coupled laser, and (d) an AC-coupled single drive MZM. In Figure 6.23a, for example, the series resistor R_s dampens the oscillations due to parasitic inductances, and provides matching to a transmission line. In the DC-coupled EAM (Figure 6.23b), the parallel resistor R_p converts the drive current into a voltage. Most EAMs are driven single ended, such as laser diodes. In the AC-coupled laser diode shown in Figure 6.23c, an RFC pulls the DC output voltage of the current-steering circuit up to the positive supply rail while presenting high impedance to the RF signal. The RF signal is AC coupled to the laser with capacitor C, which needs to be large enough to avoid baseline wander in the laser current. The bias current is injected directly into the laser with a second RFC. The detailed operation of these arrangements can be found in [116].

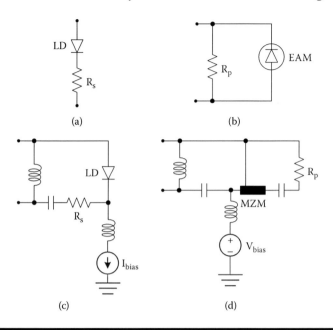

Figure 6.23 Load configuration for the current steering output: (a) DC-coupled laser, (b) DC-coupled EAM, (c) AC-coupled laser, and (d) AC-coupled single drive MZM. (*Source:* Adapted from [116].)

6.7 Back Termination Circuit

Undesirable reflections from the load end of the transmission line back into the driver may occur when using a transmission line to connect the driver to the load. Thus, the laser or modulator must be matched to the characteristic impedance of the transmission line to avoid reflections. From a power perspective, it would be best to eliminate entirely the transmission line and the matching resistor and to drive the laser directly. Three methods can be used for the laser or modulator to match the characteristic impedance: (1) open collector/drain, (2) passive back termination, and (3) active back termination. In the open collector/drain arrangement, the current-steering output stage (I_{bias}) and the laser or modulator (shown in Figure 6.22) are driven through a transmission line. This arrangement works well if the load impedance matches exactly the characteristic impedance of the transmission line. However, if there is a mismatch, a reflected wave is generated at the load and propagated back into the driver. In general, this arrangement can be used only if good matching at the load end is guaranteed.

The problem of double reflections can be resolved by incorporating a passive back termination into the driver, as shown in Figure 6.24. Here, a termination resistor R_T matching the characteristic impedance of the transmission line has been added to the current-steering output stage. This resistor prevents a second reflection

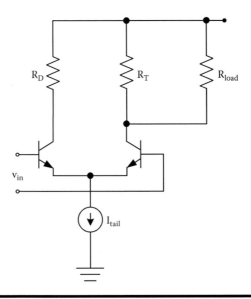

Figure 6.24 Current-steering output circuit with passive back termination. (*Source:* Adapted from [116].)

into the load and absorbs any reflected wave due to a load mismatch. The power dissipation by the driver can be reduced to some extent by choosing a back termination that is larger than the characteristic impedance of the transmission line. Despite the fact that the back matching at low frequency is degraded by this modification, the impact at high frequencies (where parasitic capacitances and inductances play an important role) may not be significant.

A method to protect against double reflections without wasting too much power can be achieved by so-called active back termination. Figure 6.25 illustrates a laser or modulator driver stage incorporating an active back termination [116]. In this arrangement, the back termination resistor R_T is connected to an AC voltage generated by a replica stage. Here, the waveform of the AC voltage under normal operation (that is, without reflections) generates a voltage drop across R_T equal to zero, which means that no power is wasted in this resistor.

6.8 Predriver

For long distance transmission, the transistors in the driver's output stage have to switch large currents of around 100 mA with bias currents of up to 60 mA. Thus, large devices are required for the last stage current switch, and their current driving capability must be made quite large [123]. As a result, their input capacitance also becomes quite large, which means that the driver may not be able to drive this large capacitance at the required speed. It must also be taken into account that the input

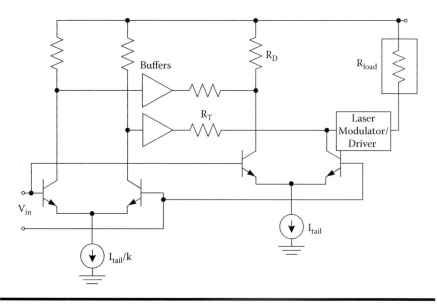

**Figure 6.25   Current steering output stage with active termination. (*Source:*
Adapted from [116].)**

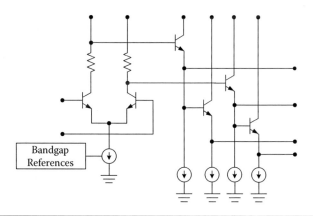

**Figure 6.26   The simplified schematic diagram of a modulator/laser predriver.
(*Source:* Adapted from [130].)**

voltage swing necessary to switch the output stage may not produce a sufficiently
large swing. To resolve this, an element called a *predriver* is generally used to drive
the output stage. Figure 6.26 provides a simplified schematic diagram of a predriver
circuit.

The predriver must be able to drive a large capacitance load while keeping its
input capacitance low. In general, the voltage swings created by the size of the

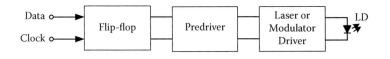

Figure 6.27　Flip-flop implementation for data retiming.

predriver and of the transistor in the output stage must be jointly optimized for best rise and fall times of the driver. Conventional laser driver amplifiers consist of an emitter coupled pair or emitter follower. When a BJT is turned OFF, there is a large amount of charge that needs to be swept out. Thus, a BJT is turned ON generally significantly faster than it is turned OFF, resulting in switching asymmetry. In addition, driving a heavy capacitive load, like the one introduced by the last stage current switch, demands a high power consumption. The first predriver has a constant-voltage swing and current levels, while a second predriver has a modulation current-dependent swing and current levels, and the output switch has Miller effect cancellation capacitors (C_M). To meet the stringent design specifications of constant rise/fall times and overshoot/undershoot specifications, a push-pull type predriver is utilized in the second stage [124].

6.9　Data Retiming

The input data signal is often retimed with a clean clock signal before being fed to the predriver for high-speed laser and modulator drivers. Data retiming is useful in eliminating the pulsewidth distortion and jitter from the input data signal. It is important that the clock source used for retiming has a very low jitter, because the jitter in the clock signal does appear undiminished at the output of the driver. To perform the data retiming function, a flip-fop is placed before the predriver, as illustrated in Figure 6.27. The retiming of a flip-flop is usually implemented with current-mode logic (CML), which is based on nested and cascaded current-steering circuits. The advantages of the CML are high speed, low sensitivity to common mode and power-supply noise, and substantially constant supply current, which minimizes power and ground bounce.

6.10　Automatic Power Control

The laser's LI characteristics are strongly temperature and age dependent. Hence, an automatic power control mechanism is usually required to stabilize the output power of the transmitter. Figure 6.28 shows a simple automatic power control circuit suitable for a continuous-mode laser driver. The automatic power control architecture of most laser driver ICs includes a photodiode that is used to monitor

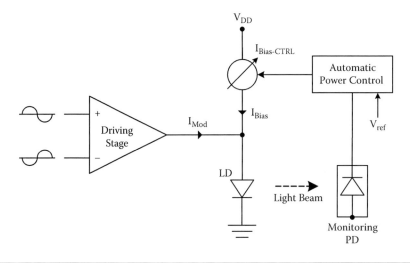

Figure 6.28 Schematic diagram of an automatic power control circuit for continuous mode laser drivers.

the output power of the LD. The monitor photodiode needs to have good temperature, age, and coupling stability to generate a current proportional to the transmitted optical power [116]. In addition, to maintain a constant optical power in the presence of temperature variations, most LD driver ICs are required to control the bias current of the VCSEL in the presence of ambient temperature variations and to have the optical power adjusted by the feedback circuit that includes the monitor PD. This conventional automatic power control is difficult to apply to 1×N or N×N multi-channel LDs. However, Kang et al. [125] have demonstrated multi-channel VCSEL driver ICs with CMOS (complementary metal-oxide-semiconductor) technology. Figure 6.29 shows the simplified block diagram of a multi-channel optical transmitter with the proposed automatic power control using a reference channel. The optical power of all channels can be adjusted by monitoring the specified node of reference-channel VCSEL. The clock signal is applied to this reference channel to extract the DC level of the specified node. The APC block consists of a buffer, a lowpass filter (LPF), and a comparator. The buffer isolates the X node and the LPF. The LPF extracts the DC voltage level of the X node. The comparator compares the extracted DC voltage level with the external reference voltage (V_{ref}) and generates the relevant voltage level to control the bias current of VCSEL. The design of the bias current control circuit that provides a fixed current independent of the X-node voltage deserves special attention in this architecture. For the given external reference voltage, as the temperature increases, the DC level of the X node decreases. This DC level extracted by the LPF is compared to the reference voltage with a comparator; and then the relevant voltage of the comparator increases the bias current to VCSEL until the X-node voltage and the reference voltage reach the same

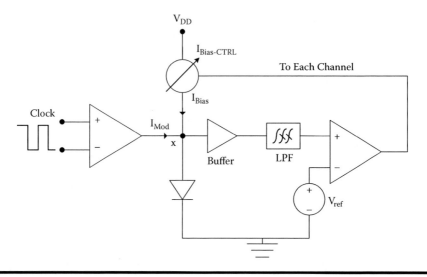

Figure 6.29 Block diagram of a multichannel optical transmitter with automatic power control. (*Source:* From [125].)

level. In this case, the APC method for multi-channel VCSEL driver ICs makes it possible to provide stable optical power over temperature variations without external additional monitor PDs.

6.11 Transmitter Linearization Techniques

The nonlinearity of the optical transmitter presents a primary limitation to the performance of an optical transceiver. This nonlinearity produces intermodulation distortion that obliges to make compromises between the modulation depth, the channel spacing, and the type of modulation scheme, which leads to a degraded bandwidth efficiency. As a result of the severe penalties incurred by the inherent nonlinearities of the optical transmitters, various linearization schemes have been proposed and implemented. These techniques can be broadly categorized into linearization in the optical domain and linearization in the electrical domain (prior to the electrical-to-optical conversion) [126]. Both of these categories offer comparable performance in linearizing an optical transmitter, yet a significant economic advantage is achieved with electronic linearization due to the lower cost associated with the necessary components. Many optical linearization techniques involve the use of duplicate lasers or optical modulators, which increases the cost of the system when compared to ordinary optical transmitters. The linearization techniques can be divided into the following four categories:

Table 6.2   Qualitative Comparison of Linearization Techniques

Linearization Method	Achievable Improvement	Bandwidth	Power Added Efficiency	Physical Size
Feedback	Good	Narrow	Medium	Medium
Feedforward	Good	Wide	Low	Large
Predistortion	Moderate	Wide	High	Small
RF synthesis	Moderate	Wide	Highest	Medium

Table 6.3   Quantitative Comparison of Linearization Techniques

Correction Technology	Correction Capability (dB)	Correction (BW)	BW Relative Cost
Feedback	10–20	<5	Medium
Feedforward	25–35	>100	High
Analog predistortion	5–10	>25	Low
Adaptive predistortion	10–20	>50	Medium

1. Feedforward
2. Feedback
3. Corrective distortion
4. RF synthesis

A survey of examples of the above categories can be found in [103]. In the context of RF power amplifiers, the above techniques have been known to have the qualitative trade-offs shown in Table 6.2 [126]. A quantitative comparison of the various linearization techniques used in cellular base stations, shown in Table 6.3, gives further insight into the benefits and shortcomings of each one [127].

Predistortion can be further divided into two subcategories: (1) digital predistortion (also called baseband predistortion) and (2) analog predistortion (also called RF predistortion). Digital predistortion is often based on look-up table (LUT) algorithms and enjoys well-established DSP hardware for its implementation. However, the major drawbacks of this technique include bandwidth constraints on the input signal imposed by DSP speed limitations, greater complexity and the necessity of digital baseband data, which may not be available in some systems. Analog predistortion, on the other hand, is relatively simple to implement

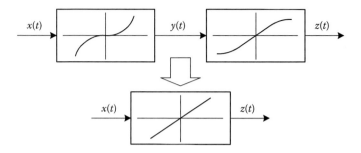

Figure 6.30 Simplified diagram illustrating the concept of predistortion.

and can handle larger signal bandwidths in adaptive architectures [128]. Analog predistortion can be performed in various ways but always involves the purposeful generation of intermodulation products with appropriate magnitude and phase to cancel the intermodulation tones created by the nonlinear transmitter. An adaptive equalizer predistortion using instantaneous digital adaptation (IDA) provides both a large signal-to-noise and distortion ratio (SNDR) and modulation depth range at the transmitter, as demonstrated by Stapleton [129]. To illustrate the concept of predistortion, consider the block diagram in Figure 6.30. Here, an input signal $x(t)$ can be seen being predistorted to produce a signal $y(t)$ that, when distorted by the transmitter, produces $z(t)$, a replica (possibly scaled) of the input signal $x(t)$. Thus, the input–output relationship from $x(t)$ to $z(t)$ is linear.

Chapter 7

Optical Wireless Receiver Design

7.1 Receiver Design Considerations

Having completed the discussion of the optical transmitter, it is now possible to discuss the optical wireless receiver. Optical wireless links operate with limited transmitter power due to eye safety considerations; and in relatively high noise environments due to ambient illumination, the performance of the optical receiver has a significant impact on the overall system performance (cf. Chapter 3). To reduce the shot noise introduced in the detector by ambient light, an optical filter is required, while the preamplifier needs to allow for shot-noise limited operation. In addition, due to link budget considerations, the receiver must have a large collection area, which, as discussed in Chapters 4 and 5, can be achieved through the use of an optical concentrator (that offers effective noiseless gain). Furthermore, as indoor and outdoor optical transceivers are intended for mass computer and peripheral markets, the receiver design is extremely cost sensitive, which makes sophisticated optical systems unattractive.

The design of an optical receiver depends on the modulation format used by the transmitter. Optical wireless receiver systems are very similar to fiber-based receiver systems. They consist essentially of a photodetector and a preamplifier, with possibly additional signal processing circuitry. Therefore, it is necessary to consider the properties of the photodetector in the context of the associated circuitry combined in the receiver because it is essential that the detector performs efficiently with the following amplifying and signal processing stages. This chapter discusses some of the key issues related to the specification and design of optical wireless receivers.

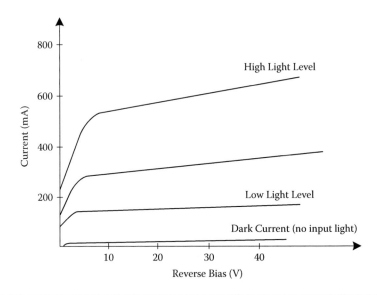

Figure 7.1   Current-voltage (CV) characteristics of a typical photodiode.

7.2 Photodetection in Reverse-Biased Diodes

Before embarking on the analysis of receiver design, it is necessary to understand the operation of the semiconductor photodiodes found in most optical fiber and wireless systems. The photodiode is similar in structure to the PN junction diode, except that its junctions can be exposed to external light. This forms a third optical "terminal" that produces a current flow that is fed to the next stage of the circuit for further amplification. Figure 7.1 shows the current output characteristic of a typical PN photodiode. A detailed analysis of these devices can be found in [101, 102, 104, 120, 131].

The front of the photodiode wafer is usually heavily doped with an opposite dopant type using ion implantation in such a way as to produce a diffusion of majority carriers across the interface of the *p* and *n* regions. This zone is important to the photodiode's performance (efficiency) because it converts photons into electron-hole pairs. When light or photons of energy impinge on the front surface of the photodiode, they penetrate into it and are absorbed by the semiconductor. If the energy of the photons is larger than the energy of the bandgap of the semiconductor material, an absorbed photon excites an electron from the valance band to the conduction band and leaves a hole in the valance band. That is, it generates an electron-hole pair. When an electron-hole pair is created inside the depletion region, these electrons and holes rapidly drift in opposite directions because of the built-in electron field [268]. This produces the photocurrent flowing in the external circuitry. The bandgap and the corresponding typical operating wavelengths for

Table 7.1   Characteristics of Various Semiconductor Materials

Material	Bandgap (eV)	Operating Wavelength (nm)
Si	1.12	500–900
Ge	0.67	900–1300
GeAs	1.43	750–850
$In_xGa_{1-x}As_yP_{1-y}$	0.38–2.25	1000–1600

Photodiode's Equivalent Circuit   Input of Amplifier

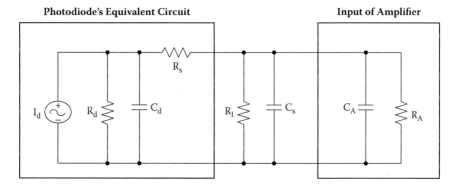

Figure 7.2   Simple equivalent circuit for PN or PIN photodetector.

several popular semiconductor materials are listed in Table 7.1. The $In_xGa_{1-x}As_yP_{1-y}$ material is commonly called indium-gallium-arsenide-phosphide, and it is particularly attractive because its bandgap can be varied by the crystal composition [132].

The equivalent circuit of a PN junction photodetector and the input of the preamplifier stage are shown in Figure 7.2. The diode shunt resistance R_j in a reverse-biased junction is usually very large ($>10^6$ Ω) compared to the load impedance R_l and can therefore be neglected. The resistance Rs represents ohmic losses in the bulk *p* and *n* regions adjacent to the junction, and Cd represents the dynamic photodiode capacitance.

One of the most important considerations when choosing optical detectors is their frequency response, that is, their ability to respond to variations in the incident intensity such as the one created by high-frequency modulation. The three main mechanisms limiting the frequency response of photodiodes are:

1. The finite diffusion time carriers produced in the *p* and *n* regions. This effect can be minimized by the proper choice of length in the depletion layer.
2. The shunting effect of the signal current by the junction capacitance Cd. This effect places an upper limit on the cut-off frequency of the photodiode.

3. The finite transit time of the carriers drifting across the depletion layer. For this reason, the photodiodes operated with a reverse-bias, which gives a larger field causing a faster drift.

Note that with regard to the junction capacitance, reducing the diode area and the doping level may reduce Cd. Increasing the reverse bias voltage as the depletion region is made wider increases the associated bandwidth, which is the opposite of what occurs in terms of transit time. There is, consequently, a trade-off between transit time and junction capacitance. Because the photodiode's quantum efficiency is also improved by lengthening the depletion region, there is a similar trade-off between quantum efficiency and transit time. A well-designed photodiode can have both high quantum efficiency and a wide bandwidth. A wider depletion region can be achieved by another type of the semiconductor PN photodiode — a PIN photodiode — which has a layer of intrinsic semiconductor material sandwiched between the *p* and *n* layers. The depletion region in this structure is almost entirely contained in the intrinsic region and it is wider than the depletion region obtained by a PN photodiode, thus giving the extra advantage mentioned above.

7.3 Choosing the Photodetector

The photodetector is an integral part of the optical front end of a wireless IR receiver because it converts an optical signal into an electrical signal. It is important for the receiver to detect low-level optical signals without introducing many errors. Also, as in digital systems, noise gives rise to bit errors, and it is important that the signal-to-noise ratio (SNR) is sufficiently large to yield an acceptable bit error rate (BER). Therefore, when considering signal attenuation along a wireless link, the system performance is determined by the detector. Thus, improvements in the detector characteristics and performance allow wider or longer optical wireless coverage. The role the detector plays demands that it satisfies very stringent requirements of performance and compatibility. The following criteria define some of the most important performance and compatibility requirements for optical wireless detectors, which, as can be observed, are similar to the requirements for optical sources:

■ High sensitivity at the operating wavelength.
■ High fidelity: to reproduce the received signal waveform with accuracy (for analog transmission, the response of the photodetector must be linear, with regard to the optical signal, over a wide range).
■ Large detection area: to offer a large collection aperture and increased effective detection field-of-view (FOV).
■ Large electrical response to the received optical signal: the photodetector should be able to produce an electrical signal as high as possible for a given amount of optical power.

- Short response time: present systems can operate at speeds of up to several GHz; thus, it is not unreasonable to suppose that future systems will operate at even higher speeds.
- Minimum noise: dark currents, leakage currents, and shunt conductance must be low. In addition, the gain mechanism within either the detector or the associated circuitry must present low noise.
- Other considerations: low cost, small size, and high stability and reliability.

Four types of photodetectors are available for the design of optical receivers: (1) avalanche photodiodes (APDs), (2) photoconductors, (3) metal-semiconductor-metal photodiodes (MSM PDs), and (4) PIN photodiodes (PIN PDs). The first three types have internal gain, whereas the PIN photodiode does not have any internal gain, which is compensated by a larger bandwidth. Because of their compliance with the requirements mentioned above, APDs and PIN PDs are two of the most popular detectors for optical receiver systems. This is explained by the fact that despite the simplicity of photodetectors and their gain, they have a low-gain-bandwidth product that makes them unsuitable for practical optical wireless systems. MSM PDs, on the other hand, present gain and bandwidth advantages when compared to other photodetectors. They can also be monolithically integrated with a preamplifier. More details about the preferred photodetectors used for wireless IR communication receivers are presented below:

- *PIN PD:* This structure offers a wider depletion region that allows operation at longer wavelengths. It also allows light to penetrate more deeply into the semiconductor material. In both device types (horizontally or vertically illuminated detectors), a depleted InGaAs layer of around 3 to 5μm is used to provide high quantum efficiency (up to 90 percent). Considerable effort was directed during the 1990s toward developing high-speed PIN photodetectors capable of operating at bit rates exceeding 100 Gbps [133–135]. A bandwidth of up to 40 GHz was demonstrated by [136] at a responsivity of 0.55 A/W (external quantum efficiency of 44 percent); and in the year 2000, such an InP/InGaAs photodetector exhibited a bandwidth of 310 GHz in the 1.55-μm spectral region [137]. Several techniques have been developed to improve the efficiency of high-speed photodiodes. In one approach, a nearly 100-percent quantum efficiency was realized in a photodiode in which one mirror of the Fabry-Perot cavity was formed using the Bragg reflectivity of a stack of AlGaAs/AlAs layers [138]. Using an air-bridged metal waveguide together with an undercut mesa structure, Tan et al. [139] demonstrated a bandwidth of 120 GHz. However, the bandwidth of current commercially available packaged detectors is usually up to 20 GHz due to limitations of the packaging.

Table 7.2   Photodetector Characteristics

	PIN			APD		
	Si	*Ge*	*InGaAs*	*Si*	*Ge*	*InGaAs*
Operating λ (μm)	0.4–1.2	0.8–1.9	1.0–1.7	0.4–1.2	0.8–1.9	1.0–1.7
Gain	1	1	1	50–500	50–200	10–40
Dark current (nA)	1–10	50–500	1–20	0.1–1	50–500	1–5
Responsivity (A/W)	0.4–0.7	0.5–0.7	0.6–0.9	80–130	3–30	5–20
Response time (ns)	0.3–1	0.1–1	0.05–0.5	0.1–2	0.5–1	0.1–.5
BW (GHz)	0.3–0.6	0.5–3	1–10	0.2–1	0.4–0.7	1–10

■ *APD:* This photodetector achieves internal gain through a more sophisticated structure that creates an extremely high electric field region. At low gain, the transit time and RC effects dominate, giving a definitive response time and hence constant bandwidth for the device. At high gain, the avalanche build-up time dominates and therefore the device bandwidth decreases proportionately with an increase in gain. Such APD operation is distinguished by a constant-gain bandwidth. Often, an asymmetric pulse shape is obtained from the APD, which results from a relatively fast rise time as the electrons are collected; and from a fall time dictated by the transit time of the holes traveling at a slower speed. Hence, although the use of suitable materials and structures give rise times of between 150 and 200 picoseconds, fall times of 1 nanosecond or more are quite common, which limits the overall response of the device. An APD employing a mesa structure has achieved a high gain-bandwidth product at 120 GHz [140]. Other researchers such as [141] have developed an asymmetric waveguide APD. This asymmetric waveguide structure is effective for achieving robustness under high input power operation and high quantum efficiencies. Quantum efficiencies of 94 percent for 1.55 mm and 90 percent for 1.31 mm wavelengths have been demonstrated. A gain-bandwidth product of 110 GHz is large enough to produce excellent performance of an APD operating at 10 Gbps. Higher gain-bandwidth products have also been demonstrated. Otani et al. [142], for example, demonstrated a gain-bandwidth product of more than 300 GHz using a hybrid approach that exhibited a bandwidth of up to 10 GHz and APD gain of up to 35 while maintaining a 60 percent quantum efficiency. Table 7.2 summarizes some typical values for PIN and APD detectors. The values given are typical for commonly used devices and do not necessarily represent fundamental limits to the performance. In many cases, it is possible to improve the performance in one specific area (bandwidth) at the cost of sacrificing the performance in others (that is, sensitivity quantum efficiency).

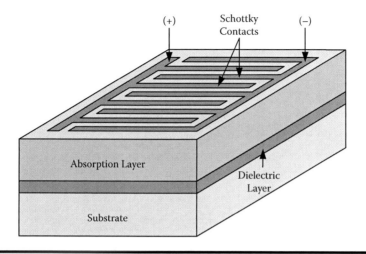

Figure 7.3   Typical structure of an MSM PD.

■ *MSM PD:* This photodetector is one of the most useful in opto-electron-ics systems, high-speed chip-to-chip connections, and high-speed sampling applications. Typically, an MSM PD structure is composed of Schottky diodes with interdigitated metallic electrodes deposited on the top surface of a light absorption semiconductor, as shown by Figure 7.3. The compat-ibility of the fabrication process with that of other electronic devices has led to research activity in this area. It has been found that they can be fabricated using a standard GaAs FET process with only minor modifications. On the other hand, the performance of the MSM diode has for a long time been inferior to that of the PIN PD, counteracting the expected advantages of monolithic integration — namely, a larger bandwidth due to a reduction in parasitic capacitance and bond inductance, and an improved sensitivity due to the reduced capacitance.

   This situation has changed lately, as recent results show that the MSM diode can achieve attractive high performances, with sensitivities of 0.1 A/W at 800 nm, and bandwidths in excess of 105 GHz [143]. For these reasons, this photodetector is becoming the preferred device for monolithic receivers [144, 145]. The key to this large bandwidth is the low capacitance (12 fF) of the diode, plus the short transit time of the carriers between the electrodes, the elimination of mounting parasitics, its low bias voltage (0.5 V), and the process compatibility with high-performance FETs [144].

For all optical receivers — fiber and wireless alike — their sensitivity is a trade-off between the photodiode parameters and the circuit noise. Applications that require a good sensitivity and a broad bandwidth will invariably use a small-area photodiode. Receivers of long-distance point-to-point fiber systems generally fall

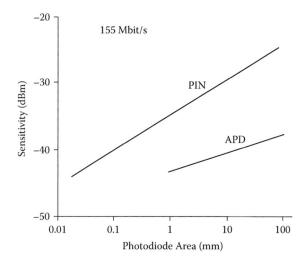

Figure 7.4 Receiver sensitivity (at 155 Mbps) in relation to photodiode type and detection area. (*Source:* From [106].)

into this category. The sensitivity improves (that is, reduces in numerical value) as the photodiode area decreases because of the correspondingly lower capacitance. However, small-area photodiodes incur a greater coupling loss due to the small aperture they present to the incoming beam. In addition, due to the high attenuation suffered by the IR signal when transmitted through air, large-area detectors are sometimes required at the receiver to compensate for this effect, which limits the receiver speed [106]. Therefore, a careful trade-off between these factors is necessary to optimize the overall performance of the receiver. Figure 7.4 shows that a receiver using an APD gives a 10-dB sensitivity advantage over one employing a PIN detector, which is consistent with observations on optical fiber receivers [106]. APD receivers, however, are more expensive and require higher operating voltages than PINs, that being the reason for which they are predominantly used in specialist systems where performance is of paramount importance. For indoor systems where economy is a priority, PIN receivers are generally favored.

7.4 Receiver Noise Considerations

Noise is an unwanted disturbance that masks, corrupts, and reduces the information content of the desired signal. A good understanding of the origins of noise is required to accurately characterize the performance of a wireless IR receiver, as the amount of noise present in a receiver is the primary factor that defines its sensitivity. It is also useful to consider the limit to the performance of a system set by the signal-to-noise ratio (SNR) at the receiver. In addition to the SNR performance,

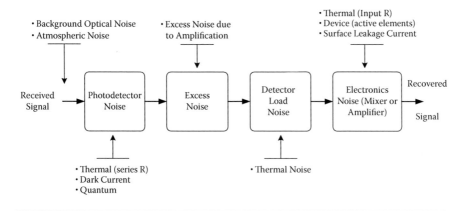

Figure 7.5   The sources of noise in a typical front-end optical receiver.

a particular receiver can also be characterized by its noise factor. In any network involving detection and amplification, the noise factor F (noise figure, $NF = 10 \log F$) is a measure of how much noise is added by the preamplifier or following the main amplifier. The noise figure is a versatile method used to determine the "quality" of such an amplifier and how much degradation of the SNR is contributed by the photodetectors. The signal power at the receiver of an optical wireless link significantly decreases in the presence of atmospheric particles such as fog or rain. Therefore, there is a maximum link distance that can be achieved before the SNR level becomes intolerable, even in the absence of other sources of noise.

Noise is the factor that ultimately limits receiver sensitivity. The noise introduced by the receiver is either signal dependent or signal independent. Signal-dependent noise, for example, results from the random generation of electrons by the incident optical power. The incident optical power level does not affect signal-independent noise. There are many sources of noise in optical communication systems. They include (1) noise from the light source, in which laser intensity noise is primarily due to spontaneous light emissions; (2) atmospheric background illumination noise (cf. Chapter 3); and (3) noise from the receiver itself. The block diagram presented in Figure 7.5 illustrates various noise sources associated with optical receiver systems. The excess noise refers to the noise associated with the photodetector's internal amplification. A comprehensive analysis of noise, especially in semiconductor photodetectors, can be found in [146–148].

The noise sources defined by points (1) and (2) above are considered noise coupled to the photodetector. The laser intensity noise, for example, is caused by intensity fluctuations due primarily to spontaneous light emissions and ambient light.

Intrinsic noise sources (3 above), such as the thermal noise found in a resistor, electronic shot noise in a junction photodiode, thermal noise in the photodiode bulk resistance, and quantum noise inherent in photodetection, on the other hand, are found in all optical receivers. Unfortunately, not much can be done to improve

(1) and (2), particularly in optical wireless receiver design. Therefore, to optimize the performance of an optical wireless system, emphasis is placed on the reduction of the sources of noise of the electronic receivers. A summary of two dominant sources of noise — thermal and shot — is presented next.

Thermal noise, also known as Johnson noise, is the spontaneous fluctuation created by thermal interaction between the free electrons and the vibrating ions in a conducting medium; and it is especially prevalent in resistors at room temperature. From Nyquist's theorem, the open-circuit emf of a resistance R, at temperature T, in a small frequency interval (bandwidth, B) is $\sqrt{4kTRB}$. A resistor thus can be used as a standard noise source. The thermal noise voltage $\overline{e}_T$, in a resistor R, can be expressed by its mean square value, $\overline{e}_T^{\,2} = 4kTBR$, where k is the Boltzmann constant, T is the absolute temperature, and B is the bandwidth in which the noise is considered [147]. The noise power, from $P = V^2/R$, can then be shown to be equal to kT over a unit bandwidth, which, at a room temperature of 290K, gives –174 dBm/Hz. Thermal noise arises from both the photodetector and the load resistor.

Shot noise is caused by current fluctuations created by the discrete nature of charge carriers. Dark current and quantum noise are two types of noise that manifest themselves as shot noise within the photocurrent. Dark current noise results from a current that continues to flow in the photodiode when there is no incident light, and it is independent of the optical signal. Thus, the dark current noise $\overline{i}_d^{\,2}$ is given by $\overline{i}_d^{\,2} = 2qI_dB$, where q is the charge on an electron and I_d is the dark current. The effect of this noise can be reduced because the dark currents in well-designed silicon photodiodes can be made very small. In addition, the discrete nature of the photodetection process creates a signal-dependent shot noise called quantum noise. This quantum noise is produced by the quantum nature of photons arriving at the detector. The noise produced is directly related to the amount of light incident on the photodetector, and it is a function of the average optical power. The mean-squared noise current is, $\overline{i}_q^{\,2} = 2qI_pB$, where I_p is the current out of the diode due to the average incident optical power and B is the electrical noise bandwidth of the measurement (typically normalized to 1 Hz [267]).

The shot noise depends on the optical signal levels, and it is far smaller for fiber links than for free-space links (depending on atmospheric and weather conditions). This is due to the fact that the optical power detected by the fiber system is merely the information signal itself. However, in wireless links the equation $\overline{i}_q^{\,2} = 2qI_pB$ describes how the detector noise varies with ambient light as well. This means that if the ambient light level increases by a factor of 4, the noise produced at the detector only doubles. This equation implies that for high ambient daytime conditions, different techniques are needed to reduce the amount of ambient light striking the detector to see a significant reduction in the amount of shot noise produced at the detector circuit. The equation $P = I^2 R_L$ can be used to convert this shot noise current into power. The shot noise power, N_s, in a 1-Hz bandwidth, is then $N_s = 2q(I_d + I_p)R_L$. For example, if the input resistance of the photodetector is 50 Ω, a photo-current of 1 mA generates a quantum noise power density of –168 dBm/Hz ($= 1.6 \times 10^{-17}$ mW).

As discussed in Chapter 3, inserting an optical filter between the lens and the light detector can reduce the effects of ambient light. However, as shown by the noise equation, the amount of light hitting the detector must be dramatically reduced to produce a significant reduction in the induced noise. Because most sun-light contains some amount of infrared light, such filters do not reduce the noise level very much. However, very narrow bandpass filters can be selected to match the wavelength of a laser diode and effectively reduce the ambient light generated by fluorescent and incandescent lamps and sunlight [269].

7.5 Bit Error Rate and Sensitivity

The performance of a receiver degrades as a result of several factors, including laser linewidth, relative intensity noise (RIN) of the source, and receiver noise. These effects have an impact on the maximum transmission distance and signal cover-age area. The performance of an optical digital link is measured by the bit error rate (BER). Conversely, in an analog optical link, the performance of the receiver is measured by the SNR or the carrier-to-noise ratio (CNR). An example of such an analog system is the optical wireless CATV system, where multiple analog or digital TV signals (or both) are combined by means of subcarrier multiplexing (SCM) into a single analog signal, which is then transmitted over an optical link. To provide good picture quality, this analog signal must have an SNR much greater than 14 to 17 dB, which is typical for non-return-to-zero (NRZ) signal. To be more precise, cable television engineers use the term "CNR" for RF-modulated signals such as the NRZ signal for SCMs systems and reserve the term "SNR" for baseband signals. For analog TV channels with AM-VSB modulation, the National Association of Broadcasters (NAB) recommends a CNR > 46 dB. For a digital TV channel with QAM-256 modulation and forward error correction (FEC), a typical CNR > 30 dB is required [149].

Receiver sensitivity is defined as the average received optical power needed to achieve a given communication rate and performance. For a digital system, the communication rate is measured by the bit rate, and the performance is measured by the BER. Both parameters — sensitivity and BER — are interrelated; and either the BER is given for a certain signal level or a minimum signal level is required to achieve a given BER. This minimum signal, when referred back to the input of the receiver, is known as the sensitivity. Sensitivity is one of the key characteristics of an optical receiver, especially of wireless IR links. It indicates the level to which

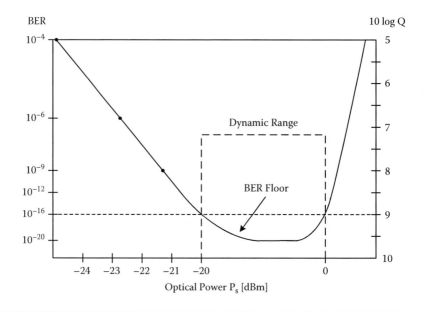

Figure 7.6 BER plot coordinate system. For small power values, the plot follows a straight line down to BER = 10^{-16}, where the sensitivity is –20 dBm.

the atmosphere can attenuate the transmitted signal and still be detected reliably by the receiver. Sensitivity can be defined in the electrical or in the optical domain [102]. The best method to characterize the optical receiver is by measuring the BER as a function of the received power and then plotting one against the other. This plot, known as a BER plot, is widely used in optical communication literature. An example of such a plot is illustrated in Figure 7.6. The BER plots do not necessarily have to be a function of received optical power P_s. They can also be a function of the input current of the preamplifier or the input voltage of the main amplifiers. A BER of 10^{-9} is sufficient in many optical communication system applications.

7.6 Bandwidth

When designing an optical receiver system, the question of how large the receiver bandwidth (BW) should be made — and more generally, what frequency response should be chosen — is not always clear. The reason for this lack of clarity is that if the receiver BW is made wide, the receiver preserves the signal waveforms without distortion, but at the same time picks up a lot of noise, which may corrupt the signal. As seen in the previous section, a high amount of noise translates into low receiver sensitivity. On the other hand, if the receiver BW is made narrow, the noise decreases and the sensitivity improves. Unfortunately, this introduces signal distortions known as intersymbol interference (ISI). Like noise, ISI also reduces sensitivity because the output swing (at the decision circuit) is reduced for certain bit sequences. Thus, there is an optimum receiver BW for which the sensitivity is optimal. A rule of thumb for NRZ receivers indicates that the optimum 3 dB BW is about two thirds of the bit rate (BW_{3dB} = 2/3B, where B is the bit rate), or similarly, the optimum BW is about 60 to 70 percent of the bit rate.

The BW of the photodetector is determined by the speed with which it responds to variations of the incident optical power, known as rise time T_r. The photodetection rise time T_r is defined as the time over which the current builds up from 10 to 90 percent of its final value when the incident optical power is changed abruptly. Clearly, T_r depends on the time taken by electrons and holes to travel to the electrical contacts. It also was discussed above that the receiver consists of a cascade of building blocks: photodetector, preamplifier, filter, main amplifier, and decision circuit. It is the combination of these building blocks that needs to have the BW_{3dB} = 2/3B. There are several strategies for assigning BWs to the individual blocks and in this way achieve the desired overall BW. Three practical BW allocation strategies [116] include:

1. To design all the receiver blocks for a BW greater than 2/3B. In this method, a precision filter is inserted, typically after the preamplifier, to control the BW and frequency response of the receiver (for example, a 4th-order Bessel Thomson filter, which exhibits good phase linearity). This technique is typically used for lower-speed receivers (less than 2.5 Gbps).
2. To design the preamplifier for a BW_{3dB} = 2/3B, and the other blocks for a BW much greater than that without using a filter. This approach has the advantage that the preamplifier bandwidth specification is relaxed, permitting a better noise performance but controlling less well the frequency response.
3. To design all blocks for a BW = 2/3B. Here, no single block is controlling the frequency response and no filter is used. This approach is typically used for high-speed receivers (greater than 10 Gbps). At these speeds, it is challenging to design electronic circuits and PDs, and they cannot be over-designed.

7.7 Signal Amplification Techniques

The current from the detector is usually converted to a voltage prior to signal amplification. The current-to-voltage converter is perhaps the most important section of any optical receiver circuit. An improperly designed circuit often suffers from excessive noise associated with ambient light, which appears as noise in the detector [269]. To optimize the collection of the optical signal received from free space, the right front-end circuitry design must be considered. An optical receiver's front-end design can be usually grouped into three preamplification categories: (1) low-impedance voltage amplifier, (2) high-impedance amplifier, and (3) transimpedance amplifier. Any of these configurations can be built using contemporary electronic devices such as bipolar junction transistors (BJTs), field effect transistors (FETs),

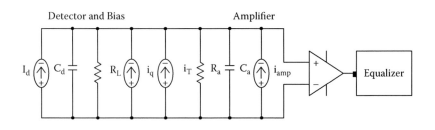

Figure 7.7 **Equivalent circuit of an optical receiver, including various sources of noise.**

or high electron mobility transistors (CMOS). The receiver performance achieved depends on the devices and techniques used in the design.

Figure 7.7 shows a full equivalent circuit for a digital optical receiver in which the optical detector is represented as a current source I_d. The noise sources ($\overline{i_q^2}$, $\overline{i_T^2}$, and $\overline{i_{amp}^2}$) and the immediately following amplifier and equalizer are also illustrated. Here, equalization compensates for the distortion of the signal due to the combined transmitter, medium, and receiver characteristics.

The design of the front end requires a trade-off between speed and sensitivity. Because using a large load resistor R_L can increase the input voltage to the preamplifier, a high-impedance front end is often used. Furthermore, a large R_L reduces the thermal noise and improves receiver sensitivity. The main drawback of the high-impedance front end is its low bandwidth, given by:

$$\Delta f = (2\pi R_L C_T)^{-1} \tag{7.1}$$

Here, an $Rs \ll R_L$ is assumed and C_T includes the contributions from the photodiode (C_d) and the transistor used for amplification (C_a). A high-impedance front end cannot be used if Δf is considerably less than the bit rate. An equalizer is sometimes used to increase the bandwidth. The equalizer acts as a filter that attenuates the low-frequency components of the signal more than its high-frequency components, thereby effectively increasing the front-end bandwidth. If the receiver sensitivity is not of concern, it is possible to simply decrease R_L to increase the bandwidth, resulting in a low-impedance front end.

Transimpedance front ends provide a configuration that has a high sensitivity, together with a large bandwidth. Its dynamic range is also improved compared with high-impedance front ends. Optical fiber receivers generally employ a transimpedance design because this presents a good compromise between bandwidth and noise, both of which are influenced by the capacitance of the photodiode. However, the large area of the photodiodes used in optical wireless receivers requires designs that are significantly more tolerant to their high capacitances. A common design in optical wireless receivers combines transimpedance with bootstrapping. The latter reduces the effective photodiode capacitance as perceived by signals [150]. This

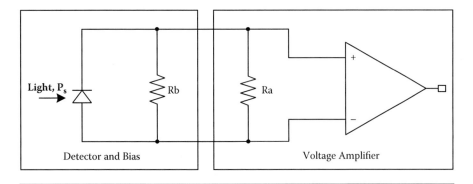

Figure 7.8   Low-impedance front end of an optical receiver with voltage amplifier.

allows for the use of a relatively high feedback impedance, which reduces noise and increases sensitivity.

7.7.1 Low-Impedance Front End

The simplest and perhaps the most common low-impedance front end is a voltage amplifier that uses an effective input resistance R_a as shown in Figure 7.8. Low-impedance front ends allow thermal noise to dominate within the receiver, which makes them impractical for optical wireless links and wideband systems.

7.7.2 High-Impedance Front End

This configuration consists of a high input impedance amplifier, together with a large detector bias resistor to reduce thermal noise. Unfortunately, this structure tends to produce a degraded frequency response, because the bandwidth is not maintained for wideband operation. In addition, it needs an equalizer at a later stage, as illustrated in Figure 7.7. Thus, it provides an improvement in sensitivity over the low-impedance front-end design but it creates a heavy demand for equalization and has problems of limited dynamic range.

A method used to convert the leakage current into a voltage; and which is shown in many published circuits [102, 148], is illustrated in Figure 7.9. This simple "high-impedance" technique uses a resistor to develop a voltage proportional to the light detector current. Unfortunately, this circuit suffers from a number of weaknesses. If the resistance of the high-impedance circuit is too high, for example, the leakage current caused by ambient light can saturate the PIN PD, thereby preventing the modulated signal from ever being detected. In addition, saturation occurs when the voltage drop across the resistor, from the photodiode leakage current, approaches the voltage used to bias the PIN device. To prevent saturation, the PIN must maintain a bias voltage of at least a few volts.

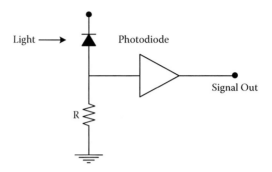

Figure 7.9   High-impedance optical front-end receiver.

Consider the following example. A PIN PD leakage current of a few milliamps may be possible in an outdoor optical wireless link operating under high background illumination conditions. If a 12-V bias voltage is used, the detector resistance would have to be less than 10,000 Ω to avoid saturation. With a 10K resistor, the conversion would then be about 10 mV for each microamp of PIN leakage current. However, to extract the weak signal of interest that may be a million times weaker than the ambient light level, the resistance should be as high as possible to get the best current-to-voltage conversion. These two needs conflict with each other in the high-impedance technique and will always yield a less-than-desirable compromise.

In addition to a low current-to-voltage conversion, there is also a frequency response penalty incurred when using a simple high-impedance detector circuit. The capacitance of the PIN PD and the circuit wiring capacitance both tend to act as frequency filters and cause the circuit to have a lower impedance when used with the high frequencies associated with light pulses. Furthermore, the high-impedance technique does not discriminate between low- and high-frequency signals. Flickering streetlights, lightning flashes, or even reflections off distant car windshields may be picked up along with the weak signal of interest. The use of high-impedance circuits is therefore not recommended for outdoor or even indoor optical wireless links.

7.7.3 Transimpedance Front End

Transimpedance amplifiers using voltage feedback operational amplifiers are widely used for current-to-voltage conversion in applications where a moderately high bandwidth and high sensitivity are required. The amplifier is termed "transimpedance" because it utilizes shunt feedback around an inverting amplifier, a technique known to stabilize an amplifier's transimpedance. The basic schematic of a transimpedance preamplifier with a resistor feedback load is shown in Figure 7.10. The resistor that converts the current to voltage is connected from the output to the input of an inverting amplifier. This front-end structure, which acts as a current converter, gives a low noise performance without presenting the severe limitation on bandwidth imposed by the high input impedance front-end design.

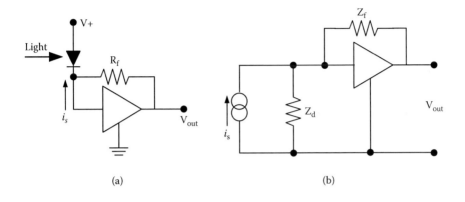

Figure 7.10 **Transimpedance front-end receiver: (a) typical circuit and (b) simplified small-signal equivalent circuit.**

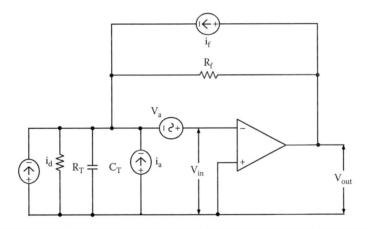

Figure 7.11 **Equivalent circuit of an optical receiver incorporating a transimpedance (current mode) preamplifier.**

This configuration largely overcomes the drawbacks of the high-impedance front end due to its low noise and to the reduction of the high impedance through negative feedback. It operates as a current mode amplifier. An equivalent circuit for an optical receiver incorporating a transimpedance front-end structure is shown in Figure 7.11. In this equivalent circuit, the parallel resistances and capacitances are combined into R_T and C_T, respectively. It provides a far greater bandwidth without equalization than the high-impedance front end, and it has a greater dynamic range, which makes it the preferred option in most wideband optical-fiber communication receivers. The amplifier acts as a buffer and produces an output voltage proportional to the photodiode current. The most important improvement the transimpedance amplifier presents over the simple high-impedance circuit is its

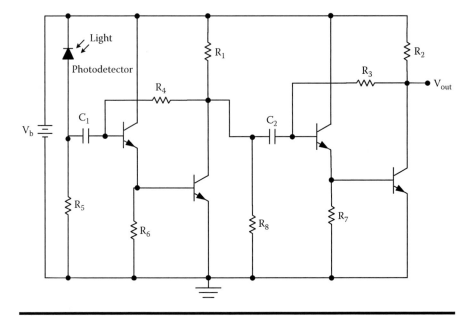

Figure 7.12   Transimpedance front end with direct coupling collector feedback bias and Darlington pair configuration.

canceling effect of the circuit wiring and diode capacitance. The effective lower capacitance allows the circuit to work at much higher frequencies. However, as in the high-impedance method, the circuit still uses a fixed resistor to convert the current to a voltage and is thus prone to saturation and interference from ambient light, which is highly problematic for outdoor receiver applications. An example of a transimpedance front-end receiver with direct coupling collector feedback bias and Darlington pair is shown in Figure 7.12.

7.7.4 Bootstrap Transimpedance Amplifier

As discussed in Chapters 4 and 5, an optical wireless receiver is required to have a large collection area due to power consumption considerations. One of the main sources of noise in wideband preamplifiers employing large-area detectors is the noise due to the lowpass filter formed by the detector capacitance and the input impedance to the preamplifier. Typical commercial large-area photodetectors have capacitances of around 100 to 300 pF, compared to 50 pF in their optical-fiber link counterparts. Hence, techniques to reduce the effective detector capacitance are required to achieve a low noise and wide bandwidth design.

In any photodetector application, capacitance is one of the main factors that limits the response time. Decreasing the load resistance improves this situation, but at the expense of sensitivity. In the amplifier presented in this subsection, positive feedback can be used with caution. Here, it is possible to combine the effective

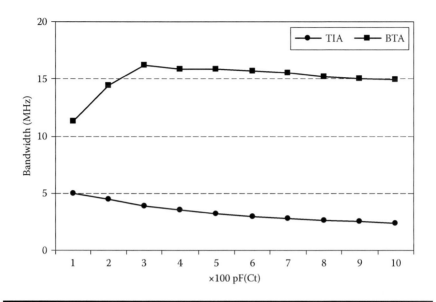

Figure 7.13 **Frequency response presented by a BTA and a transimpedance amplifier (TIA) with varying total receiver capacitance C$_t$.** (*Source:* **From [152].)**

stability of negative feedback with the desirable features of the positive one. In addition, the input capacitance, in effect, constitutes part of the feedback network of the op-amp and hence reduces the available loop gain at high frequencies. In some cases, a high input capacitance can cause the circuit to have a lightly damped or unstable dynamic response. Lag compensation by simply adding feedback capacitance is generally used to guarantee stability; however, this approach does not permit full exploitation of the full gain-bandwidth characteristic of the op-amp. An alternative approach to reduce the input capacitance is the bootstrap transimpedance amplifier (BTA) reported by [150, 151], which was originally intended for receiver bandwidth enhancement. This technique offers the usual advantages of the transimpedance amplifier, together with the effective capacitance reduction technique for optical wireless detectors cited above. Figure 7.13 illustrates the bandwidth enhancement provided by the BTA compared to a conventional transimpedance amplifier (TIA) for very large input capacitance [152].

Active techniques for reducing the input capacitance (for example, bootstrapping) have been previously reported and described. Successful examples and analysis of transimpedance circuits using bootstrapping can be found in [153, 154]. The basic bootstrapping principle is to use an additional buffer amplifier to actively charge and discharge to input capacitance as required. By doing so, the effective source capacitance is reduced, enabling the overall bandwidth of the circuit to increase. There are four possible bootstrap configurations (series or shunt bootstrapping modes, with either floating or grounded sources) that can be applied to the basic circuit. The series configuration has been reported in [151, 155], and the

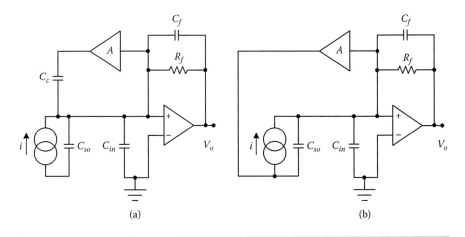

Figure 7.14 Equivalent circuit for shunt BTA: (a) grounded source and (b) floating source. (*Source:* From [153].)

shunt technique can be found in [153]. The series BTA configuration has an additional buffer amplifier in series with the main transimpedance amplifier for the generation of a suitable "forcing voltage." The shunt bootstrap circuit has an additional buffer/voltage amplifier in parallel with the main transimpedance amplifier to allow the generation of a suitable forcing voltage to keep the AC voltage drop across the input capacitance virtually zero. The two possible configurations for this circuit are shown in Figure 7.14, where the current source represents photocurrent. In Figure 7.14a, the source is connected to ground. The capacitance associated with the source is shown in parallel as C_{SO}. This source is connected to a transfer resistance circuit that consists of an amplifier of gain A with feedback components R_f and C_r in parallel. The output of the bootstrapping amplifier A is fed back via a coupling capacitor C_c to keep the AC voltage drop across the source capacitance C_{SO} virtually zero. The floating source configuration is shown in Figure 7.14b. Here, assuming appropriate bias conditions, it can be connected directly to the source.

The increase in bandwidth exhibited by the bootstrap circuit can be attributed to this reduction in feedback capacitance. It was demonstrated in [156] that the 3-dB cut-off frequency of an optical link based on an InP PIN photodiode has been enhanced from about 3 GHz to 7 GHz with a Monolithic Microwave Integrated Circuit (MMIC) BTA.

The BTA also improves receiver sensitivity through a circuit configuration such as the one shown in Figure 7.15, where the photodiode used is an Si APD. The preamplifier without bootstrapping incurs a 3-dB sensitivity degradation with respect to the bootstrap case [157]. A large number of recent optical wireless receivers incorporate bootstrapping front ends. An example of such a receiver can be found in [158]. Here, a direct detection optical receiver for a 155-Mbps terrestrial line-of-sight free-space application with good sensitivity and high dynamic range is demonstrated.

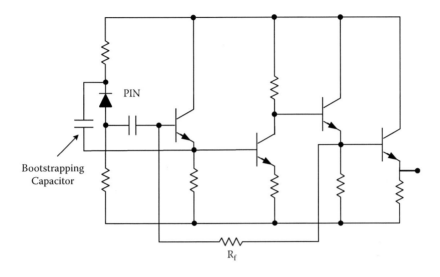

Figure 7.15 **Simplified schematic of a PIN BJT bootstrapped common collector preamplifier. (*Source:* From [157].)**

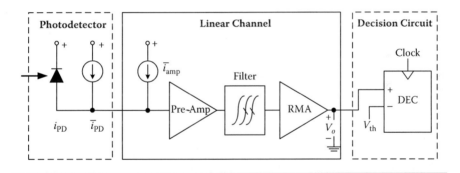

Figure 7.16 **Optical receiver model with RMA and decision circuit blocks.**

7.8 Receiver Main Amplifier (RMA)

The purpose of the RMA or post-amplifier is to amplify the small signal from the preamplifier to a level that is sufficiently high for reliable operation of the clock and data recovery circuit, as illustrated in Figure 7.16. The RMA can be designed as a stand-alone part (with normal 50-Ω transmission line impedance at the input or output) or it can be integrated with the preamplifier or the photodetector in the same chip. There are two main types of RMA: (1) the limiting amplifier (LA) and (2) the automatic gain control (AGC) amplifier. Depending on the application, it may or may not be acceptable for the RMA to introduce nonlinear distortion. If low

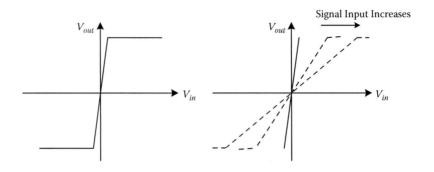

Figure 7.17   Transfer function of (a) an LA and (b) an AGC.

distortion is mandatory, the AGC amplifier must be used. If nonlinear distortion can be tolerated, the simpler LA design is generally preferred. For small input signals, most amplifiers display a fairly linear response. For large signals, however, nonlinear effects may distort the output signal. Specifically, a differential stage with constant tail current has limited output swing; when all the tail current is switched through one of the two transistors, several distortions in the form of clipping set on.

An LA is an amplifier with no special provisions to avoid clipping or limiting the output signal. An AGC amplifier consists of a variable-gain amplifier and an automatic gain control mechanism that keeps the output swing constant over a wide range of input swings. Here, where the LA would start to distort for large input signals, the AGC reduces its gain and manages to stay in the linear regime. This is shown in Figure 7.17, where the idealized DC transfer function of an LA and AGC amplifier is illustrated. The DC transfer function of an AGC amplifier depends on the input signal. Nevertheless, for very large signals, the AGC amplifier cannot reduce its gain any further and limiting eventually occurs. The system designer must make sure that the input dynamic range of the AGC amplifier is sufficiently wide to avoid this situation.

Because an LA does not have a gain-control mechanism, it is generally easier to design than an AGC amplifier. Furthermore, its power dissipation, bandwidth, and noise are often superior to those of an AGC amplifier realized in the same technology. The linear transfer function of an AGC amplifier preserves the signal waveform and permits analog signal processing to be performed on the output signal. Examples of such signal processing tasks are equalization, slice-level steering, and soft decision decoding. The LA severely distorts when operating in the limiting regime, which causes much of the input information signal to be lost. However, the zero-crossings are preserved as long as the LA is free of offset and memory. Hence, it is sometimes said that LA performs the slicing function.

The preamplifier specifications determine the primary performance of the optical receiver, such as the sensitivity and the overload limit, whereas the RMA specifications have a smaller impact. However, not giving the right importance to the

RMA specifications may lead to a degradation of the entire optical receiver performance. The main specifications considered when designing an RMA are the voltage gain, the bandwidth, the group delay variation, the noise figure, the input dynamic range, the input offset voltage, and the low-frequency cut-off. These specifications are explained in detail in a number of contemporary electronics communication references (see, for example, [159, 160]). The typical values of the RMA bandwidth are often made much larger than the desired receiver bandwidth. For this reason, another term used for RMA is "broadband amplifier." The RMA has little impact on bandwidth. The receiver bandwidth then is mainly controlled by the front-end stages, that is, the photodetector, the preamplifier, or the filters.

The gain is usually specified for small input signals. In the case of an LA, the signal must be small enough so that it stays in the linear regime of the transfer function in such a way that gain compression or limiting does not occur. In the case of the AGC, the signal must be small enough so that the maximum gain is realized, which means that the AGC mechanism does not reduce the gain. The AGC range is defined as the ratio of the maximum to the minimum gain. For example, a minimum gain of 10 dB in a particular AGC and a maximum gain of 50 dB give an AGC range of 40 dB. The noise generated by an RMA adds to the total receiver noise and thus degrades the receiver sensitivity. Common RMAs are built in multistages to increase the total gain, which might increase the total noise, especially the thermal noise associated with the load resistor. Therefore, RMA performance is usually represented by its noise figure (*NF*), which is the ratio of the total output power to the fraction of the output noise power due to the thermal noise of the source resistance.

The input dynamic range of the RMA describes the minimum and maximum signal for which the RMA performs a useful function; for example, for which the bit error rate (BER) is sufficiently low. The maximum input signal depends on the type of amplifier. An AGC amplifier operates linearly and therefore maximum input swing is reached when the amplifier starts to distort. A commonly used criterion is the 1-dB gain compression point. For an LA that operates in the nonlinear regime, on the other hand, the maximum input swing is reached when the amplifier produces so much pulsewidth distortion and jitter that the specified BER cannot be maintained. For example, in BJT implementations, a large input signal can cause the base-collector diodes to become forward biased, thus leading to such distortion [161].

The input offset voltage of an RMA is the input voltage for which a zero output voltage results. In an AGC amplifier, a small offset voltage is less significant because it can be compensated at the output of the amplifier. In particular, if slice level steering is used after the AGC amplifier, the offset voltage is eliminated automatically. However, a nonzero offset voltage in an LA results in a slice level error, which in turn first causes pulsewidth distortion because of the finite rise and fall times of the received signal and then more bit errors because it becomes more likely that the noisy signal crosses the off-center slice level.

Virtually all RMAs are structured as multi-stage amplifiers because this topology permits the realization of very high gain-bandwidth products. Furthermore, broadband techniques such as series feedback, emitter peaking, cascading, transimpedance load (shunt or active feedback), negative Miller capacitance, buffering, scaling, inductive load (shunt peaking), inductive inter-stage network, and distributed amplifier are applied to the gain stages to improve their bandwidth and to shape their roll-off characteristics. Some RMAs also permit the introduction of a controlled amount of offset to adjust the slice level. This feature is useful to optimize the BER performance in the presence of unequal noise distributions for "0" and "1." The AGC amplifier consists of a variable-gain amplifier (VGA), an amplitude detector, and a feedback loop that controls the gain in such a way that the output amplitude remains constant. Most RMAs include an offset compensation circuit to reduce the random offset voltage below critical value, such as the one presented by Schick [161], who reported attenuation compensation techniques in distributed SiGe HBT amplifiers using highly lossy thin-film microstrip lines for operation in a 40-Gbps fiber-optic communication system. The amplifier chip with the extended unit cell is shown in Figure 7.18. It can be seen that, in terms of speed, the LAs present advantages over the more complex AGC amplifier.

7.8.1 Optimization Decision Circuit

In a general optical receiver chain for digital format, after the RMA there is a binary decision circuit with a fixed threshold V_{th}, as illustrated by Figure 7.16. The decision circuit determines the output voltage from the linear channel of the RMAs (whether a bit is zero or one) by comparing the sampled output voltage V_{th}, which is located at the midpoint between the zero and one levels. The comparison in the decision circuit is triggered by a clock signal, which is typically provided by a clock-recovery circuit. This receiver block is nonlinear, and other techniques used to optimize the overall performance of the receiver (included in this model) are adaptive equalizer, jitter, decision threshold control, and forward error correction.

The optimum frequency response of the receiver depends on the shape of the received pulses. An adaptive equalizer can be used to cancel intersymbol interference (ISI) and automatically respond to varying received pulse shapes. For digital modulation schemes, such as non-return-to-zero (NRZ), the linearities of the receiver and transmitter are of secondary importance; however, for analog modulation, like the one used in CATV applications, linearity is critical. ISI and noise appear both in the frequency and in the time domains, where they are known as data-dependent jitter and random jitter, respectively. In systems with APDs (photodetection with internal amplification), the optimum decision threshold is below the halfway point between the zero and one levels. In this situation, either a manual slice-level adjustment or an automatic decision point steering scheme can be used

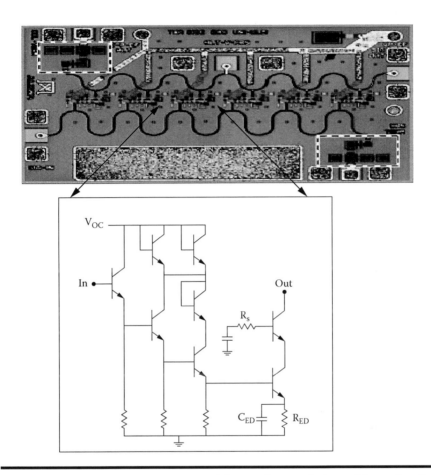

Figure 7.18 **Distributed SiGe HBT amplifiers with extended amplifier unit cell.**
(*Source:* **From [161].**)

to optimize the performance. Also, forward error correction (FEC) can be used to improved BER performance. An advanced type of FEC decoder requires the receiver to have a multilevel slicer for the average binary decision circuit.

The most popular type of equalizer is the adaptive decision-feedback equalizer (DFE), which consists of two finite impulse response filters (FIRf), one feeding the received signal to the decision circuit and other one providing feedback from the output of the decision circuit illustrated in Figure 7.19. The DFE is a nonlinear equalizer because the decision circuit is part of the equalizer structure. In contrast to the simpler linear feed-forward equalizer (FFE), which consists of only the first FIRf, the DFE produces less amplifier noise. Full treatment of the DFE and FFE can be found in [162, 163]. Note that the RMA must be linear (AGC amplified) to prevent nonlinear signal distortions of the input of the equalizer.

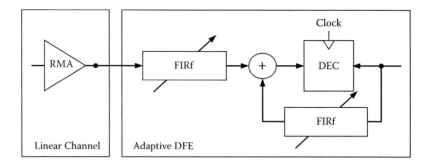

Figure 7.19   Linear RMA channel followed by an adaptive decision feedback equalizer.

The equalizer of an optical wireless receiver can be implemented in the digital or the analog domain. For digital implementation, the biggest challenge is the AD converter, which samples and digitizes the signal from the linear channel. For a 10-Gbps NRZ receiver, one needs a converter with about 6 bits of resolution sampling at 10 GHz. For analog realization, the delays can be implemented with cascades of buffers, and the taps with analog multipliers and current summation nodes. The challenge here is to achieve enough BW and precision over the process, supply voltage, and temperature. An analog 10-Gbps DFE with eight precursor taps and one post-cursor tap is described in [155]. When implanting a high-speed equalizer, two parallel decision circuits are used: first is slicing, in case the previous bit is a zero and the other one in case the previous bit is a one. Then, a multiplexer in the digital domain selects which result to use.

7.8.2 Loss of Signal Detector (LSD)

In the event of an accidental blockage of the link path by a moving object (which usually occurs in indoor optical wireless links) or a failure of the laser or its power supply, the receiver loses the signal. The receiver must detect this loss condition to give an alarm and allow the possibility of restoring connectivity automatically. Some commercial RMAs include LSD on the chip to support this function. The block diagram of the LSD of an optical wireless receiver chain is illustrated in Figure 7.20. An amplitude detector monitors the signal strength at the RMA output. The measured amplitude is compared with a threshold voltage V_{th} with a comparator circuit. The V_{th} is set to the amplitude that just meets the BER requirements. The comparator usually exhibits a small amount of hysterisis to avoid oscillations in the LOS output signal. Finally, a timer circuit suppresses short loss of signal events. For example, the SONET standard requires that only the loss of signal events that persist for Δt_{LOS} longer than 2.3 μs are signaled [164].

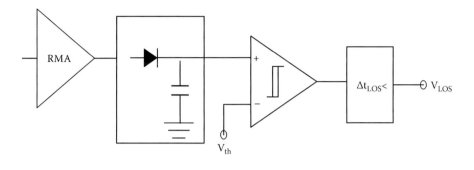

Figure 7.20 Schematic diagram of an LSD, where $\Delta t_{LOS} > 2.3$ μs for SONET.

7.9 Transceiver Circuit Implementation Technologies: Hybrid and Monolithic Integration

As seen in Figures 6.1 and 6.2, optical transceivers consist of many functional blocks. In early optical transmission systems, optical transceivers were often built on printed circuit boards using discrete electronic components and commercially available integrated circuits. However, as transmission speeds and bandwidths increased, it became necessary to integrate the optical transceiver as much as possible. This explains the rapid growth of research work in the area of implementation technology; especially for optical fiber-based applications where the high bandwidth of the single-mode fiber can be exploited. Some of the benefits of transceiver integration in optical-fiber systems include high performance, high speed, high reliability, small size, and potentially low cost. Simultaneously, the listed advantages endeavor for the optical free-space system, especially for optical wireless transceiver mobile units.

Optical receivers and transmitters can have two types of integration: (1) electronic and (2) opto-electronic. Electronic integration refers to the case where the optical detector or source is still a discrete device, while all the electronic circuitry in the preceding optical receiver or transmitter block is integrated in either one or several integrated circuits (ICs).

Opto-electronic integration, on the other hand, refers to the case where the detector is monolithically integrated with the preamplifier (or the light source integrated with the laser driver) in an opto-electronic integrated circuit (OEIC) at the front end of the system. Thus, the optical receiver or transmitter consists of an OEIC preceded by either one or several ICs. Opto-electronic integration can be applied to both the receiver and the transmitter to form the single OEIC of an optical transceiver module.

The broadband circuits of the optical transceiver blocks are realized in a wide variety of monolithic integration technologies. The design principles discussed in the previous section, for example, can be implemented using a wide variety of technologies and employing different types of transistors, such as bipolar junction transistors (BJTs), heterojunction bipolar transistors (HBTs), heterostructure field-effect transistors (HFETs), metal-semiconductor field-effect transistors (MESFETs), and complementary metal-oxide-semiconductor transistors (CMOS). Bipolar transistors consistently outperform field-effect transistors in switching applications. However, field-effect devices in the form of high electron mobility transistors (HEMTs) fabricated from InP hold most of the high-profile transistor speed records, such as the unity gain cut-off frequency and the maximum frequency of oscillation [116]. For medium- and low-speed applications, standard silicon technologies such as MOSFETs or BJTs (or both) are preferred due to their cost advantage. However, for high-speed applications, silicon-germanium (SiGe), gallium-arsenide (GaAs), or indium-phosphide (InP) technologies, which offer fast heterostructure transistors in the form of HFETs and HBTs, become necessary. This issue is well explained by Sackinger [116]. Further information on this subject can be found in [130, 165, 166]. Some examples of current technology developments for optical transceiver sub-circuits are presented below.

Over the past decade, most modulator drivers have been fabricated with compound semiconductors owing to their high-breakdown and high-speed characteristics [167–170]. However, some of them are inefficient, both with respect to the die area and in terms of power consumption. To achieve low-cost system integration with other digital functional blocks, the driver circuit must contain high-speed transistors with lower breakdown voltages. That is why 10- to 14-Gbps SiGe HBT drivers with output swings over 3 V_{pp} and 10-Gbps CMOS drivers with 2.5-V_{pp} output swings have been reported [171, 172]. However, it is difficult for single-transistor topology to generate a voltage swing larger than 3.5 V_{pp}. That is why, to solve this problem, a series-connected voltage balancing (SCVB) topology has been proposed to double the breakdown voltage [173]. This was the idea of Mandegaran and Hajimiri [174], who modified the SCVB and implemented the driver in 0.18-μm SiGe BiCMOS technology to give a differential output swing of 8 V_{pp}.

Li and Tsai [175] reported in 2006 the effective implementation of 10-Gbps modulator drivers using silicon-based process technology. Here, an intrinsic collector-base capacitance feedback network (ICBCFN) was incorporated into the conventional cascade and SCVB circuit configurations at the output stage, to implement drivers in 0.35-μm SiGe BiCMOS technology, whereas an intrinsic drain-gate capacitance feedback network (IDGCFN) modified from ICBCFN was used to implement drivers in 0.18-μm CMOS technology. Thanks to the advantages offered by ICBCFN, the power consumption was greatly reduced (from 2 W to 1 W).

Kucharski et al. [176] claimed a higher speed (20 Gbps) and lower power consumption but they demonstrated only a small output swing of less than 1 V_{pp}. These performances and comparisons are shown in Table 7.3.

Table 7.3 Comparison of Modulation Drivers with Respective Implementation Technology

Technology	Freq. (Gbps)	Power (W)	Supply Voltage (V)	O.p. Swing (Vpp)	Die Area (mm2)	Ref.
AlGaAs/InGaAs pHEMT (0.2 μm)	10	1	5.2	6	7.5	[167]
AlGaAs/InGaAs pHEMT (0.15 μm)	10	2.6	5/–3.3	17	2.1	[168]
InP/InGaAs DHBT	12	1.6	60/60	7	1.1	[169]
InP SHBT	12.5	1.4	20/18	6	0.91	[170]
Si HBT	14	2.2	5.2	7.2	n.a.	[171]
CMOS (0.18 μm)	10	0.7	1.8	5	n.a.	[172]
InP HEMT (0.1 μm)	10	3.5	3.4/–7.3	7.2	3.4	[173]
SiGeBiCMOS (0.35 μm)	10	3.6	6.5	8	n.a.	[174]
SiGe BiCMOS (0.35 μm)	10	1	5	6	0.80	[175]
CMOS (0.18 μm)	10	0.6	1.8	8	0.68	[175]
CMOS (0.18 μm)	20	0.12	2.5	<1	n.a.	[176]

Note: n.a. = not available.

Finally, on the receiver side, the pre- or main amplifier and the following subcircuits have been implemented using a wide range of technologies, which similar to the case of the optical transmitter block, include MESFETs, HFETs, BJTs, HBTs, BiCMOS, and CMOS. Examples of amplifiers or preamplifiers implemented with these technologies can be found in [177–182]. Currently, researchers are trying to create 50-Gbps (and faster) amplifiers, as well as working on reducing circuit implementation cost. Usually, heterostructure devices such as HBTs and HFETs, based on compound materials such as SiGe, GaAs, and InP, are used to reach this goal [183, 184]. The following subsection discusses further each implementation technology used for higher integration of the optical transceiver.

7.9.1 OEIC for Higher Integration

Opto-electronic integration is under intense investigation by many laboratories around the world. The simplest form of an OEIC is the PIN-FET, which monolithically combines a PIN photodetector with a FET on the same substrate. An alternative to OEIC consists of using a flip-chip detector, which can be mounted on top of a circuit chip. In this way, the detector and the circuit technologies can be chosen and optimized independently.

Some of the advantages of the OEIC technology include reduced size, smaller interconnect parasitics, and lower packaging cost. Bringing the photodetector and the preamplifier on the same chip, for example, reduces the detector's total capacitance at the summing node of the preamplifier. Thus, the OEIC approach is particularly attractive for high-speed receivers. However, it is still a less-mature technology than electronic integration. In addition, OEIC receivers are often less sensitive than their hybrid counterparts because compromises must be made when fabricating the photodetector and the transistors on the same substrate and with the same set of processing steps. An expert panel at the 2001 GaAs IC Symposium, with representation from Lucent, Nortel, JDS, and others, unanimously concurred that the solutions in the near term will require hybrid rather than integrated approaches. Each individual component will be chosen based on its own price–performance ratio. The cost savings from integration cannot yet overcome the performance penalty inherent in mixing several functions on the same die. Hence, no single technology is likely to dominate on the transceiver module [185]. However, with the increasing use of erbium-doped fiber amplifiers (EDFAs), the loss in sensitivity may not be a major drawback.

Since the pioneering work by Yariv et al. [186, 187] on monolithic integration of optical and electronic components on the same chip, numerous contributions have been reported in the literatures of this field [188–193]. Thus far, both GaAs and InP materials have been used to demonstrate different types of OEICs. The high electron drift velocity and high electron mobility of these systems, as well as the ability to form heterojunctions exhibiting high quantum efficiency, combine to make these materials highly desirable for high-speed electronic and opto-electronic devices.

Lattices matched to InP are capable of supporting almost all the functions in the receiver or transmitter module, including the photodiode, the high-speed digital electronics, the high-bandwidth analog components, the laser, and even the modulator. Some of the functions can even be integrated on the same substrate. Another advantage of the InP-HBT technology specific to optical communication applications is that it permits the integration of long-wavelength opto-electronic devices [270]. For example, PIN photodiodes sensitive in the 1.3- to 1.55-μm wavelength range may be integrated on the same chip by reusing the base-collector diode [116]. Narrower bandgap semiconductors such as InGaAs are direct bandgap absorbers of 1.3- and 1.55-μm near-IR radiation, which make excellent photodiodes. Because of the much lower surface recombination velocity in InP than in GaAs, InP-HBTs can be scaled to submicron dimensions, resulting in record power-delay performance [185]. However, a drawback of the InP technology is the present lack of large substrates.

In some MOSFET and HFET circuit technologies, a high-quality junction photodiode may not always be available; and a metal-semiconductor-metal (MSM) photodetector, which consists of two back-to-back Schottky diodes may not offer a solution. Silicon circuit technologies offer a medium-speed photodiode that is sensitive to near-IR radiation at wavelengths around 0.85 μm. In CMOS technologies, for example, the junction between the *p* drain/source diffusion and the *n*-well can serve as a photodiode [194].

As explained in the previous subsection, optical wireless applications require fast and sensitive photodetectors with large light-sensitive areas. Large photodetector areas, for example, are useful for optical wireless communication systems [195] and for optical distance measurement sensors [196]. Unfortunately, the high junction capacitance of a photodiode with a large diameter limits both the speed and the sensitivity of the TIAs connected to the photodiode. This can be partly compensated for by using a photodiode with a low-capacitance vertical PIN structure. Noise analysis has shown that in BiCMOS technology, the noise is dominated by the base series resistance of the first transistor stage of the TIA. Compared to a photoreceiver using a photodiode with a diameter of 700 mm, which attains a maximum data rate of 20 Mbps and a sensitivity of 20 dBm, Phang and Johns [195] and Hein et al. [197] have achieved a data rate of 300 Mbps and a sensitivity of 24.7 dBm with a receiver using a photodiode with a diameter of 400 mm. Much faster OEICs are possible, even with large-area photodiodes. An optical receiver with a PIN photodiode of diameter 300 mm, for example, achieves a data rate of 2.5 Gbps and a sensitivity of 20.1 dBm [193] at the cost of a high power consumption of 70 mW, whereas that of Hein et al. [197] consumes less than 6 mW. Other recent reports on high-speed, large-area OEIC receivers can be found in [145, 194, 198]. Most of the receivers presented in these reports have photodiodes with areas between 0.6 and 0.85 μm.

Intense work has also been carried out on realizing transmitter subcircuitry in OEIC technology by combining the laser diode, the monitor photodiode, and the driver circuit on the same chip. The vertical cavity surface emitting laser (VCSEL) offers another opportunity for opto-electronic integration using InP. The most promising VCSELs at 1.3 to 1.5 μm are formed with lattice-matched materials on InP substrates; and, despite the possibility of integration with electronics exists, there have been no demonstrations until now. Yet, the VCSEL offers a considerable improvement over edge-emitting lasers by allowing multiple sources to be integrated in a package-friendly format [144]. It is a challenge to combine laser and circuit technologies effectively into a single one due to the significant structural differences between lasers and transistors. Lasers, for example, require mirrors or gratings for their operation. As a result, transmitter OEICs are not as far advanced as receiver OEICs. An alternative to the above-mentioned monolithic OEIC is the integration of the laser and the driver by means of flip-chip technology. An important advantage of this flip-chip OEIC approach is that the technologies for the laser chip and the driver chip can be optimized independently, thus avoiding the compromises of monolithic OEICs.

7.10 Summary and Conclusions

Chapters 6 and 7 presented optical transmitters and receivers, which are usually packed together as an optical transceiver, as well as their roles in optical wireless communication systems. This included a detailed description of the transmitter and receiver design requirements, as well as an introduction to the circuit implementation in hybrid and monolithic technologies. In addition, a brief discussion of the design limitation and further development of optical transceiver design, specifically focused on optical wireless communication links, was presented.

The design of the optical transceiver for a wireless IR system must cater to a number of issues that are very different from the case of an optical-fiber system. The atmospheric turbulence across the path (cf. Chapter 2), for example, leads to variations in the received power level (scintillation) and beam angle of arrival (AOR) at the receiver's optical antenna. To accommodate these fluctuations, a photodiode of diameter equal to 1 mm may be required for an outdoor link, compared to the 50 pm required for an optical-fiber receiver. Thus, this larger photodetection area results in a larger capacitance and can lead to a serious degradation in terms of receiver noise performance, bandwidth, and gain. For this reason, several researchers [150, 151, 153] suggest the use of bootstrapping techniques to reduce the effect of large input capacitance. The free-space optical receiver must also have a large dynamic range to allow for the long-term variations in received power level resulting from weather attenuation and for the short-term variations or scintillation. Unlike a fiber system where the BER performance of the link is governed by the Gaussian noise statistics associated with the receiver, the BER of a free-space optical link is determined by both the receiver sensitivity and the strength of the scintillations across the path. Other than that, the system designer must cater to a number of optical wireless transceiver design limitations, such as eye safety, power budget limitations (especially for distance outdoor links), uninterrupted line-of-sight (LOS) propagation, and transmitter beam divergence angle.

Some future trends and research directions of the optical wireless transceiver design include:

1. Improved system capability and reliability: research in this area is meant to overcome the above-described system design limitations, such as the limitations to the receiver sensitivity due to noise and other hindrances. The increase in the diameter of the reception lens influences the mass and dimensions of the system proportionally. For this reason, lens sizes of no more than 150 mm are generally used, with much smaller sizes needed for mobile units.

2. Another research field is the one related to the design of safer optical wireless links for indoor use. The optical wireless hardware currently used can be classified into two broad systems that operate near 0.85 and 1.55 μm [271]. Laser beams at 0.85 μm are near-infrared and therefore invisible, which is considered a retinal-hazard wavelength region when the light focuses onto a tiny spot on the retina. Even at longer wavelengths ($\lambda > 1400$ nm), the surface of the eye can be burned. It is possible to design eye-safe laser transmitters at 0.85 μm. However, due to the aforementioned biophysics, the allowable laser power is about 50 times higher at 1.55 μm. This factor of 50 is important to the communication system designer because a higher optical beam power can be used, while still being safe for humans. In outdoor links, the additional laser power allows the system to propagate over longer distances or to compensate for a higher attenuation, as well as to support higher data rates.

3. Large bandwidth and higher speed transceiver systems. Historically, every new generation of optical telecommunications equipment has been 4 times faster than the previous generation. In the world of data communications, the steps taken are even larger; every new generation has been 10 times faster than the previous one. Work on the technological aspects of 40-Gbps ICs has reached the stage where cost-effective commercial products are being developed. Competing technologies in this area include InP HBT, InP HEMT, GaAs p-HEMT, and SiGe BiCMOS. The interface speeds of telecommunications and data communications systems, such as Ethernet, have increased by a factor of 4 and 10, respectively. This means that if current trends continue, interface speeds will reach 160 Gbps for the former and 100 Gbps for the latter in the next generation.

   InP-based ICs are promising candidates for high-bit-rate optical transmitters and receivers because of their excellent high-frequency response, high breakdown voltage, and the possibility of monolithic integration with 1.5-μm photodetector and semiconductor optical modulators. A number of research groups are currently aiming at 100-Gbps speeds and beyond [199]. To reach such speeds, fast optical front-end receivers and fast laser/modulator drivers (possibly with integrated MUX and DEMUX) must be designed. Currently, laser technology is developing in such a way that it allows increasing transfer factors and the link distances, which is particularly relevant for long-distance, high-speed communication systems. However, restricted operation in poor weather conditions in outdoor free-space optical links constitutes a common problem for the majority of these systems.

4. Higher circuit integration (cf. Section 6.6). Despite the many potential advantages of the OEIC, to date OEICs have not outperformed hybrid circuits performing similar functions. The reason for this has been generally attributed to the challenges presented by certain materials with respect to the integration of very different devices on a single chip. However, the OEIC circuits that monolithically integrate optical and electrical components on a single semiconductor chip represent a device technology with potential to meet a broad range of future telecommunication and computing systems needs. As for the case of integrated electronics, monolithic integration offers significant advantages over hybrid circuits in compactness, reliability, possible performance improvements resulting from reduced parasitics, and potentially significant reductions in cost (particularly in the case of arrays). OEICs, for example, offer the advantage of having reduced size, which is particularly significant in applications where small opto-electronic components are required or where the high-speed capabilities of opto-electronic devices allow for a significant reduction in interconnect lines and improved noise immunity. In addition, in terms of reliability, once the fabrication process has been established, the reliability of the finished integrated components is superior to the one presented by their hybrid counterparts. This is due to the fact that the number of wire bonds and related mechanically weak points in the circuit are minimized. Furthermore, reliability is also enhanced in terms of noise immunity.

5. Cost effective for mass production; one of the most important research areas in optical wireless transceiver design concerns the development of low-cost systems for mass production and system implementation for public use. The optical wireless access point and mobile units for indoor use must be very low cost to be able to compete with traditional RF devices. Both the mobile units and the access points should be low power to minimize the size and the cost of the battery. Expensive laser emitters, complex optical focal systems, and high-precision automatic mechanisms (required for pointing and tuning the source with a small beam divergence) make optical wireless transceivers expensive. A collimated beam, for example, is desirable in outdoor systems to compensate for the high energy losses (cf. Chapter 2). This explains why special pointing tools and automatic tuning devices are sometimes used. However, the use of these elements increases the cost of the system. In large-angle systems, on the other hand, the special pointing optical devices and the automatic tuning systems are not required, which makes the system cheaper. Unfortunately, their power requirement is higher. For this reason, increasing efforts are directed toward finding optimum optical front ends at the receiver and the transmitter.

Chapter 8

Modulation, Coding, and Multiple Access

8.1 Introduction to Modulation and Multiple Access Techniques

The selection of the right modulation scheme is of paramount importance when designing an optical wireless system. The modulation technique employed defines important parameters such as bandwidth and power efficiency, which affect the overall performance of the system. A power-efficient modulation scheme, for example, is generally required in wireless infrared (IR) systems where the transmitted power level is limited due to battery consumption restrictions. A bandwidth-efficient modulation scheme, on the other hand, can be used in applications where the bandwidth must be optimized, but where power consumption is not critical. Some of the most popular modulation schemes used for optical wireless communication systems are pulse position modulation (PPM), on-off keying (OOK), and subcarrier modulation [24].

Multiple access refers to the way in which the IR medium can be shared among different users. A number of electrical and optical multiple access techniques have been explored in the past with this purpose. Some of the most important techniques include TDMA (time division multiple access), WDMA (wavelength division multiple access), and FDMA (frequency division multiple access). This chapter presents some of the most important characteristics of these and other multiple access techniques, as well as the most popular modulation and coding schemes used for wireless IR communications.

8.2 Modulation

A number of analog, digital, and pulse modulation techniques and their use in optical communication systems have been explored through the years with the intention of improving the performance of wireless IR links. Modulation in these systems occurs at two different levels: (1) the electrical signal that contains the message modulates a carrier, and (2) the electrically modulated signal is used at the driver of the optical source to convert the electrical signal into optical power, which means that the amplitude of the modulated electrical signal is used to modulate the IR radiation emitted by the transmitter.

At the optical transmitter end, one of the most popular forms of modulation is *intensity modulation* (IM), wherein the waveform of the information is used to modulate the instantaneous power of the transmitted energy at the desired wavelength. In this technique, the electrical data signal is used as the input of the transmitter driver. At the receiver end, the information is recovered (down-converted from an optical into an electrical signal) using a technique called *direct detection* (DD), in which the photodetector generates an electrical signal according to the instantaneous power of the received optical signal. This transmission-detection technique, which has been widely discussed by Barry and Kahn [21, 47], presents a fundamental characteristic: in an IM/DD channel, the input power at the receiver cannot be negative.

8.2.1 Analog Modulation

One of the modulation techniques that has been proposed and used in the past for wireless IR communication systems is analog modulation. Examples of analog modulation include amplitude modulation (AM) and frequency modulation (FM). The latter is preferred over AM for optical wireless communications due to the fact that amplitude modulation suffers from additive noise [66]. FM, on the other hand, has been proposed for audio signal applications in indoor environments because some of the effects of ambient illumination can be minimized by shifting the frequency of the signal from the spectrum of unwanted background illumination through a high frequency carrier. A further benefit of this modulation technique is that, by using different frequency carriers, multiple users can access the wireless IR system (and the optical channel) using frequency division multiplexing (FDM). An example of how frequency modulation can be combined with intensity modulation is illustrated in Figure 8.1. Here, the message amplitude (which in the system cited above is an audio signal) is used to modulate a carrier in frequency; and the resulting electrically modulated signal is then used to drive the emitting source and therefore transmit the information using intensity modulation (IM). An example of an IR audio system using this modulation scheme was presented by Akerman [66, 201].

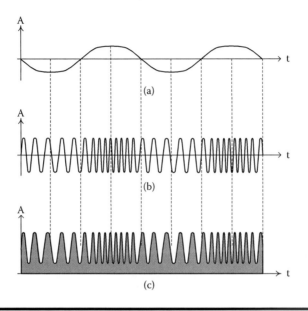

Figure 8.1 FM/IM modulation scheme: (a) variation of the message amplitude with time, (b) variation of the FM modulated signal with time, and (c) variation of the light intensity of the transmitted signal with time.

8.2.2 Pulse Modulation (PM)

Alternative ways to modulate analog information such as voice include the use of pulse position modulation (PPM), pulse amplitude modulation (PAM) (which for the same reasons explained above — additive noise — is generally not favored for wireless IR communication systems), and pulse duration modulation (PDM) [66]. Modulation using any of these techniques is achieved by sampling the analog signal and then transmitting it using digital pulses that are modulated in amplitude, position, or duration according to the amplitude of the analog signal. In PAM, for example, the amplitude of the carrier is modulated according to the characteristics of the analog signal. In an L-PAM scheme (a PAM scheme with L levels), the bits of data are grouped and transmitted as a single pulse. The amplitude of such a pulse is selected from a symbol alphabet with a number of normalized levels given by [20]:

$$L = 2^M \tag{8.1}$$

where M is the number of bits. The pulse amplitude $p(t)$ can then take any of the next possible levels [20]:

$$p(t) \in \left\{0, \frac{1}{L-1}, \frac{2}{L-1}, \dots, 1\right\} \tag{8.2}$$

If the analog signal is sampled and quantized, techniques such as pulse code modulation (which is one of the most common techniques to convert an analog signal into a digital one in a process also known as analogue-to-digital or A/D conversion) can be used. This pulse modulation technique (also known as PCM) consists of sampling the analog signal in time and rounding off (quantizing) its value at each sampling time to the closest quantized level. The size of each quantized level is obtained by dividing the amplitude of the analog signal (which in the case of Figure 8.2 goes from n_l to $-n_l$) by the number of quantized levels (L) as follows:

$$\Delta_v = \frac{2n_l}{L} \tag{8.3}$$

The larger the number of samples (the sampling rate) and the number of quantized levels used, the more accurately the original signal can be reproduced. This is illustrated in Figure 8.2a, where an analog signal has been sampled and quantized. Figure 8.2b shows the equivalent quantized signal for the number of quantized levels (which in this example is $L = 20$) and sample rate presented in Figure 8.2a.

Once the quantized level size has been established, the amplitude of each sample is assigned to the closest midpoint value between quantized levels. Each sample

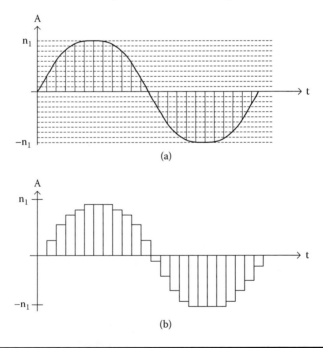

Figure 8.2   Pulse code modulation (PCM): (a) original signal being sampled and quantized, and (b) quantized signal being digitized, where a binary code is assigned to each of the quantized levels and samples.

can then be assigned to one of the digital values corresponding to each quantized level in a process also known as digitization. The resulting signal is denominated *L*-ary according to the number of quantized levels.

Due to their simplicity, binary signals are usually favored to represent the values of the quantized signals over other digital forms. Binary representation is achieved through a coding process where each quantized level is assigned a binary code according to its amplitude. If 16 quantized levels (16-ary signal) are used, for example, each one of these levels can be represented by four binary numbers. To transmit the final digital signal, each digit (or bit) must be represented by a distinctive pulse shape. The simplest way of doing this is to assign a positive pulse for a "1" and a negative pulse for a "0".

Once the information has been modulated using PCM, it is then possible to use a coding scheme to represent the digital information and drive the transmitter source to emit intensity-modulated pulses of IR radiation. There are a number of coding schemes that can be used to represent PCM binary signals. A good example is Manchester coding, which offers good timing extraction but requires doubling of the transmission pulse rate. The way in which PCM and Manchester coding can be used together in an optical wireless communication link is illustrated in Figure 8.3.

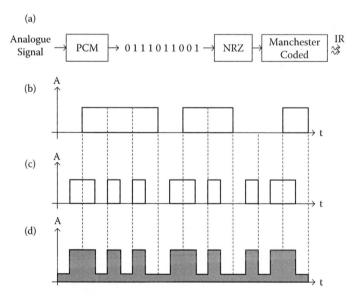

Figure 8.3 PCM scheme with Manchester coding: (a) sequence of conversion from an analog signal to a train of IR pulses using PCM and Manchester coding, (b) conversion of the PCM data into NRZ, (c) conversion of the NRZ data into Manchester coded data, and (d) light pulses generated from the Manchester coded data.

Here, it is shown how the data generated by the pulse code modulator is represented by NRZ and then converted into Manchester coded data, which is then used to drive the source of IR radiation that transmits the IM data.

PPM is another pulse modulation technique widely used in wireless IR communications. This technique was originally developed to provide high power efficiency for long-distance, point-to-point optical-fiber communications, and its use for wireless IR communications has been recently explored [17, 202, 203]. Here, the temporal position of the pulses is varied according to the characteristics of the modulating signal. In this technique, the symbol intervals are divided into L chips or subintervals. The symbol intervals have a duration given by [204]:

$$T = \frac{\log_2 L}{R_b} \tag{8.4}$$

where R_b is the bit rate. Similar to PAM, where the amplitude of the pulse can be selected from an alphabet with a number of levels given by 2^M; in L-PPM, the number of positions available (denoted by L) is related to the number of bits M by [20]:

$$L = 2^M \tag{8.5}$$

The transmitter emits an optical pulse during only one of the chips, which have a duration T/L. A PPM signal satisfying the condition that the optical power emitted by a wireless IR transmitter is positive (but lower than the power constrain of the transmitter) can be represented as [48]:

$$x(t) = LP \sum_{k=0}^{L=1} c_k p\left(t - \frac{kT}{L}\right) \tag{8.6}$$

where T is the duration of the symbol interval, $p(t)$ is a rectangular pulse of unit height and duration T/L as described above, L is the number of subintervals, and $[c_0, c_1, c_2, ..., c_{L-1}]$ is the PPM codework. All the signals are equidistant, with a minimum distance given by [48]:

$$d_{\min}^2 = \frac{2LP^2 \log L}{R_b} \tag{8.7}$$

In terms of power efficiency, if the number of levels L is larger than 2, this modulation scheme exhibits a higher efficiency than OOK. Unfortunately, an increase in L (and in power efficiency) also implies an increase in the bandwidth requirement. The relationship between the average power requirement of PPM with respect to OOK is given by [48]:

$$\frac{P_{PPM}}{P_{OOK}} = \frac{d_{OOK}}{d_{min}} = \sqrt{\frac{2}{L \log_2 L}} \qquad (8.8)$$

and the bandwidth requirement of PPM (for a given bit rate R_b) can be calculated as the inverse of one chip duration as follows [48]:

$$B_{PPM} = \frac{L}{T} = \frac{LR_b}{\log_2 L} \qquad (8.9)$$

L-PPM is illustrated in Figures 8.4, 8.5, and 8.9. In Figure 8.4, the position of the transmitted light pulses varies according to the fluctuations in amplitude of an audio signal. L-PPM is also illustrated in Figure 8.5, where the pulse conveying the 4 bits of a 4-PPM scheme is represented. This figure also shows a temporal guard that is allocated to prevent interference between frames.

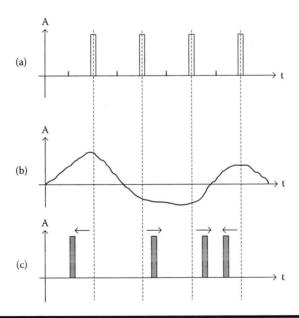

Figure 8.4 PPM scheme: (a) original position of the pulses, (b) variation of the audio signal with time, and (c) variation of the position of the light pulses according to the amplitude of the audio signal.

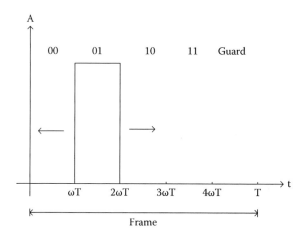

Figure 8.5   L-PPM scheme with L = 4, M = 2, and a temporal guard band used to minimize interference between frames. (*Source:* From [20].)

PPM is generally favored in any of its variants due to its high power efficiency. Some of the different versions of PPM include differential pulse position modulation (DPPM), wavelength shift keying pulse position modulation (WSK-PPM), and digital pulse interval modulation (DPIM) [205]. DPPM, for example, improves the power or the bandwidth efficiency (in low bit rate applications based on the directed-LOS topology) by introducing a simple alteration to the traditional PPM. In addition, when compared to conventional PPM, DPPM offers the advantage of not requiring symbol synchronization. An example of a system based on DPPM is the 100-Mbps networking technology presented by Singh et al. [24].

WSK-PPM is a technique that combines the features of WSK and PPM to provide bandwidth and power efficiency. This means that it is sometimes used as an alternative to conventional PPM to compensate for poor bandwidth efficiency.

Further transmission capacity can be achieved using DPIM [205]. In this technique, each symbol is initiated with a pulse and therefore no symbol synchronization is necessary. Moreover, as each symbol is transmitted as a pulse of one time slot duration, this modulation scheme eliminates the unused time slots within a symbol, which improves the transmission capacity when compared to PPM [20]. Another advantage of DPIM is that average power efficiency can be improved at the cost of increasing the complexity of the system. This modulation scheme is more complex than OOK because the receiver needs slot and symbol-level synchronization.

Another modulation technique that compared to DPIM and conventional PPM requires less transmission bandwidth and achieves higher bit rates is called dual header-pulse interval modulation (DHPIM). This modulation scheme presents the advantages of providing low power consumption, reliability, high speed, and low cost.

The advantages offered by PPM have motivated its use in applications based on power-limited intensity modulation with direct detection (IM/DD). This explains why the IEEE 802.11 has adopted it for the Physical Layer of its IR standard [24]. The main disadvantages of PPM are that it does not offer a high bandwidth efficiency, and that it is more susceptible to intersymbol interference (generated by multipath propagation). For this reason, a number of researchers have investigated ways to mitigate this problem. Some of the solutions proposed include the use of equalization, trellis-coded modulation, and maximum likelihood sequence detection (MLSD) [24].

Among the different modulation techniques proposed in the past for optical wireless communications, one of the most attractive is the one called rate-adaptive transmission. This scheme has been proposed for different coding and decoding schemes. The purpose of this technique is to maintain a high-enough SNR by allowing a progressive degradation of the bit rates under adverse operation conditions and for a reasonable transmitter power, rather than allowing a complete disruption of the communication link [24, 206].

Schemes, such as L-PPM and OOK, suffer from ISI (due to multipath distortion) for data rates above 100 Mbps [24]. Therefore, even if L-PPM scheme improves power consumption, it is not necessarily the most appropriate technique for high data rate applications (above 50 Mbps). This is not only due to the fact that the bandwidth requirement is high, but also because the f^2 noise in the preamplifier increases [20].

Frame and slot synchronization is one of the most important issues related to PPM. Despite the fact that the frame rate presents spectral components, it is not possible to use them directly to achieve frame synchronization because the pulse position varies from one frame to another [20, 207]. Frame synchronization has also been identified as one of the factors that affects the sensitivity performance in PPM. For this reason, different researchers have worked on finding ways to improve the frame and slot synchronization in this modulation scheme. Wilson et al. [207], for example, consider that this can be done by improving the parameters of the PLL used for slot synchronization.

Note also that the short-term DC component of PPM (NRZ and RZ) fluctuates as a function of the bit-stream pattern (this effect is also known as unbalance [20]), which implies that the detected signal is subject to baseline wander due to the fact that the preamplifier of the detector is AC coupled [20]. This problem can be solved using Manchester coding, which is widely used for wireless IR communications because it has zero spectral weightings at DC, but a similar one to RZ at high frequencies. Another way to reduce the baseline wander created by the AC coupling of the detector's preamplifier is to use baseline restoration techniques (at the cost of increasing the complexity of the system).

A variation of PPM is called multiple pulse position modulation (MPPM). In this case, as in traditional PPM, each symbol interval is partitioned into n chips. The symbol interval has a duration given by Equation (8.4), and each chip has a duration given by T/n. The difference here is that the transmitter, rather than sending a single chip as in traditional PPM, sends an optical pulse during w of these chips. The MPPM transmitted signal can be represented as [48]:

$$x(t) = a \sum_{k=0}^{n-1} c_k \phi\left(t - \frac{kT}{L}\right) \tag{8.10}$$

where $\phi(t)$ is a rectangular pulse of unit energy and duration T/n given by:

$$\phi(t) = p(t)\sqrt{\frac{n}{T}} \tag{8.11}$$

and $[c_0, c_1, c_2, \ldots, c_{n-1}]$ is the MPPM codework of weight w. The constant a is chosen so that the average optical power satisfies the following expression [48]:

$$a = \left(\frac{P}{w}\right)\sqrt{nT} = d_{OOK}\frac{\sqrt{n\log_2 L}}{2w} \tag{8.12}$$

It is sometimes desirable to use only a fraction L of the $\binom{n}{w}$ binary n-tuples of weight w. The bandwidth, which is roughly the inverse of the chip duration n/T, satisfies:

$$\frac{B_{MPPM}}{R_b} = \frac{n}{\log_2 L} \tag{8.13}$$

and the average power requirement, compared to OOK, is given by [48]:

$$\frac{P_{MPPM}}{P_{OOK}} = \frac{2w}{\sqrt{nd\log_2 L}} \tag{8.14}$$

It is interesting to notice the similarities between Equations (8.8) and (8.14). From this comparison one can conclude that PPM is a particular case of MPPM, with specific values of n, d, and w.

A further variation of PPM is offered by overlapping pulse position modulation (OPPM), which is considered a special case of MPPM. In OPPM (as in PPM and MPPM), each symbol interval (for which the duration is given by Equation (8.4)) is divided into n chips, where each one of the chips has a duration of T/n. In this case also, as in MPPM, the transmitter, rather than sending a single chip as in traditional PPM, sends an optical pulse with a duration equal to w of these chips. The difference between MPPM and OPPM is that there is a restriction in the latter:

the w ones must be consecutive (with the purpose of decreasing the bandwidth). Unfortunately, the cost of reducing the bandwidth is a reduced alphabet size. The rectangular pulse transmitted consists of w chips that begin at any of the $M = n - w + 1$ chips. In OPPM, the transmitted waveform can be represented by [208]:

$$X(t) = \sum_k P_{peak} p_{T_D}^{l(k)}(t - kT_c) \tag{8.15}$$

where $p_{T_D}^{l(k)}(t) = 1$ for $l(k)T_c \leq t \leq [l(k)+w]T_c$, and $p_{T_D}^{l(k)}(t) = 0$ elsewhere. In this case, the relation between the peak optical power P_{peak} and the average optical power is given by:

$$P_{peak} = \frac{nP_{avg}}{w} \tag{8.16}$$

In addition, the pulse duration T_D is equal to wT_c. For OPPM, the bandwidth is given by $n/(wT)$ so that [208]:

$$\frac{B_{OPPM}}{R_b} = \frac{\frac{n}{w}}{\log_2(n - w + 1)} \tag{8.17}$$

Comparing Equation (8.17) and Equation (8.9), it can be concluded that because L is larger than n/w, the bandwidth of OPPM is smaller than that of conventional PPM. In terms of power, the requirement of OPPM when compared to OOK is given by [48]:

$$\frac{P_{OPPM}}{P_{OOK}} = \frac{2w}{\sqrt{2n\log_2(n - w + 1)}} \tag{8.18}$$

Here, as in MPPM, it can be observed that for specific values of w and n, Equation (8.18) reduces to Equation (8.8). In this case, the specific values are $n = L$ and $w = 1$.

8.2.3 Digital Modulation

In the context of wireless IR communications, digital modulation is generally defined as the techniques used to convert a digital set of data into a format appropriate for optical transmission. The NRZ information produced by PCM and illustrated in Figure 8.3b, for example, can be used to modulate a new electrical carrier. This new signal can then be used as the input to the driver of a wireless IR transmitter.

Pahlavan [66] encompasses in this category the non-amplitude modulating schemes. Digital modulation techniques used for wireless IR communications include frequency-shift keying (FSK) and phase-shift keying (PSK). These techniques are favored for optical transmission due to the fact that transmission of the binary information produced in a number of applications can be easily implemented. In single-subcarrier modulation (SSM) schemes, the original signal is modulated into a radio-frequency subcarrier, which is then used to drive the transmitter, where the instantaneous power of the IR transmitter is proportional to the modulated signal. Modulation schemes such as FSK, ASK (amplitude-shift keying), and PSK are good examples of SSM. In them, the spectrum of a baseband signal is shifted to facilitate the transmission of the information signal. In the current context, the modulating signal (the binary signal) modulates a sinusoidal carrier of a higher frequency, and the resulting signal is used to drive the transmitter source. It is worth noting that a DC bias is generally added to make sure that the subcarrier, which is a sinusoidal waveform that can have positive or negative values, always has a positive value.

If the variations in amplitude of the carrier occur in proportion to variations in the amplitude of the modulating signal, then the modulation scheme is called ASK or on-off keying (OOK). If, on the other hand, the representation of the binary signal is carried out by shifting the phase of the sinusoidal carrier, the modulation scheme is called PSK. This is illustrated in Figure 8.6 where a binary set of data is first converted into NRZ form. The new set of data is then used to modulate a carrier using PSK as described above. If, rather than the phase or amplitude of the carrier, it is its frequency that is varied in proportion to the amplitude of the modulating signal (the binary code), then the modulation scheme is called FSK. This is

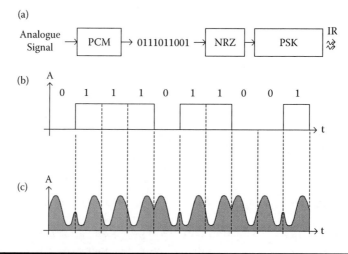

Figure 8.6 PSK modulation: (a) sequence of conversion from an analog signal to a train of IR pulses using PSK, (b) NRZ set of data, and (c) PSK light pulses generated from the NRZ data.

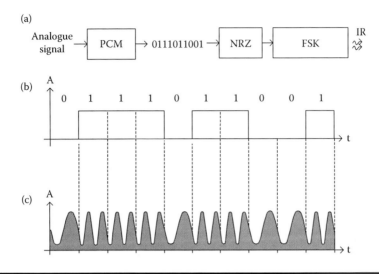

Figure 8.7 FSK modulation: (a) sequence of conversion from an analog signal to a train of IR pulses using FSK, (b) NRZ set of data, and (c) FSK IR pulses generated from the NRZ data.

illustrated in Figure 8.7. Depending on the bandwidth efficiency required in the channel bandwidth use, more complex versions of these modulation techniques can be employed at the cost of increased complexity. Examples of systems based on PSK and FSK can be found in [10, 11]. Here, Gfeller et al. shown that if the environment is diffuse and there are different paths for the transmitted energy, then PSK presents synchronization problems. For this reason (and because of its flexibility to data rate variations), FSK is preferred over PSK even if both of them present similar performances in terms of bit error rate. Another example of a system based on PSK is the 48-kbps networking system developed by Fujitsu [35] (cf. Chapter 1).

The directed-LOS networking system described by Minami et al. from Fujitsu [34], the 1-Mbps networking technology described by Nakata et al. from Hitachi [36], and the 45-Mbps networking system developed by Matsushita [40], on the other hand, are examples of systems based on FSK.

Contrary to SSM where only one signal is modulated onto an electrical subcarrier, multiple-subcarrier modulation (MSM) consists of modulating not just one, but several digital or analog signals onto different electrical subcarriers that are then modulated onto a single optical carrier using IM [49]. One of the main attractions of MSM is that it permits asynchronous multiplexing of different information streams and the receiver can demodulate only the information stream desired. In addition, MSM is attractive because it provides immunity to fluorescent light noise near DC and because it minimizes ISI in multipath channels. Unfortunately, MSM presents poor optical power efficiency due to the fact that, to avoid clipping the negative part of the electrical signal, a DC bias must be added to the MSM electrical signal to compensate for the negative values of the modulated sinusoids. This DC bias increases with the number of subcarriers, thereby reducing the power efficiency of the technique. Therefore, the number of carriers is generally kept to a minimum, especially in applications that involve portable units where small power consumption is required. Clipping, on the other hand, introduces interference and clipping noise.

One way to reduce the optical average power consumption of MSM (when using digital modulation — QPSK or BPSK — with rectangular pulses in a system based on intensity modulation/direct detection, IM/DD) is by using a DC bias that, rather than being fixed, varies on a symbol-by-symbol basis. In this method, proposed by Yu et al. [49], this is achieved using just enough bias during each symbol period to prevent negativity. This time-varying signal is baseband PAM and uses root-raised-cosine pulses (that is, pulses whose Fourier transform is the square root of a raised cosine).

8.2.3.1 On-Off Keying

On-off keying (OOK) is one of the preferred modulation techniques employed in wireless IR communications due to its good bandwidth efficiency and because it is the easiest scheme to implement. A set of OOK data can be represented by return-to-zero (RZ) or non-return-to-zero (NRZ) pulses. In a number of wireless IR communication systems, the preferred modulation-signaling scheme combination is NRZ-OOK, which has been widely explored for indoor applications employing IM/DD [21, 47].

The normalized transmitted pulse of an NRZ-OOK scheme is $p(t) = 1$ for $0 \leq T_b$, and $p(t) = 0$ elsewhere [20]. Here, clock recovery results complicated due to the fact that the spectra of NRZ pulses do not contain discrete spectral components. Therefore, line coding or nonlinear clock recovery must be introduced. The transmitted pulse of RZ signaling (with a 50 percent duty cycle), on the other hand, is of the form $p(t) = 1$ for $0 \leq T_b/2$ and $p(t) = 0$ elsewhere. The spectra of this form of signaling do contain discrete components at the bit rate, which simplifies clock recovery. Unfortunately, the bandwidth efficiency of RZ signaling with 50 percent duty cycle is also reduced by 50 percent. This makes it necessary to find a compromise between the complexity of the clock recovery extraction and the bandwidth efficiency.

The waveform of an OOK modulation scheme with NRZ can be represented as [208]:

$$X(t) = \sum_k a_k P_{peak} p_{T_b}(t - kT_b) \qquad (8.19)$$

where $p_{Tb}(t) = 1$ for $0 \leq t \leq T_b$ and $p_{Tb}(t) = 0$ elsewhere, and $a \in \{0,1\}$. Here, the peak optical power P_{peak} is related to the average optical power P_{avg} as follows: $P_{peak} = 2P_{avg}$.

As discussed above, the difference between OOK-RZ and OOK-NRZ is that in the former, the transmitted signal is "on" (transmits a "1") only during a fraction of a bit duration γT_b (where γ represents the duty cycle) while in the latter it is "on" during the entire bit duration. In this case, the waveform of the OOK modulation scheme with RZ can be represented as [208]:

$$X(t) = \sum_k a_k P_{peak} p_{\gamma T_b}(t - kT_b) \tag{8.20}$$

where $a \in \{0,1\}$, $P_{\gamma Tb}(t) = 1$ for $0 \le t \le \gamma T_b$, and $p(t) = 0$ elsewhere. Here, the relationship between the peak optical power P_{peak} and the average optical power is given by:

$$P_{peak} = \frac{2P_{avg}}{\gamma} \tag{8.21}$$

8.3 Modulation Techniques Comparison

As explained above, two of the most important parameters related to the performance of the different modulation techniques used for wireless IR communications are the power efficiency and the bandwidth efficiency. For this reason, several authors have carried out comparisons of a variety of modulation schemes to determine which ones present advantages with regard to power or bandwidth efficiency and under which circumstances [21, 47, 48, 204]. An example of a modulation scheme comparison can be found in [47], where the characteristics of a variety of modulation techniques for an IM/DD channel assuming additive white Gaussian noise are presented. Here, the power requirements and bandwidth efficiencies of different modulation schemes have been normalized to the NRZ-OOK case, which is used as a reference. The results are presented in Figure 8.8.

A similar analysis is presented in Table 8.1 [21], where the normalized bandwidth and the normalized average power requirements of several modulation schemes are compared to those of OOK. This comparison also assumes an ideal channel with additive white Gaussian noise. OOK was chosen as a reference because it represents a good compromise between the bandwidth and power requirements. It can be observed in this table that when the pulses used by OOK are changed from NRZ to RZ, the average power requirement decreases at the cost of increasing the bandwidth by the factor $1/\gamma$ (where $0 < \gamma < 1$). As noted above, the good compromise presented by OOK with RZ pulses makes it the preferred option for a number of applications. However, other schemes such as PPM are preferred over OOK with RZ when the duty cycle of the latter is too small.

As explained in Subsection 8.2.2, one of the modulation schemes that offers the best power efficiency for wireless IR communication systems is L-PPM. This is illustrated in Figure 8.8, where it is observed that the larger the L value, the better

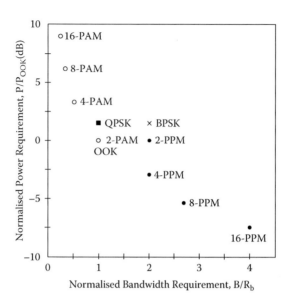

Figure 8.8 Bandwidth and power efficiency comparison of different modulation schemes. This comparison assumes an IM/DD channel and has been normalized to the OOK scheme. (*Source:* From [20].)

the power efficiency becomes. Unfortunately, this is done at the expense of reducing the bandwidth efficiency. Figure 8.8 also shows that, in terms of bandwidth efficiency, the best modulation scheme is L-PAM. However, it presents a high power consumption. It is therefore important to take into account that when choosing a modulation scheme, a compromise must be made between the power and the bandwidth efficiencies. For applications based on IM/DD that do not require a high data rate, but that need to optimize the power efficiency, 16-PPM appears to be the best option, with a power efficiency improvement of around 8 dB over OOK. If, on the other hand, power consumption is secondary, but bandwidth optimization is the priority, then 16-PAM offers the better choice.

Figure 8.8 shows that quadrature phase-shift keying (QPSK) and binary phase-shift keying (BPSK) do not offer any significant advantage over OOK with regard to power or bandwidth efficiency; but, on the contrary, both schemes present a lower power efficiency (of approximately 1.5 dB) than OOK. Moreover, the bandwidth requirement of BPSK is about twice that required for OOK. Despite this, these modulation schemes are used for some applications due to the fact that, as explained in Subsection 8.2.3, they minimize the noise introduced by background illumination (fluorescent lamps).

Yu et al. [49] have analyzed multiple-subcarrier modulation techniques due to the fact that MSM allows the use of frequency division multiplexing, with intensity modulation and direct detection, and because transmission at high bit rates can

Table 8.1   Modulation Schemes Comparison

Modulation Scheme	Normalized Average Power Requirement (in optical dB)	Normalized Bandwidth Requirement
OOK with NRZ	0	1
OOK with RZ	$5\log_{10}\gamma$	$\dfrac{1}{\gamma}$
L-PPM (soft decision)	$-5\log_{10}\left(\dfrac{L\log_2 L}{2}\right)$	$\dfrac{L}{\log_2 L}$
L-PPM (hard decision)	$-5\log_{10}\left(\dfrac{L\log_2 L}{4}\right)$	$\dfrac{L}{\log_2 L}$
L-DPPM (hard decision)	$-5\log_{10}\left(\dfrac{L\log_2 L}{8}\right)$	$\dfrac{L+1}{2\log_2 L}$
N BPSK	$1.5+5\log_{10}N$	2
N QPSK	$1.5+5\log_{10}N$	1

Note: The variable γ represents the duty cycle [21].

be achieved through MSM, while maintaining a low intersymbol interference (cf. Section 8.6).

Park et al. [48] presented another comparison of a number of modulation schemes for indoor applications. Their results, which are in agreement with the information presented by other authors, show that low duty-cycle modulation schemes present high power efficiencies at the expense of reduced bandwidth efficiencies.

In the comparison by Kah et al. [21] mentioned above, the modulation schemes analyzed were PPM, differential PPM (DPPM), multiple subcarrier modulation (MSM) using binary phase-shift keying (BPSK) or quaternary phase-shift keying (QPSK), and OOK with NRZ and RZ pulses.

Compared to OOK, PPM presents a decrease in the average power requirement at the cost of increasing the bandwidth requirement. In addition, PPM presents a much better immunity to the near-DC noise from fluorescent lamps; this is the reason why this modulation scheme is employed for indoor wireless IR applications. Furthermore, for a given bit rate, the bandwidth that L-PPM requires is given by the factor $L/\mathrm{Log}_2 L$, which means that for a large L, the power requirement is much lower than that of OOK.

The fact that the average power requirement of an L-PPM scheme decreases as L increases is true if multipath distortion does not exist or if it is not taken into account. The low power requirement of the L-PPM scheme makes it a favorite modulation scheme for portable devices in mobile applications. However, it must be taken into account that the improvement in power consumption is made at the cost of increasing the bandwidth requirement. In addition, the chip and symbol-level synchronization requirement implies that more complexity is needed. Furthermore, the peak power requirement of PPM, when compared to OOK, is higher (Figure 8.9).

When transmitted over multipath channels, L-PPM may suffer from intrasymbol interference (interference in chips within the same symbol) and intersymbol interference (interference between adjacent symbols). However, the use of a large L with its corresponding low power consumption compensates for this situation. This is true for low bit rates and for small delay spreads. However, if the delay spreads increase, PPM suffers a faster power requirement increase than OOK, where the larger the L of the L-PPM scheme, the faster the increase in the power requirement [21].

While QPSK requires the same bandwidth as an OOK signal, binary phase-shift-keying (BPSK) requires twice the bandwidth needed by it. In terms of optical power, QPSK and BPSK require 1.5 dB more optical power than OOK (on a distortionless channel). It must be noted that, compared to OOK transmission over an ideal channel, MSM suffers from a higher degradation when transmitted over a multipath channel. This is due to the fact that the subcarriers may suffer from ISI and interference from other subcarriers, and that the subcarriers also suffer from an attenuation that increases as the frequency of the subcarrier increases. The interference between subcarriers can be reduced by using a large number of them, but this is not always practical and possible due to the increased optical power requirement at the transmitter, which increases with the number of subcarriers.

Other interesting comparisons between different modulation schemes have been performed by Park et al. [48] and Ho-Quang et al. [208]. They presented a comparison of the power and the bandwidth requirements of the PPM scheme and its variations (MPPM and OPPM) with respect to OOK. Some of the values obtained by Park et al., for example, are presented in Table 8.2 (the values presented are for an AWGN channel). From this table it can be observed that, as in previous comparisons and analysis on PPM, as L increases, the power requirement decreases (at the cost of increasing the bandwidth). It is interesting to note that, in general, the performance of MPPM is better than that of PPM, with the bandwidth requirement decreasing as the weight increases. In this case, MPPM with a weight of 2 exhibits better performance than PPM; and MPPM with a weight of 8 exhibits better performance (with regard to the bandwidth requirement) than MPPM with a weight of 2 (for the same power requirements). The drawback in using MPPM with weight 8 is that, to achieve the appropriate power efficiency, a large number of chips are required. It also can be observed in Table 8.2 that the bandwidth efficiency of overlapping PPM with duty cycle $\alpha = \frac{1}{2}$ is very good; but as the duty

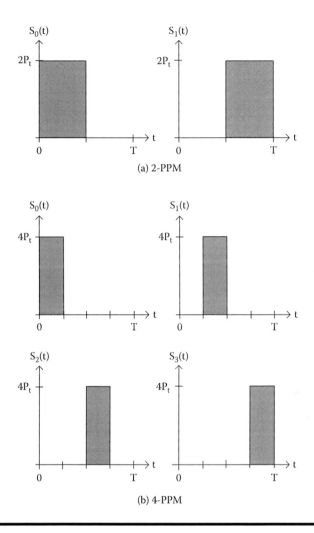

Figure 8.9 L-PPM modulation scheme waveforms: (a) 2-PPM, and (b) 4-PPM. (*Source:* Adapted from [21])

cycle is changed to $\alpha = \frac{1}{4}$, the power efficiency increases at the cost of increasing the bandwidth requirement.

Park et al. [48] also performed a comparison of the modulation schemes presented in Table 8.2 on a quantum-limited channel. Their results show a significant difference in performance with respect to the same schemes in an AWGN channel, especially with regard to the power requirements. In general, the modulation schemes on an AWGN channel are much more power efficient than their quantum limited channel counterparts. Some of the bandwidth requirement values are presented in Table 8.3.

Table 8.2   Bandwidth Requirement (B/R_b) Comparison of Different Modulation Schemes (Corresponding to a Variety of Power Requirements P/P_OOK Given in dB (for an AWGN Channel))

Power Requirement	PPM	MPPM (w = 2)	MPPM (w = 8)	OPPM	TCM 8– OPPM
2 dB	0.63	—	1.82	0.63 (α = 1/2)	1
1 dB	0.98	1.88	1.5	0.98 (α = 1/2)	1 (d_c = 4)
0 dB	2 (2-PPM)	1.66	1.28	2 (2-OPPM)	1
–1 dB	1.95	1.54	1.18	1.95	1
–2 dB	1.9	1.54	1.18	1.9	1
–3 dB	2 (4-PPM)	1.6	1.28	2 (4-OPPM)	1
–4 dB	2.2	1.78	1.41	2.2 (α = 1/4)	1 (d_c = 16)

Source: Data obtained from [48].

Table 8.3   Bandwidth Requirement (B/R_b) Comparison Corresponding to a Variety of Modulation Schemes (for a Power Requirement P/P_OOK Given in dBs and a Quantum Limited Channel)

Power Requirement	PPM	MPPM (w = 2)	MPPM (w = 8)	OPPM
4 dB	—	1.88	1.65	1.28
3 dB	2 (2-PPM)	1.72	1.4	2 (2-OPPM)
2 dB	1.95	1.56	1.27	1.95
1 dB	1.9	1.5	1.19	1.9
0 dB	2 (4-PPM)	1.55	1.19	2 (4-OPPM)
–1 dB	2.34	1.7	1.27	2.34
–2 dB	2.85 (8-PPM)	2.08	1.5	—

Source: Data obtained from [48].

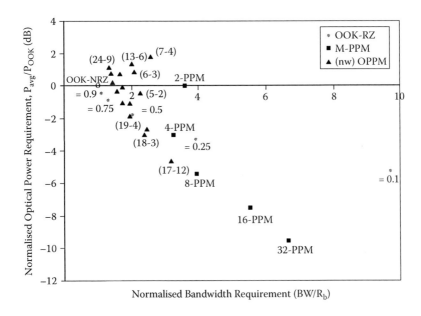

Figure 8.10   Bandwidth and power requirements comparison for different modulation schemes. (*Source:* From [208].)

Ho-Quang et al. [208] defined the power requirement P used to compare the different modulation schemes as the average optical power required by the system to achieve a given bit error rate (BER) and bit rate R_b.

Some of the results obtained by Ho-Quang et al. compare different modulation schemes in terms of bandwidth and power requirements are shown in Figure 8.10. From this graph it can be observed that the modulation schemes that require less optical power are the MPPM, where the larger the M becomes, the lower the power requirement (at the cost of increasing the bandwidth requirement). It can also be observed that OOK-NRZ presents one of the lowest bandwidth requirements, but this is achieved at the cost of increasing the power consumption. Note that although 2-PPM seems to require more bandwidth than OOK-NRZ (for the same power requirement), as the M value of the MPPM scheme increases, the power requirement is improved at the cost of increasing the bandwidth requirement. It is also interesting to notice that in the specific case of 4-PPM, both the power and the bandwidth requirements appear to be better than the values presented for 2-PPM. On the other hand, 8-PPM appears to be a good compromise for medium-speed wireless systems, where a trade-off between the power and bandwidth requirements can be achieved. Portable low-speed systems, where power consumption is of paramount importance, benefit from the use of a scheme such as 16-PPM.

The (n, w)-OPPM modulation schemes offer, in general, a lower bandwidth requirement than their MPPM counterparts, but also a higher power requirement.

This can be observed in the comparison between (18 – 3)-OPPM and 8-PPM. Here, the latter presents a lower bandwidth requirement than the (18 – 3)-OPPM scheme, but also requires a higher bandwidth. Compared to 4-PPM, for which the power requirement is practically the same, (18 – 3)-OPPM offers a lower bandwidth requirement. A good compromise between the bandwidth and the power requirements can be obtained from either (9 – 2), (18 – 3), or (17 – 2)-OPPM.

The modulation scheme analysis for outdoor systems is also of interest. For this reason, researchers such as Zhang [204] have also performed comparisons between a variety of modulation schemes in terms of power and bandwidth efficiency. For his analysis, Zhang considered OOK-NRZ, L-PPM, and DPIM. His comparison shows that from these three modulation schemes, the one that requires minimum transmission power for a given BER is L-PPM, where the larger the L, the more power efficient the PPM scheme. DPIM also offers an improved power efficiency when compared to OOK but its power efficiency is not as good as that of L-PPM. Zhang's analysis also shows that in terms of bandwidth efficiency (from these three), the modulation scheme that presents the lowest bandwidth requirement is OOK, followed by DPIM and L-PPM. Therefore it can be concluded that, just as for indoor applications, the selection of one modulation scheme over another depends on whether power or bandwidth efficiency has priority for a specific application. As none of these schemes offers at the same time optimum power and bandwidth, a trade-off between them must be found.

8.4 Modulation Schemes in the Presence of Noise

As explained in Chapter 3, background illumination is the main source of noise in a wireless IR receiver. This noise can be reduced using optical or electronic filtering and by restricting the FOV of the receiver. Electronic ballasts introduce low-frequency components at multiples of the line frequency, and the effect of these ballasts can be highly reduced or even eliminated using subcarrier modulation (with a subcarrier frequency that is high enough).

The problems associated with background illumination noise (especially those related to indoor environments) have prompted a number of researchers to investigate their effect on different modulation schemes [65, 209, 210]. Narasimhan et al. [65], for example, have analyzed the effect of fluorescent lamps on the optical power requirement and the BER of a variety of modulation schemes.

Their results reveal that the OOK modulation scheme is significantly affected by fluorescent illumination when the optical power from the fluorescent lamps is high. This is the case even when an optical filter is used. BPSK, on the other hand, presents a response that is basically independent from the optical power of the fluorescent lamps. This is due to the fact that this modulation scheme can employ a high-frequency subcarrier to make it immune to florescent illumination. In addition, the different PPM schemes analyzed suffer from less power penalties than OOK.

Moreira et al. [211] have also performed a modulation scheme comparison under background illumination. They analyzed three modulation schemes: NRZ-OOK, L-PPM, and BPSK. Their results show that optical wireless systems depend significantly on the channel characteristics. For an environment with no interference from artificial illumination, PPM is the modulation scheme that presents the best power efficiency. When interference from background illumination (created from fluorescent lamps using electronic ballasts) is present, both OOK and L-PPM suffer high power penalties. Fortunately, as mentioned in Chapter 3, the effects from background illumination can be significantly reduced by using filtering techniques and by reducing the FOV of the receiver.

The type of detection appears to be relevant to the power efficiency of the different modulation schemes operating under background illumination. In this respect, the power penalty of L-PPM with maximum likelihood detection (ML) is determined by the amplitude and the bandwidth of the interference signal. In L-PPM with threshold detection (TH), on the other hand, the power penalty is dictated by the maximum and minimum values of the interference signal. One also can conclude from the modulation scheme comparison of these authors that the effect of interference is more significant at low data rates than at higher ones. In the case of OOK, the optical power required for a given BER is almost independent from the bit rate; however, for L-PPM systems with ML detection, the optical power requirement is closely related to the bandwidth of the interference (and to its amplitude). Table 8.4 shows the power penalty introduced by the OOK and the L-PPM modulation schemes under different sources of illumination [211] (the filter considered for the OOK scheme with NRZ was highpass with cut-off frequency at 4 kHz). The required optical power shown in this table has been normalized to that required to achieve a BER of 10^{-5} in a channel without interference. Here, it can be observed that, for OOK-NRZ, the major penalty is introduced by incandescent illumination with no electronic filter. When a filter is used, this power penalty is significantly reduced. It can also be observed that the power penalty introduced by florescent lamps using electronic ballasts (operating at 400 Hz) is higher than the one introduced by fluorescent lamps without ballasts.

The power penalty of L-PPM with ML operating under incandescent illumination is very small. This is also the case when operating under fluorescent illumination when no electronic ballast is used. However, if an electronic ballast is used to drive the fluorescent lamps, the power penalty increases. This effect can be reduced using high-pass electronic filters. The effect of two different filters is also shown in Table 8.4. Here, one of the filters has a cut-off frequency at 200 kHz, while the other has its cut-off frequency at 600 kHz.

The penalty introduced by a 16-PPM system using a threshold detector is also presented in Table 8.4. Here it can be observed that incandescent illumination introduces a high power penalty when no electronic filter is used. It can also be observed that when an electronic filter is used, the penalty introduced by incandescent illumination is highly reduced. In the case of fluorescent lamps driven by

Table 8.4 Penalty (indBs) Induced by Background Illumination in OOK and L-PPM Schemes with ML and TH, at 1 Mbps

	BER = 1e-01	BER = 1e-02	BER = 1e-03	BER = 1e-04	BER = 1e-05
OOK-NRZ No interference	−5	−2.5	−1.2	−0.5	0
OOK-NRZ Incandescent, unfiltered	24	24.4	24.4	24.4	24.4
OOK-NRZ Incandescent, filtered	−4	−1	0	1.5	2.5
OOK-NRZ Fluorescent with ballast	16	16.6	16.9	17	17.1
OOK-NRZ Fluorescent interference	−5	−2	−0.5	0.6	1.8
16-PPM/ML No interference	−11	−9.3	−8.5	−7.8	−7.2
16-PPM/ML Incandescent, unfiltered	−10	−8.4	−7.5	−6.8	−6.4
16-PPM/ML Incandescent	−10	−8.4	−7.5	−6.8	−6.4

16-PPM/ML Fluorescent, no filter	−10.8	−9	−8	−7.4	−6.9
16-PPM/ML Elec., unfiltered	4.2	5.6	5.8	5.9	6
16-PPM/ML Elec. (600 kHz)	−7	−5	—	—	—
16-PPM/ML Elec. (200 kHz)	−3	−1.5	0	0.9	1.5
16-PPM/TH No interference	−8.8	−7.5	−6.8	−6	−5.6
16-PPM/TH Incandescent	15	15.3	15.3	15.3	15.3
16-PPM/TH Incandescent, filtered	−8.8	−7.1	−6.1	−5.6	−5
16-PPM/TH Fluorescent, filtered	−8.8	−7.5	−6.7	−5.8	−5.4
16-PPM/TH Elect (100 kHz)	4.2	5	5.5	6.3	6.7

Note: The required optical power is normalized to that required to achieve a BER of 10^{-5} in a channel without interference.

Source: Data obtained from [211].

electronic ballast, the penalty is also high, but in this case too, the penalty can be reduced using appropriate electronic filtering. Compared to OOK, PPM offers a more effective mitigation of background illumination due to the fact that higher cut-off frequencies can be used.

Chan et al. [210] also presented a comparison of the performance of OOK and PPM under background illumination conditions. Their reasons for comparing the performance of these two modulation schemes in particular was due to their popularity for wireless IR communications and the good sensitivity of PPM when compared to OOK. They demonstrated that background illumination noise reduced the sensitivity of a wireless IR receiver, especially when the receiver is positioned underneath a very directive spotlight; and, from these modulation schemes, the sensitivity of OOK is more strongly affected to background illumination noise than PPM.

8.5 Modulation Schemes in the Presence of Multipath Distortion

Different researchers have also analyzed the effect of multipath dispersion on a variety of modulation schemes [46, 210]. A good example of this type of analysis is the one presented by Chan et al. [210] in relation to OOK and PPM. They modeled the multipath dispersion by calculating the power level and the time needed for every ray traveling from the transmitter to the receiver. The time was calculated by considering the distance traveled by every ray and the velocity of light (the transmission in this analysis was assumed to be diffuse and from a Lambertian source). In addition, by dividing the ceiling and the walls (which were considered to have 100 percent reflectivity) around the communication plane into small elements, they took reflections into account. The size of the reflective surface elements depended on whether the reflections were first or second order.

The experiments carried out by Chan et al. on a 10-Mbps transmission using two different modulation schemes demonstrate that the sensitivity performance of a wireless IR receiver deteriorates under multipath dispersion. It was also found that the sensitivity performance of PPM is more strongly affected by multipath dispersion than OOK due to the fact that the pulse width in a PPM scheme is smaller than in its OOK counterpart.

Park et al. [46] also performed an evaluation of the effect of multipath dispersion on different modulation schemes. In their case, the schemes analyzed included multiple pulse position modulation (MPPM), PPM, and OOK. Their report included upper boundaries for the probability of error of PPM-based schemes with ML and unequalized receivers under intersymbol interference. For numerical comparisons, they considered a first-order lowpass filter channel with bandwidth W. Their results showed that all the PPM versions are susceptible to multipath dispersion. In addition, they demonstrated that the susceptibility presented by the PPM schemes is higher than that of OOK with maximum likelihood detection. Furthermore, their results showed that unequalized OPPM is very sensitive to intersymbol interference (ISI) and that equalization is beneficial at high bit rates. Another conclusion derived from the work of Park et al. is that, despite the fact that the power efficiency of PPM schemes is high when compared to OOK on an ideal ISI-free channel, their power efficiency drops when ISI is present.

8.6 Multiple Access Techniques

Multiple access techniques refer to the way in which the IR medium can be shared among a number of users. These techniques define the transmission capacity allocated to different users when a number of them are trying to access the system. This is particularly relevant in indoor systems where the same device may be required to exchange information with a number of other units, as compared to outdoor systems that tend to operate in a point-to-point fashion. The current section refers to physical-layer multiplexing techniques, which are based on some of most important characteristics of the IR medium. These characteristics include [21]:

- Optical filters can be used to discriminate unwanted wavelengths and to allow only the relevant ones to reach the detector.
- It is possible to use the same wavelength in different rooms (in a building or a house) due to the fact that IR radiation does not penetrate walls.
- Intensity modulation (IM) favors the use of waveforms with short duty cycles because they improve power efficiency. IM also favors the use of time division multiple access (TDMA).
- It is possible to achieve high angular resolution (in an angle-diversity receiver) due to the fact that the wavelengths corresponding to the IR part of the spectrum are short, which is important for space division multiplexing.

There are two possible ways to achieve multiple access at the physical layer level in a wireless IR communication system: (1) electrically and (2) optically. Examples of electrical multiplexing techniques include code division multiple access (CDMA) and time division multiple access (TDMA), the latter being probably the most popular form of electrical multiplexing for optical wireless applications. Examples of optical multiplexing techniques, on the other hand, include wavelength division multiple access (WDMA) and space division multiple access (SDMA).

A good way to provide multiple access while making use of the simplicity of IM/DD is by using multiple-subcarrier modulation (MSM). This technique consists of modulating a number of independent bit streams onto subcarriers of different frequencies. The modulated signals can then be added and used to drive the transmitter. Multiple bandpass demodulators can then be used to recover the original bit streams at the receiver end. Unfortunately, compared to SSM, MSM requires more optical power at the transmitter (on the order of $5\log_{10} N$ for a transmission with N subcarriers [21]) to ensure that $X(t)$ is non-negative (cf. Section 8.2.3).

8.6.1 Electrical Multiple Access Techniques

Electrical multiplexing is a reliable form of providing multiple access between a number of users. As discussed above, two of the electrical modulation techniques widely used in wireless IR communications are TDMA and CDMA. TDMA, for example, consists of dividing the time allocated to each user in a synchronized manner, which requires some degree of system complexity. In TDMA systems, the transmitters emit energy according to their allocated slots of times while the receivers monitor the same channel. Examples of systems based on TDMA include the telephone technology developed by Kotzin et al. from Motorola [38] and the telephone system presented by Poulin et al. from MPR Teltech [41].

One of the advantages of TDMA is that, due to its low duty cycle, high power efficiency can be achieved, which makes it the preferred option in applications involving portable devices where power consumption is critical. Unfortunately, the fact that high coordination is required to avoid loss of synchrony means that a relatively high complexity is required for the system.

Frequency division multiple access (FDMA) allows simultaneous transmission of information within the same physical space using different frequencies. Compared to TDMA, FDMA has a low power efficiency; and, as the number of subcarriers increases, the power efficiency decreases even further.

In the case of CDMA, different users can have access to the same channel, thanks to the use of different orthogonal or quasi-orthogonal code sequences. Because different transmitted waveforms (with different duty cycles) can be used with CDMA, it is not possible to draw conclusions regarding the efficiency of this multiplexing technique. Kahn et al. [21] performed a comparison of different multiple access techniques (in a cellular system with spatial reuse) in terms of their optical average power efficiency as the radius of the cells varied. For this analysis, they assumed a diffuse system where a number of terminals communicated through an optical satellite connected to an optical-fiber backbone such as the one proposed by Gfeller et al. [9–11] (Figure 1.1). For this comparison, they also assumed that WDMA (which is explained below) was used to separate the downlink from the uplink in such a way that signals from the downlink and the uplink did not interfere with each other. For this reason, only the downlink — the bandwidth of which was partitioned to implement CDMA, TDMA, or FDMA — was considered. The different cells within the room were illuminated by different downlink transmitters, which had equal bandwidth allocated to them. Another important assumption for this comparison was that non-adjacent transmitters could reuse the same bandwidth (a *reuse factor* was therefore defined according to the number of bandwidth partitions). In this way, the interference between cells was kept to a minimum (in reality, no interference was created by adjacent cells when using FDMA and TDMA, but some interference was created when using CDMA).

The signal to co-channel interference ratio (SIR) — which depended on variations of the channel DC gain and on the ratio of the reuse distance of the cell radius — could be increased by either augmenting the reuse factor or increasing the cell radius. Unfortunately, an increase in the reuse factor implied a loss of capacity, reason for which the reuse factor was kept as small as possible.

Some of the values obtained by Kahn et al. [21] in relation to the theoretical performance of six fixed reuse schemes in a diffuse system are presented in Table 8.5. The multiple access schemes and the modulation techniques considered in this comparison include TDMA with 2-PPM, TDMA with 4-PPM, TDMA with OOK, CDMA with Optical Orthogonal Codes (OOC), CDMA with *m*-sequences, and FDMA with BPSK. The cells considered in this comparison are hexagonal, with a throughput in each one of 10 Mbps and a reuse factor of 3. The factor γ, which is the evaluation factor presented in Table 8.5 for the different multiple access techniques with their respective modulation schemes, is proportional to the square of the transmitted optical power, and equal to the SNR for unit optical path gain. According to Marsh et al. [21, 22], a 10-dB variation in the factor γ corresponds to a 5-dB variation in the transmitted optical power. Therefore, a small γ *value* is advantageous because it represents a small optical power at the transmitter.

One of the first things observed in Table 8.5 is that the value of γ increases for a cell radius less than 2 m. The reason for this is that when the cell size is too small, the co-channel interference increases, which makes it impossible to maintain the BER value. As the radius of the cells increases, the co-channel interference becomes less critical, which allows a better evaluation of the different multiple access techniques and modulation schemes in terms of their power requirement. Thus, it can be concluded that while FDMA using BPSK is the multiple access technique that presents the largest power requirement, the multiplexing technique that requires less power at the transmitter to achieve a target value of BER = 10^{-9} is TDMA with 4-PPM, which has the lowest duty cycle. TDMA with 2-PPM requires a higher transmitter power because the duty cycle is also higher.

In applications where multipath ISI and hard-decision detection are minimal, differential pulse position modulation (DPPM) offers advantages over PPM in terms of power or bandwidth efficiency. The difference between 4-PPM and 4-DPPM, is that, while in 4-PPM one of the four chips constituting a symbol is "high" and the other three are low at every single moment; in 4-DPPM, the "low" bits that would follow the "high" chip are omitted (which implies that the symbols have different durations).

Table 8.5 Multiple Access Techniques Comparison on a Cellular System with a Reuse Factor of 3

Cell Radius (m)	TDMA/4-PPM	TDMA/2-PPM	TDMA/OOK	CDMA/OOC	CDMA/m-seq.	FDMA/BPSK
1.5	123	129	135	—	122	143
2	116	122.5	124	132	124	134.5
2.5	117.5	123	124	122.5	127	135
3	119.3	125.5	125.5	124	130	137.5
3.5	122.5	128.5	128.5	127	132.6	140
4	126	132	132	130	136	143
4.5	129	135	135	134	140	147
5	132.5	138.5	138.5	137	142.7	150

Note: The values presented in this table represent the γ required (in electrical dB) to achieve a BER of 10^{-9}.

8.6.2 Optical Multiple Access Techniques

Optical multiplexing refers to techniques in which multiple users can share the IR medium by making use of its optical properties. As discussed above, two of the optical multiple access techniques used in the past for wireless IR communication systems are wavelength division multiple access (WDMA) and space division multiple access (SDMA). WDMA defines the way in which a number of users can transmit information simultaneously within the same physical space using narrow-spectrum emitters with different wavelengths. Systems based on WDMA are simpler than their TDMA counterparts because they make use of some of the inherent characteristics of the IR medium noted above. In its simplest form, a WDMA transceiver can be easily implemented using a single laser diode with a narrow linewidth, and a receiver consisting of a photodetector and a narrowband thin-film optical filter, each operating at a different wavelength to achieve simultaneous and uncoordinated transmission and reception. Unfortunately, in applications that require the use of several terminals, each operating at different wavelengths, the process becomes more complicated and costly. This is because for WDMA to work effectively, the emitter sources and the detectors need to be tuned to exact wavelengths. In addition, depending on the application, multiple terminal systems based on WDMA may require the use of tunable lasers at the transmitter and complex tunable bandpass filters at the receiver, which may or may not be possible to couple with the desired optical concentrator. Furthermore, tunable LDs are expensive and require complex techniques to tune them to exact wavelengths. For these reasons, rather than using single LDs and single detectors that can be tuned to different wavelengths, it is easier to implement transmitters that use different sources at different wavelengths and receivers with a number of detectors tuned to different wavelengths through thin-film bandpass optical filters, which increases the cost of the system.

An example of an application based on the concept of WDMA is the system developed by Gfeller et al. [9] described in Chapter 1, in which a number of terminals are clustered within a room where they exchange information with an optical satellite attached to the ceiling. In this application, the optical satellite transmits data at a specific wavelength (or range of wavelengths), while the terminals transmit at a different wavelength. This is possible due to the fact that direct communication between computer terminals is not required, which simplifies the number of wavelengths to only two — one for the uplink and another one for the downlink.

SDMA offers an alternative to WDMA. This technique relies on the use of angle-diversity receivers (ADRs) to discriminate signals according to the direction from which they come. SDMA can be implemented using a variety of ADRs combined with different configurations of quasi-diffuse transmitters to optimize the performance of the link (a description of ADRs is given below).

There are a number of techniques available to detect and process the signals received by the different structures of an ADR. The specific advantages and characteristics of an ADR system depend on the specific technique used. These techniques include equal-gain combining (EGC), maximum-likelihood combining (MLC), selection diversity (SD), and maximum ratio combining (MRC). MLC, for example, provides simultaneously high gain and large FOV while mitigating noise, multipath dispersion, and co-channel interference. EGC, on the other hand, offers the advantages of being capable of using only one preamplifier and avoids the need of having channel and noise estimation. SD and MRC, just as MRC, provide high gain and large FOV simultaneously. In addition, both techniques mitigate noise and co-channel interference; but while SD mitigates multipath distortion always, MRC mitigates it only under specific circumstances [21].

Maximum-likelihood combining (MLC) is considered the optimum reception technique in applications and configurations that suffer from high multipath distortion. Here, each one of the signals generated by the different detectors is processed by a different continuous-time matched filter. The outputs are then sampled and combined as illustrated in Figure 8.11, with each sample weighted in inverse proportion to the power spectral density of the noise. The weighted sum of the different samples can be used by the receiver to perform maximum likelihood sequence detection (MLSD). An ADR consisting of J elements produces a reception $Y(t)$ as follows [21]:

$$Y(t) = RX(t) \otimes h_j(t) + N_j(t) \quad j = 1, 2, \ldots, J \tag{8.22}$$

where $N_j(t)$ represents white Gaussian noise and $h_j(t)$ represents the impulse response of channel between the transmitter and the jth receiver. This expression assumes that there is no fluorescent light or co-channel interference present in the system. If this is the case, the matched filter at the exit of each detector can be represented by $h_j(-t)$.

The drawback of MLC is its complexity, which sometimes surpasses the requirements of a number of applications. If this is the case, simpler alternative techniques such as EGC, MRC, or SD are usually preferred. They all have in common that the $Y_j(t)$ receptions are combined in a memoryless linear combiner and that the weighted sum is filtered using a single continuous-time filter. What makes these techniques different from each other is the way in which the weights are chosen.

SD is a technique that sacrifices gain in an effort to improve the SNR. The latter is achieved due to the fact that only the signal with the best SNR is used, which means that the transmitted signal can be separated from the noise more easily. In addition, a significant reduction in the multipath effect can be achieved through the directional elements when the receiving structures have an FOV narrower than 50° [21, 212]. Unfortunately, when compared to non-angle-diversity receivers, SD presents a higher complexity due to the fact that the SNR must be estimated in each of the receiving elements.

In applications where multipath reflections and ambient light noise arrive from directions far away from where the information signal is arriving, MRC offers a net

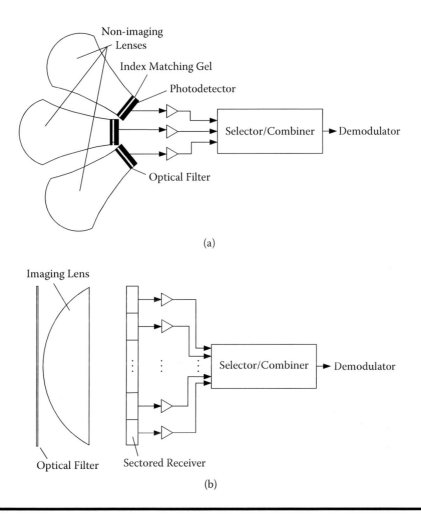

(a)

(b)

**Figure 8.11   Angle-diversity receiver (ADR): (a) array of DTIRC-detector struc-
tures and (b) imaging lens with sectored photodetector.**

decrease in multipath distortion (compared to a wide FOV receiver using a single
photodetector). If, on the other hand, the ambient light noise and the multipath
reflections are arriving from directions relatively close to the direction from where
the information signal is arriving, this may result in an increase in multipath distortion. Another important characteristic of MRC is that it reduces the transmitted
optical power requirement by between 4 and 6 dB in diffuse applications operating
at low data rates [21, 213]. Compared to SD, MRC requires between 1 and 2 dB
less optical power [21, 214]. In addition, MRC maximizes the SNR because the
sum of the $Y_j(t)$ reception is carried out according to the weights, which in this case
are proportional to the ratio of the signal current to the noise-PSD. The drawback

of this technique is that it also needs to estimate SNR in each one of its receiving elements, which, when compared to non-angle-diversity receivers, increases the complexity of the system.

Equal-gain combining (EGC) increases the FOV of the receiver but it cannot separate the information signal from co-channel interference or from noise. In this technique, as its name indicates, the signals from the different receiving elements are added together with equal weights. This reduces the complexity of the system because a single preamplifier can be used for the signals of all the photodetectors and because the SNR does not have to be estimated in any of the receiving elements of the ADR. The drawback of this technique is that it cannot be used for high data rates because an increase in the FOV also means an increase in multipath distortion. Examples of this technique include the 4-Mbps diffuse system presented by Gfeller et al., where speed was sacrificed to provide maximum coverage and robust operation [215].

Space division multiple access (SDMA) can use imaging or non-imaging ADRs with either selection diversity (SD) or maximum ratio combining (MRC). Potential applications of SDMA were presented by Kahn [21]. One possible scenario where Kahn proposes the use of this multiple access technique in a wireless IR LAN employing imaging ADRs and quasi-diffuse transmitters illuminating different spots on a reflecting surface (such as a wall or ceiling), as shown in Figure 8.12. In this type of application, small co-channel interference is achieved when the imaging ADRs detect energy from a reflective spot (originating from the transmitter in the reflecting surface) without receiving energy from another transmitter that may be creating an overlapping spot with the first. In mobile applications, the probability of spot overlapping becomes higher as the number of users increases. Therefore, it is necessary to use additional electrical multiplexing to attain a robust system. Another application where the use of SDMA has been proposed is in a wireless LAN hub such as the one described in Figure 1.1, where the optical satellite (or hub) provides access to a number of terminals (in a computer cluster) within a room.

8.6.2.1 Angle-Diversity Receivers

Angle-diversity receivers (ADRs) consist of multiple receiving elements oriented in different directions. They are used as an alternative to single-element receivers in a variety of configurations. ADRs can be created using arrays of photodetectors with optical concentrators mounted on them and pointed in different directions. If this is the case, the photodetector-concentrator structures can be arranged in such a way that their FOVs barely overlap, which provides an overall wide FOV and a high gain (as the FOV of each non-imaging concentrator can be reduced, the optical gain can be increased). ADRs provide design flexibility because the specific FOV requirement of a receiver can be achieved by combining the FOV of each individual concentrator. In addition, due to the fact that each concentrator-detector

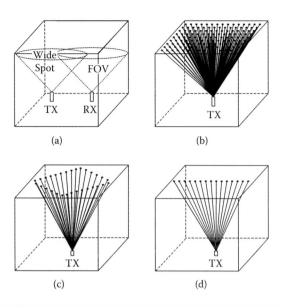

Figure 8.12 **Different spot-diffusing configurations: (a) diffuse configuration, (b) uniform multi-spot diffusing system, (c) diamond multi-spot diffusing system, and (d) line-strip multi-spot diffusing system. (*Source:* Adapted from [221])**

structure can have a narrow FOV, this solution improves the sensitivity of each receiving element or improves the speed of the system by allowing the use of smaller photodetectors (which is possible thanks to the gain of the concentrators). Further advantages of ADRs include the reduction in co-channel interference, background illumination noise, and multipath dispersion [21]. Background illumination noise, for example, can be minimized due to the fact that the transmitted signal is rarely isotropic, which means that it illuminates the receiver mainly from some specific directions (which frequently coincide with the orientation of the transmitter and the direction of the emitted signal). This is also true for background illumination, which, just as the transmitted signal, is rarely isotropic. Therefore, it illuminates the detector mainly from some directions. This means that ADRs can exploit the directional nature of the transmitted signal and of the background illumination to increase the SNR. An ADR comprised of DTIRCs mounted on a number of photodetectors is illustrated in Figure 8.11a.

An alternative way of creating an ADR involves the use of an imaging lens on top of a segmented photodetector (or an array of photodetectors) in such a way that changes in the position of the focal point created by the angular variations in the incident energy are detected by a different sector of the photodetector (or on a different photodetector), depending on the angle with which the energy is imping-ing on the lens. One of the advantages of using an imaging lens in an ADR, when compared to their non-imaging counterparts, is that the photodetector array can be planar, which facilitates the use of a large number of pixels (or receiving elements). Each one of these pixels or sectors can have its own preamplifier, which means that each of them can act effectively as an independent detector. If this is the case, the high power consumption associated with a large number of preamplifiers can be minimized by deactivating the preamplifiers that are not receiving the desired signal.

The ADR technique based on the use of an imaging lens (concentrator) at the entrance of a photodetector array is called an imaging angle-diversity receiver. This is illustrated in Figure 8.11b, where the cross-sectional view of an imaging lens and a sectored photodetector connected to a selector or to a combiner and leading to a demodulator is shown. Examples of imaging ADRs can be found in [44, 216–219]. The research group of Yun and Kavehrad [219], for example, was one of the first to investigate the use of ADRs for wireless IR communications. They investigated this technique with the intention of improving the power budget of the system, and called their transmitter "spot-diffusing" and their receiver "fly-eye." Their "fly-eye" receiver could be implemented using a ball lens or an imaging lens with photodetectors placed in different positions. This way, variations in the angle of incidence (and therefore variations in the focal point position) made the energy of the signal impinge on a different photodetector.

Imaging ADRs present the additional advantage (compared to non-imaging ADRs) of requiring only one concentrator, independently of the number of sectors of the photodetector. In addition, with regard to the fabrication of the sectored photodetector, a single planar array simplifies the manufacturing process. This is due to the fact that the photodetector array can be fabricated monolithically, which facilitates the use of a large number of detector segments [21]. Unfortunately, the FOV of imaging diversity receivers is narrower than one of their non-imaging versions because, in the latter, each receiving element can be oriented individually, giving the designer more control over the desired overall FOV.

Another example of an imaging ADR is found in [218]. Here, Djahani et al. described an ADR consisting of an imaging lens placed on top of J equal-sized hexagonal pixels, where the spot generated by the imaging lens falls on no more than three pixels at every single time. The pixels in the photodetector array presented no gaps between them, and their number was defined as [218]:

$$J = 2 \sum_{i=1}^{n} (n-i-1) - 2n + 1 \tag{8.23a}$$

or as:

$$J = 3(n^2 - n) + 1 \tag{8.23b}$$

where n is an integer greater than or equal to 2 and less than 20. In this method, the area of an individual pixel is given by [218]:

$$A'_{IMG,i} = \cfrac{d^2_{IMG}}{\left(\cfrac{1 + \sqrt{1 + 4\dfrac{J-1}{3}}}{2} + \left[\cfrac{-1 + \sqrt{1 + 4\dfrac{J-1}{3}}}{4} \right] \right)^2} \left(\frac{3\sqrt{3}}{8} \right) \qquad (8.24)$$

where d_{IMG} is the diameter of the image created by the imaging lens, and the average signal power received by the ith pixel is given by:

$$P_{IMG,i} = f_i(\Psi,\theta)P_{IMG} \qquad (8.25)$$

where $f_i(\psi,\theta)$ is the overlap factor that indicates the fraction of the total power received by a specific pixel (when the signal image spot overlaps more than one of them); ψ and θ are the polar and the azimuthal angles of incidence, respectively, and P_{IMG} is the average optical signal power received by the ADR.

ADRs can be used in conjunction with spot-diffusing transmitters (SDTs) (also called quasi-diffuse transmitters), which replace conventional diffuse transmitters and compensate for the high path loss of the diffuse configuration. This is possible due to the fact that SDTs can emit multiple narrow beams pointed in different directions, effectively simulating a diffuse link but making optimum use of the transmitted power. Further information on SDTs is presented in Subsection 8.6.2.2.

A possible configuration for an ADR, called the pyramidal fly-eye diversity receiver, was presented by Al-Ghamdi et al. [220, 221]. This technique consists of using narrow FOV receivers mounted on the sides of a pyramidal structure with the aim of minimizing directional interference through an optimized FOV. Another version of an ADR can be found in [222]. Here, Valadas et al. proposed a receiver conformed by a variety of sectors, with each sector consisting of an optical collector, its associated front-end, and its control circuit. Each sector (which they called diversity branches) had a specific area, FOV, and orientation; and the total FOV of the overall receiver was created by the addition of the individual FOV of each sector. In these sectored receivers, the optimum signal could be chosen, depending on which sector presented the best SNR. In addition, a resulting signal could be obtained by the addition of the signals of the different sectors.

8.6.2.2 Quasi-Diffuse Transmitters

Spot-diffusing or quasi-diffuse transmitters employing multiple transmitting elements with a narrow emission field have been proposed as a way to minimize multipath dispersion and optimize power consumption in a configuration that simulates a diffuse link.

One way to achieve a spot-diffusing transmitter involves the use of a narrow-beam optical source (such as a laser) in combination with a holographic diffuser. The resulting emitter generates multiple beams that illuminate different points on a reflective area. Particular intensity distributions and spot intensities can be achieved using a CGH (computer-generated hologram) such as the one described in Chapter 3. Examples of different configurations achieved using CGHs include uniform wide-angle, line-strip, and diamond multibeam transmitters. Al-Ghamdi et al. [221] have analyzed different spot-diffusing techniques, finding that compared to other geometries, line-strip multibeam systems (LSMS) offer improved performance (an SNR improvement of about 20 dB) when combined with ADRs consisting of three narrow-FOV detectors oriented in different directions.

Another way of creating a multibeam transmitter involves using arrays of vertical-cavity surface emitting laser (VCSEL) arrays combined with diffractive elements that allow the emitted beam to be shaped or steered, as presented by Karpinnen et al. [223]. In the compact multibeam transmitter they proposed, the Gaussian beam of a VCSEL could be converted into a flat-top beam and directed with the desired angle by a diffractive element. The transmitters could be made to emit concurrently or selectively (which provided a better power efficiency and less multipath dispersion), depending on the degree of complexity required at the transmitter. In the specific example presented by Karpinnen et al., the transmitter achieved a total emission angle of ±50° through the addition of the emission angle of individual elements, which provided an emission angle of 10° each (in one dimension). The diffractive elements were created through an optical map transformation and fabricated by electron-beam lithography using fused silica.

Yun and Kavehrad [219] proposed a multi-spot diffusing system (MSDS) in 1992. It consisted of multiple transmitters, each one emitting energy within a narrow beam and oriented in a different direction. The transmitted beams pointed to a ceiling that exhibited high reflectivity. Al-Ghamdi et al. performed a comparison of a multi-spot diffusing system with a conventional (diffuse) system. The diffuse system was based on a reflector. They found that the uniform multibeam transmitter improves the received optical power when compared to the conventional diffuse system. One of the MSDS configurations used in their comparison was a diamond array. In this case, a diamond shape on the ceiling (or the reflecting surface) was achieved using four line strips consisting of 20 spots separated 10 cm from one another. The analysis of the diamond multi-spot diffusing system shows that despite its less uniform power distribution (compared to its uniform counterpart), and the fact that the collected power at the center of the room was lower, the collected power level near the side walls was higher. Another array used for this comparison was the line-strip multibeam system (LSMS). The LSMS used consisted of an array of 80 × 1 beams with equal intensities and with a distance between adjacent spots of 10 cm. The beam was oriented in such a way that it formed a line in the middle of the ceiling. The three configurations proposed by Al-Ghamdi et al. are illustrated in Figure 8.12.

A comparison of the different configurations indicates that the power received from the multi-spot diffusing transmitters is better than the power received from a conventional diffuse transmitter. In terms of the delay-spread performance, the uniform multibeam configuration using a single wide FOV receiver presents a larger delay spread than a conventional diffuse system. This is also true for the line-strip and the diamond multibeam configurations that use a single wide FOV receiver. The uniform multibeam transmitter presents the higher delay, followed by the diamond multibeam configuration and the line-strip multibeam transmitter. A comparison of the different multi-spot diffusing configurations also indicates that the delay spread can be considerably reduced if angle diversity receivers are used in combination with MSDS. If this is the case, the spread of the multi-spot diffusing configurations can be reduced to values below those of the conventional diffuse configuration (in most cases). The lowest delay spread (of all the configurations) is presented by the LSMS combined with an angle diversity receiver, where most of the total received power is contained within the first-order reflections. Here, high-order reflections contain only a small fraction of the total power, which allows a lower pulse spreading and a better receiver power [219].

A receiver configuration that can be used with a line-strip multi-spot transmitter is the pyramidal fly-eye receiver described above in Subsection 8.6.2.1). This type of receiver can be configured in such a way that the SNR is optimized for an LSMS, while the probability of blockage of the multi-spot array is minimized. In addition, this transmitter-receiver combination can be used to minimize a large amount of background illumination noise, which is particularly the case when very directive receivers are used. This is due to the fact that the noise is more directive that the signal, which means that the rate of noise reduction is higher than the signal power reduction when the FOV of each receiver decreases. In addition, reducing the FOV of each receiver minimizes the pulse spread by limiting the range of accepted rays. Unfortunately, this is done at the cost of reducing the received power also. The FOV of an optical concentrator can be set to a value that allows for a reduction in the background illumination, yet maintains a convenient number of direct paths between the diffusing spots and the ADR.

8.7 Summary and Conclusions

This chapter has underlined the importance of selecting the right modulation scheme when designing an optical wireless system. The modulation technique used defines important parameters such as bandwidth and power efficiency. For this reason, different modulation schemes such as pulse position modulation (PPM), on-off keying (OOK), and subcarrier modulation [24] have been analyzed by different researchers, both for indoor and outdoor applications.

Choosing a modulation scheme with a high power efficiency is important in applications involving portable devices, for which power consumption is critical.

From the comparisons presented in this chapter, it can be concluded that one of the modulation schemes that offers the highest power efficiency for wireless IR communication systems is L-PPM. Here, the larger the L value, the better the power efficiency. Unfortunately, this is done at the expense of reduced bandwidth efficiency. In addition, this modulation scheme requires a higher level of complexity due to the fact that chip and symbol-level synchronization are required.

In terms of bandwidth efficiency, one of the most efficient modulation schemes is L-PAM, but in this modulation scheme too, this is done at the expense of sacrificing another parameter — power efficiency in this case. Therefore, if bandwidth is the most important parameter for a specific wireless IR application, and power consumption is not too critical, then 16-PAM offers a very attractive option. If, on the other hand, power consumption is the most important parameter and bandwidth efficiency is not too critical, then 16-PPM presents clear advantages over other modulation schemes.

Note that the performance analysis of a variety of modulation schemes carried out by different researchers is generally made with respect to OOK. This is due to the fact that OOK offers a good compromise between bandwidth and power efficiency. This, and its simplicity of implementation, make OOK one of the most attractive modulation schemes for a variety of applications. OOK is generally used with NRZ pulses. In cases where the power efficiency of OOK must be optimized, it can be used with RZ pulses at the cost of increasing the bandwidth requirement.

With regard to the performance comparison of different PPM schemes (that is, PPM, MPPM, and OPPM), it can be concluded that, with regard to bandwidth efficiency, the performance of MPPM is better than PPM (in general), with the bandwidth requirement decreasing with increments in weight. MPPM with a weight of 2, for example, performs better than traditional PPM, while MPPM with a weight of 8 performs a lower bandwidth requirement than PPM or than an MPPM scheme with a lower weight. The drawback in MPPM when using a high weight is that, in this case, a large number of chips are required.

OPPM also presents a good bandwidth efficiency when the duty cycle is $\alpha = \frac{1}{2}$, but it is reduced when the duty cycle is changed from $\alpha = \frac{1}{2}$ to $\alpha = \frac{1}{4}$ (that is, when the power efficiency is improved by reducing the duty cycle).

From the comparison of different modulation schemes, it can also be concluded that quadrature phase-shift keying (QPSK) and binary phase-shift keying (BPSK) do not present an obvious advantage in terms of power or bandwidth efficiency when compared to OOK (when operating in environments that do not have background illumination). However, when background illumination noise is present, both modulation schemes help to alleviate the problem. This applies, in particular, to indoor environments. Fluorescent lamps employ electronic ballasts that introduce low-frequency components at multiples of the line frequency. The effect of these electronic ballasts can be reduced significantly (or eliminated) using subcarrier modulation with a high enough subcarrier frequency. For this reason, choosing a modulation scheme such as BPSK with an appropriate subcarrier frequency is attractive.

From the comparison of the different modulation schemes under fluorescent lamp illumination, it can be concluded that OOK is one of the schemes most strongly affected, even when a high pass filter is employed at the receiver. PPM, on the other hand, presents a much better immunity to near-DC noise from fluorescent lamps than OOK. This is why PPM is favored for indoor applications where the detector is subject to illumination from fluorescent lamps.

This chapter also presented different physical-layer multiple access techniques used for wireless IR communications. These techniques define the way in which the IR medium is shared by a variety of users, which is especially important in indoor LANs where one terminal is required to exchange information with a number of other terminals or devices. The physical-layer multiplexing techniques presented in this chapter are based on some of the characteristics of the IR medium.

The two ways of achieving physical-layer multiple access are (1) electrically and (2) optically. Electrical multiplexing techniques include code division multiple access (CDMA) and time division multiple access (TDMA). Optical multiplexing techniques include wavelength division multiple access (WDMA) and space division multiple access (SDMA).

TDMA consists of dividing the time allocated to each user in a synchronized manner. This multiplexing technique offers high power efficiency at the cost of requiring a relatively high complexity, which is necessary to achieve the high coordination required. FDMA, on the other hand, grants access to multiple users by allowing them to transmit information simultaneously (within the same physical space) using different frequencies. TDMA presents lower power efficiency than FDMA. In the other electrical multiple access technique called CDMA, different users can have access to the same channel, thanks to the use of different orthogonal or quasi-orthogonal code sequences.

Optical multiplexing techniques make use of the optical properties of the IR medium to grant access to a number of users. In WDMA, for example, a number of users achieve simultaneous transmission using different wavelengths. This can be easily implemented using transmitters with narrow linewidths and receivers that incorporate narrow bandpass optical filters. Compared to TDMA, WDMA offers simplicity. Another optical multiple access technique is SDMA, which uses ADRs to discriminate signals, depending on the direction from where the energy emanates. ADRs consist of multiple receiving elements oriented in different directions. They are used as an alternative to single-element receivers and can be used in a variety of configurations. The receiving elements of ADRs are generally a combination of an imaging or non-imaging concentrator and a photodetector. They provide flexibility of design because, by using a number of detectors with reduced FOV each, any overall FOV desired can be achieved. The narrow FOV of each detector allows a better background noise rejection and a higher gain.

Angle-diversity detection is advantageous because through it, a reduction in background illumination noise, multipath dispersion, and shadowing (created by objects) can be achieved. This technique also improves the received optical power level. ADRs are particularly useful in non-directed configurations. Here, the conventional large FOV of a single-element receiver can be replaced by a number of detectors orientated in different directions (where each one of them has a narrow FOV). This presents the advantage of minimizing multipath dispersion as each one of the directive elements receives only a small fraction of a signal that has undergone multiple reflections (Carruthers and Kahn, for example, reported that the multipath dispersion in an experimental system operating at 70 Mbps and incorporating nine single receiving elements with a FOV = ±22° each was basically insignificant [217]). Furthermore, the use of this technique offers the possibility of rejecting unwanted sources of light by eliminating the unwanted energy impinging on specific detectors, which otherwise would be received by the wide FOV of a single-element detector. An example of an angle-diversity configuration is the fly-eye diversity receiver, which consists of a pyramidal base containing a narrow FOV receiver attached to its walls.

The angle-diversity technique can be combined with spot-diffusing or quasi-diffuse transmitters for maximum benefit. Spot-diffusing transmitters employ multiple transmitting elements with a narrow emission field. The spots generated can have a number of different configurations, of which one of the most effective is the line-strip multibeam system. The combination of line-strip multibeam transmitters and angle-diversity receivers has not only been reported to reduce the delay spread created by multiple reflections from walls and ceiling, but also to produce a significant improvement in the SNR due to the reduction of background illumination noise. Thus, the combination of these two techniques implies that a system based on them can benefit from the advantages offered by the directed and the diffuse configurations at the same time.

Note that if the complexity of a system based on an angle-diversity receiver and a multibeam transmitter needs to remain a minimum, configurations such as the pyramidal fly-eye receiver (based on a small number of detectors) can be used.

Chapter 9

Infrared Data Association (IrDA) Protocols

9.1 Wireless Protocol Standards

A protocol is a set of rules that defines the way in which communication between systems and devices is carried out. A number of wireless communication protocol standards are currently in use for radio and infrared. IEEE 802.11, for example, is one of the most popular standards for RF wireless communication. Equipment based on this standard operates at the thus-far unregulated frequencies of 2.4 GHz and 5 GHz. Another popular low-range and low-power consumption RF wireless communication protocol is Bluetooth, which also makes use of the 2.4-GHz unregulated frequency bands.

With regard to infrared, one of the most popular standards is the one created by the Infrared Data Association (IrDA). This chapter presents general information about the IrDA, as well as an overview of the IrDA standard and technology. In addition, basic features of the different layers of the IrDA protocol are introduced. The information presented in this chapter (and in Chapter 10) is based on information available in the IrDA Web site [17], on the description of the protocol by Knutson et al. [224], and on the specifications of the standard [225–228]. Further details and information about the IrDA and its protocol can be found in these sources.

9.2 The Infrared Data Association

In 1993, an international organization called the IrDA (Infrared Data Association) [17] was formed to provide interoperability standards for the software and hardware used in wireless IR communication systems worldwide. This organization has more than 160 members, which include major manufacturers of software, hardware, components, peripherals, systems, and communication systems. It also includes automobile and service providers, as well as telephone and cable companies. The idea behind the standard was to create a simple, cost-effective,* and low-power transceiver that could be implemented in all sorts of devices. According to Mike Watson, ex-president of the IrDA, more than 600 million electronic devices are already based on the IrDA standards, with a growth rate of more than 20 percent annually [229]. The devices that currently use the IrDA transceiver include notebooks, digital cameras, desktops, printers, public phones and kiosks, PDAs, palm PCs, electronic books, electronic wallets, cellular phones, pagers, watches, toys, and other mobile devices.

The IrDA consists of an Architecture Council and several Special Interest Groups (SIGs) who are responsible for developing the protocols and specifications used to define the way in which data is transferred between IrDA-enabled devices. The SIGs are [17]:

- *IrFM*, which develops the specifications for worldwide wireless proximity payment for commercial transactions at the point of sale (credit card payments, for example). They also develop specifications that allow compatibility with other wireless technologies.
- *IrTM*, whose function is to develop the specification for transportation telematics and payment systems.
- *IrBurst*, which develops the specifications for digital information transfer at speeds up to 100 Mbps.
- *UFIR*, whose function is to modify the digital information transfer specification produced by IrBurst to increase the speed to up to 500 Mbps
- *UFIR Marketing*, which develops vendor support for the design and manufacture of IrDA products based on the UFIR specification.

The IrDA also has a number of Committees, whose objectives are to develop new markets, resolve technical issues, and provide tools and services that ensure interoperability of devices and compliance to specifications. These Committees include [17]:

- *Technical Committee*: responsible for maintaining the existent standards and for identifying and resolving technical problems related to both hardware and software. It is also in charge of supervising the creation of new protocol specifications for the session, application, and core protocol layers.

---

* The current cost of an IrDA transceiver is approximately US$1.00 [229].

- *Marketing Committee:* responsible for the development of new markets. It is also responsible for the strategic direction of the IrDA. Its four subcommittees include Extensions, Market Development, Interoperability, and New Markets.
- *Test and Interoperability Committee:* responsible for creating and promoting standards for the interoperability of IrDA-based devices. It is also responsible for producing documentation to assist in testing the compliance of products with the IrDA standards specifications, for creating certification standards, and for protecting the association's trademarks.

9.3 IrDA Standard Overview

The first standards of the IrDA were released in 1993 and 1994. The increasing use and acceptance of the IrDA standards have stimulated the adoption of the IrDA specifications by different standards organizations. In 2000, it was estimated that the use of optical wireless devices based on the IrDA standard was growing at 40 percent annually, with a global installed based of over 150 million units [230].

The IrDA-Data protocol defines a standard for an interoperable universal two-way wireless IR transmission data port. This protocol is organized in a layered fashion, with the physical layer at the base (just as in the OSI Model). When the IrDA physical detector receives data, the data packet is directed to the lowest layer of the protocol stack, which obtains the control data required and transfers the remaining information to the next layer up in the model. Each stack extracts the required control data and transfers the rest to the upper layer until the data payload is received by the application. Information created by an application follows the inverse process. The information is managed by the protocol of the application layer, which takes the required data, appends control information to it, and sends this new set of information to the next lower layer. The process terminates when the information is transferred from the lowest-level protocol to the physical level that will transmit the information via IR radiation.

The IrDA basic protocol stack layers are [224]:

- *IAS.* The "Information Access Service" is responsible for detecting services on other devices, and for supplying the appropriate mechanisms for advertising to remote devices.
- *TinyTP.* This layer is responsible for per-channel flow control, segmentation, and reassembly.
- *IrLMP.* The "Link Management Protocol" provides multiplexing of the IrLAP layer as well as service registry and discovery mechanisms.
- *IrLAP.* The "Link Access Protocol" is responsible for establishing a link between two devices (which includes device discovery, connection, and negotiation) and for maintaining it.

■ *Framer/Driver.* This layer is responsible for satisfying the driver requirements and for performing packet framing according to the requirements of the physical layer.
■ *Physical layer.* This layer specifies the physical hardware for the IR link, as well as other characteristics such as the wavelength of operation, the range, the BER (bit error rate), and the angular capabilities of the transmitter and the receiver. The components of the physical layer are illustrated in Figure 9.1.

The six core protocol stack layers are illustrated in Figure 9.2. Here, the optional stack layers OBEX (Object Exchange) and IrComm (IR Communication) are also indicated.

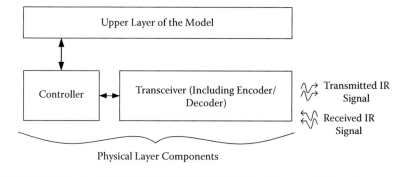

Figure 9.1   IrDA Protocol physical layer components.

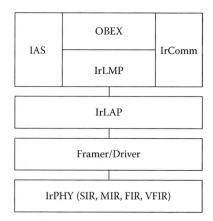

Figure 9.2   IrDA Protocol stack layers [2].

As explained in Chapter 1, IR and radio technologies can operate in a complementary manner; and one may be preferred over the other, depending on the application. In terms of standards, Bluetooth has specified a session protocol called IrOBEX that makes it possible for applications to use either the Bluetooth radio technology or the IrDA infrared technology for applications over short distances [18, 230]. Originally, the Object Exchange protocol (OBEX) was developed to exchange data objects over an IR link. However, it can appear above other transport layers such as RFComm. IrOBEX is based on the IrDA data standard, therefore linking it to current IR standards.

A comparison between the implementation costs of IrDA and its radio counterpart Bluetooth is also of interest. According to an analysis from *Extended Systems* presented in 2000 [230], device manufacturers estimated at the time that they could get a complete solution for about US$1.00, and that the cost of integrating IR into a device could be as little as US$2.00. This estimation has become true; and according to Knutson et al. [224], the current cost of an IrDA transceiver is approximately US$1.00. This has been achieved thanks to the fact that implementing IrDA in consumer devices is simple, and IrDA controllers are simple also. The estimations from *Extended Systems* also suggested that the cost of implementing Bluetooth in a device would be about US$20.00 for first-generation devices, with future devices targeted to about US$5.00.

Another advantage of IrDA devices over components that support other forms of wireless communication is the low power consumption of its transceivers. This can be observed in Table 9.1 [224], where a comparison between IrDA, Bluetooth, and WiFi is presented. It can be seen here that the power consumption of IrDA transceivers is smaller than that of Bluetooth while providing higher data rates.

Table 9.1   Comparison of IrDA, Bluetooth, and WiFi Wireless Technologies

Type of Wireless Technology	Data Rate	Range	Power Consumption
Low-power IrDA	4 Mbps	20 cm	60–120 μA
Normal IrDA	16 Mbps	1 m	0.9–1.2 mA
WiFi 802.11b	11 Mbps	100 m	480–700 mA
WiFi 802.11a	54 Mbps	100 m	500–600 mA
Bluetooth Class 3	723 kbps	1 m	0.3–1.2 mA
Bluetooth Class 2	723 kbps	10 m	75–100 mA
Bluetooth Class 1	723 kbps	100 m	220–350 mA

Source: From [224].

It is also smaller than that of WiFi on its different classifications. The low power consumption of IrDA transceivers is particularly relevant when wireless communication must be implemented in portable devices where power consumption is critical. A good example of this is wristwatches. The low power consumption required by these devices motivated the creation of the low-power IrDA specification. Unfortunately, reducing the power consumption in a device is generally made at the cost of also reducing its range. A further advantage of the IrDA standard when compared to Bluetooth is the minimum bit error rate (BER) specified by each standard. While for IrDA the worst value of BER allowed is 10^{-8}, for Bluetooth the worst BER is only 10^{-3}, which makes the latter more unreliable.

IrDA transceivers employ digital intensity modulation as the way to convey information over the near-IR part of the spectrum (cf. Chapter 8). The wavelength at which these transceivers operate is 880 nm (equivalent to 300 THz), with current data rates up to 16 Mbps. In terms of the architecture, the IrDA Protocol defines point-to-multipoint connection from a single primary unit to up to 16 devices. Here, the primary unit assigns time slots to the secondary units, controlling in this way the data traffic. The specific algorithm to implement this connection can be dictated by each manufacturer.

Despite the fact that the transmission range specified by the IrDA is 1 meter, the actual range achieved by a device based on this specification varies depending on the background illumination conditions under which the transceiver operates. The reason why the distance is specified as only 1 meter has to do with power constraints. The specification also defines a ±15° emission cone at the transmitter, and a wide FOV at the receiver (achieved through a hemispherical lens mounted on top of the photodetector).

There are different data rates supported by IrDA devices. This is the result of the evolution of the technology, which covers data rates between 2.4 kbps and 10 Mbps. The specific transmission rates of Serial Infrared (SI), which was the first physical layer specification, are 9.6, 19.2, 38.4, 57.6, and 115.2 kbps. These data rates are supported by all types of IrDA devices. A further specification, commonly known as Medium Infrared (MIR), defines faster bit rates based on an encoding scheme similar to the one used for SIR (standard IR). Its speeds are 0.576 and 1.152 Mbps. A later version of the specification, commonly known as Fast Infrared (FIR), is based on PPM and defines an improved data rate of 4 Mbps. The latest speed improvement defined by the IrDA physical layer specification supports a transmission speed of 16 Mbps. This is known as Very Fast Infrared (VFIR) and it is based on run-length limited encoding [231].

One of the preferred operation modes of the IrDA devices is the *point-and-shoot* usage model. This means that to establish the link, the user must aim his device at the appropriate target device, which in turn can detect and authenticate the other device and establish a communication link. The user's device can, at that moment, "shoot" (send) some data to the device of interest.

9.4 The Physical Layer Protocol

The Physical Layer Protocol defines the low-level physical components used in a wireless IR link to transmit and receive information. It includes the characteristics of the transmitter (such as the type of source, which in this case is an LED), and the characteristics of the receiver (which contains an Si PIN photodiode). Other characteristics, such as the optical, electrical, and physical characteristics with which manufacturers associated with the IrDA need to comply, are also specified in the Physical Layer Protocol. These characteristics include wavelengths of operation, data rates, the optical components required to provide the appropriate FOVs for emission and detection, rise and fall times, bit error rates (BERs), pulse duration, range, etc.

The first specification of the physical layer, which supported a maximum speed of 115.2 kbps, was released in April 1994 and supported speeds up to 115.2 kbps. Since then, technology has improved and newer versions of the physical layer specification have been released. Version 1.4, for example, was approved in 2001 and specifies the necessary components to support data bit rates up to 16 Mbps.

9.4.1 IrDA Transmitters

The main physical layer devices, such as the LED (light-emitting diode) and the photodetector, are encapsulated in a single component called a *transceiver*. The LED is controlled by an LED driver that receives information from an encoder (whose function is to convert data bits into an appropriate electrical signal). The output driver modulates the output power of the LED in such a way that the electrical pulses are converted into equivalent optical pulses in a process known as intensity modulation (IM) (cf. Chapter 8). Roughly speaking, this type of modulation consists of switching the LED on and off at the rate specified by the protocol. The specific encoding scheme defines whether the pulse represents a single or several bits. The characteristics of the pulse are important to achieve the right quality of information transmission and the right transmission speed. The transmitter characteristics that must be taken into account to guarantee an appropriate quality of transmission include the emitted power intensity, the LED's response time (which consists of the rise and fall times of the component and defines the speed at which the LED can be switched on and off), and the time the LED emits at full power. The raise time is the time it takes the LED to go from 10 to 90 percent of its peak value, and the fall time is defined as the period of time it takes the LED to go from 90 percent of its peak value to 10 percent of it (that is, the time it takes the LED to go from a high emission power to a practically switched off state). The period of time that an LED is emitting energy defines the duration of the transmitted pulse; and the pulse duration is defined as the length of time between the moment that the LED reaches 50 percent of its steady peak value during rise time to the moment

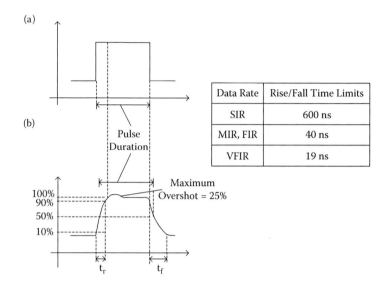

Data Rate	Rise/Fall Time Limits
SIR	600 ns
MIR, FIR	40 ns
VFIR	19 ns

Figure 9.3   IrDA's infrared pulse limits: (a) LED current and (b) radiated light intensity.

when the emitted energy reaches 50 percent of the peak value during fall time (this is illustrated in Figure 9.3). The pulse duration depends on the encoding scheme and ranges from 1.41 to 88.55 μs for slow infrared to between 28.3 and 45 ns for very fast infrared. The complete specification of pulse lengths for different modulation schemes is presented in Table 9.2.

When the LED is switched on, it is not unusual for the radiated energy to exceed its typical value of peak intensity. This effect is called *overshoot*, and it is generally just a transitory effect that occurs before the component can reach its stable or steady state. Another effect whose limit is specified by the IrDA standard is the one commonly known as *edge jitter*. This is defined as abrupt and unwanted variations in the position of the pulse edge with respect to where it should be. In the case of slow IR, it refers only to the leading edge of the pulse, and its limit must be within ±6.5 percent of the width of the pulse to comply with the protocol. For fast and very fast IR, edge jitter refers to both the leading and trailing edges of the pulse. For medium and fast IR speeds, the edge jitter limit is ±4 percent of the width of the pulse; and for medium IR speeds, the limit is set at ±2.9 percent of the pulse's width.

9.4.2 IrDA Receivers

The receiver side of an IrDA transceiver consists of a photodetector and a decoder. The function of the photodetector is to convert the received optical pulses into

Table 9.2   IrDA Pulse Duration Specification for Different Modulation Schemes

Signalling Rate	Modulation Scheme	Minimum Pulse Duration	Nominal Pulse Duration	Maximum Pulse Duration	Rate Tolerance % of rate
16 Mbps	HHH (1, 13)	38.3 ns	41.7 ns	45.0 ns	±0.01
4 Mbps (double pulse)	4-PPM	240 ns	250 ns	260 ns	±0.01
4 Mbps (single pulse)	4-PPM	115 ns	125 ns	135 ns	±0.01
1.152 Mbps	RZI	147.6 ns	217 ns	260.4 ns	±0.01
0.576 Mbps	RZI	295.2 ns	434 ns	520.8 ns	±0.01
115.2 kbps	RZI	1.41 μs	1.63 μs	2.23 μs	±0.87
57.6 kbps	RZI	1.41 μs	3.26 μs	4.34 μs	±0.87
38.4 kbps	RZI	1.41 μs	4.88 μs	5.96 μs	±0.87
19.2 kbps	RZI	1.41 μs	9.77 μs	11.07 μs	±0.87
9.6 kbps	RZI	1.41 μs	19.53 μs	22.13 μs	±0.87
2.4 kbps	RZI	1.41 μs	78.13 μs	88.55 μs	±0.87

Source: From [224, 235].

electronic pulses that convey the information to pass to the decoder. The decoder then converts the received electrical pulses into bits of data. The fact that the emitted pulse generated by the transmitter is well defined facilitates the identification of the transmitted signal and the discrimination of radiation from unwanted sources of illumination (that is, noise) at the receiver.

The receiver also incorporates a gain control system that facilitates accommodating the different intensities of illumination detected by the PIN diode. These gain variations arise from the fact that the distance between the transmitter and the receiver in a given application at a specific time may vary between 1 cm and 1 m, which, taking into account that the transmitted signal attenuates exponentially, results in very different power intensities at the photodetector. Automatic gain control (AGC) is the process by which gain is automatically adjusted in a specified manner as a function of a specified parameter, which in this case is the received optical power. This gain control is implemented electronically, for example, by varying the transimpedance of the receiver's preamplifier [224].

The receiver in an optical IR link must be able to cope with different levels of incident energy regardless of its distance with respect to the transmitter as long as both the transmitter and the receiver are within the range defined by the protocol. That is why the IrDA specifies limits to the maximum and minimum irradiance values of the receiver for the range specified by the protocol.

9.4.3 Transceiver Specifications

A hardware component called a *controller* controls the IrDA transceivers. This controller is the interfacing element that allows reliable information exchange between the upper layers of the model and the transmitter or receiver.

There are four different types of transceiver defined by the IrDA protocol. They are the result of the evolution of the protocol; and, for this reason, different speeds are supported by each of them.

Faster transceivers, for example, require a more complex controller than their low-speed counterparts. The names of the different transceivers are [224]:

- SIR — standard infrared (low speed) — its driver performs its own framing and assigns bytes to a controller from which the bytes are taken by the transmit encoder.
- MIR — medium infrared (medium speed) — its controller uses Direct Memory Access (DMA) to drive the encoder/decoder, do the framing, and relocate big quantities of raw data.
- FIR — fast infrared (high speed) — its controller uses a DMA like the one used in the MIR transceiver.
- VFIR — very fast infrared (high speed) — its controller uses a DMA like the one used in the MIR transceiver.

Note that the position (that is, the physical location) of the IrDA transceivers is not defined by the standard. This is the reason why different electronic devices such as mobile phones, PDAs, and laptop computers have the infrared port situated in a different location. This is possible due to the fact that, given the characteristics of the physical layer components in terms of FOV, transmitted power, distance, etc,, the devices must be easy to use regardless of the position of the port in the device. Unfortunately, this is not always the case. The infrared port is sometimes placed in locations that make it difficult for the user to use the point-and-shoot feature of the IR device (an example of this situation is when the IR port is located at the base of a device, making it difficult for the user to establish communication with another device while seeing the instructions on the screen at the same time). The mounting of the transceiver during the manufacturing process is also crucial for the correct operation of a wireless IR device. If the component is not positioned properly, it may suffer from shadowing or from a deformation of the transmitted beam, deteriorating the quality of the communication.

Devices incorporating IrDA transceivers use a protective window in front of them. This window is also specified by the IrDA standard as it not only provides physical protection for the component, but also acts as an optical filter at wavelengths between 850 and 900 nm (cf. Chapter 3). It is important to take into account that even if the window is transparent at the IR wavelengths of interest, it still presents some absorption. Therefore, to conform to the IrDA compliance certification, a 20 percent power loss must be taken into account when designing the system.

The shape and position of the plastic window is very important for the proper transmission and detection of IR beams. If the window is not completely perpendicular to the axis of the transmission/reception beams, or if the window is bent, it may deviate the IR rays in an unwanted way. The overall effect should be such that the IrDA transceiver with the window in place emits a beam of IR energy of at least ±15°, as specified by the physical layer standard. It is also important that the window is completely flat because concave or convex windows will have the effect of either diverging or collimated the beams of IR radiation.

9.4.4 Other Issues Related to the Physical Layer

With regard to the characteristics of the configuration defined by the IrDA, the protocol is based on a point-to-point topology (although IrLAP can be used for point-to-multipoint communication) that is achieved by a directional and narrow emission beam and a wide field-of-view (FOV) at the receiver. This configuration is line-of-sight and directed. Multipoint-to-multipoint connectivity is not currently available.

Compared to a wired system that employs Carrier Sense Multiple Access/Collision Avoidance to detect and avoid data collision, the IrDA Protocol does not specify a collision detector system. Collision avoidance is achieved by taking into account that the user has control over the physical space illuminated by the IrDA device. This physical space is well defined by the range and the emission angle of the IrDA transceiver; therefore, it is possible to assume that this space is used by the primary device (and maybe a few secondary ones) only.

The narrow emission beam defined by the IrDA physical layer specification is generated by an LED operating in the near-IR part of the spectrum at 880 nm with a half emission angle within the range ±15 to ±30° with respect to the axis of the component (the optical axis). As mentioned above, this well-defined angular emission allows space reuse and avoids data collision. The emission distance of the wireless IR link specified by the IrDA is 1 meter, which means that, from geometrical considerations, a spot of energy of approximately between half a meter and a meter in diameter can be generated at the specified range. This range not only further supports data collision avoidance and space reuse, but also reflects the fact that power efficiency was taken into account when defining the protocol (power consumption

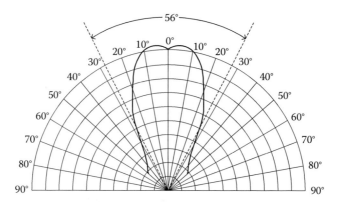

Figure 9.4   Typical plot of angle of radiation versus relative intensity for the LED of an IrDA transceiver. (*Source:* Adapted from [224].)

increases with the square of the transmission distance). Figure 9.4 shows a typical plot of angle of radiation versus relative intensity. As can be seen, the emission pattern is not exactly a cone, even if the angle of emission is defined in that way at the exit of the LED. This effect must be taken into account at the moment of aligning the device for ranges in excess of 0.5 m.

Figure 9.5 illustrates the characteristics of the LED specified by the IrDA. It contains information regarding the minimum and maximum emission angles of the beam as well as the acceptable wavelength range and the maximum allowable emission power. The power level limitations of these devices are related to their emission angles. The power restrictions are given in terms of their irradiance, which is defined as power per unit solid angle. The maximum allowable power defined by the IrDA for its transmitters is 500 mW/sr. If the transmitter emitted energy with a power intensity beyond the maximum value specified by the standard, the transmitted energy could saturate the photodetectors at close range.

There is also a minimum emitted intensity defined by the standard. The definition of this minimum value is necessary to ensure that the maximum range is achieved and that the signal is properly received by the photodetector. Contrary to the maximum power intensity, which as mentioned above depends on the emission angle, the value of the minimum intensity depends on the version of the transceiver (which can be "low power" or "standard") and on the encoding scheme used. Low power transceivers, for instance, need to produce 3.6 mW/sr at standard IR speeds. For higher speeds (MIR, FIR, and VIR), the minimum emission intensity becomes 9 mW/sr. Standard transceivers need larger minimum signal intensities than their low power versions. The minimum intensity specified for standard (low) speed transceivers is 40 mW/sr. It is 100 mW/sr for the rest of the speeds. The minimum and maximum signal intensities at the transmitter are summarized in Table 9.3 [data obtained from 224, 232].

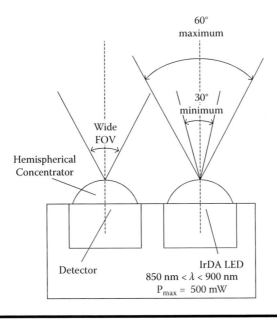

Figure 9.5 Optical characteristics of an IrDA transceiver.

Table 9.3 Minimum and Maximum Signal Intensities and Irradiances for IrDA-Based Transmitters and Receivers

Transceiver Part and Speed	Power Type	Minimum Limit	Maximum Limit
Transmitter SIR	Low power	3.6 mW/sr	500 mW/sr
Transmitter SIR	Standard	40 mW/sr	500 mW/sr
Transmitter MIR, FIR, VFIR	Low power	9 mW/sr	500 mW/sr
Transmitter MIR, FIR, VFIR	Standard	100 mW/sr	500 mW/sr
Receiver SIR	Low power	9 mW/cm	500 mW/cm
Receiver SIR	Standard	4 mW/cm	500 mW/cm
Receiver MIR, FIR, VFIR	Low power	22.5 mW/cm	500 mW/cm
Receiver MIR, FIR, VFIR	Standard	10 mW/cm	500 mW/cm

With regard to the rise and fall times defined by the IrDA for the transmitters, the limits depend in this case on the speed of the device. For slow IR, the rise and the fall times must be 600 ns maximum; while for medium and fast IR, their maximum value is 40 ns. For very fast IR, the rise and fall time is defined as 19 ns. The rise and fall times for each case are illustrated in Figure 9.3. The IrDA also defines limits to the overshoot values of the transmitter. The standard establishes that the percentage of overshoot optical emitted power must not exceed 25 percent of the typical peak emission value. This limit applies to all the encoding schemes.

Note that that consumer products using the optical sources specified by the IrDA must comply with the IEC 60825-1 eye safety regulation when fabricated in Europe. This is due to the fact that, even if IR LEDs are diffuse sources that cannot be focused onto the retina and currently do not pose any risk to the human eye (cf. Chapter 3), assumptions are being made by the European Committee for Electrotechnical Standardisation CENELEC [233] that this may not be the case in the future. These assumptions have been made taking into account that LED technology is undergoing constant development and that the increasing efficiency of these devices or their combination with optical concentrators or collimators may pose potential threats to eye safety in the not too distant future. For this reason, consumer products must be tested for eye safety and certified as Class 1 when manufactured in Europe.

It is worth noting that the real values obtained from an IrDA transceiver sometimes vary from the values specified by the protocol. The range, for example, may be longer or shorter, depending on the background illumination conditions of the place where the system is operating. In addition, the radiance obtained from an LED when embedded into a product can be reduced by up to one fifth of its original value due to the absorption of the plastic windows that are sometimes used in front of the transmitter as a casing to improve the appearance of the product or to protect the component.

The beam angle value specified for the IrDA LEDs also differs frequently from the real value. In practice, this value may range between 20 and 25° (half-angle), compared to the ±30° emission angle specified by the protocol. In this case too, the manufacturing casing and the component position within the casing may be responsible for this variation.

Another value that differs from the one specified by the protocol is the wavelength. While the standard defines an emission wavelength of 880 nm, the real value may be within the range 850 to 900 nm [224].

When the emitter of an IrDA transceiver is activated and data is being transmitted, the receiver of this (its own) transceiver is deactivated. Once the transmitter has finished sending data, the detector becomes active. The receiver requires some time to stabilize to its normal state of high sensitivity. This time lapse is commonly known as *receiver latency* and it is defined as the time lag between a change in input concentration and the output level reaching 95 percent of its final value. The receiver latency defined by the IrDA protocol is 0.5 ms for standard, medium, and fast infrared speeds at low power levels, and 10 ms for SIR, MIR, and FIR speeds at standard powers. The receiver latency value becomes even smaller for very fast IR speeds, where for the low and standard power options the value is 0.1 ms.

The reason why the emitter and the detector are active one at a time is to avoid interference from the same device. If both the transmitter and the receiver of an IrDA transceiver were allowed to operate at the same time, the energy emitted by the transmitter could potentially be detected by its own detector due to the fact that both are contained within the same package and at a very short distance from each other. This makes it necessary to deactivate one when the other is activated. This operation mode, where transmission and reception do not occur simultaneously, is known as half-duplex. If, on the contrary, both the transmitter and the receiver are active at the same time and are allowed to operate simultaneously, the operation is classified as full-duplex. The IrDA Protocol defines half-duplex operation only.

Chapter 3 described how background illumination introduces noise into wireless IR receivers. The typical mechanisms used to reduce or eliminate this unwanted energy were also discussed in the same chapter. Because IrDA devices are generally employed in places where they are exposed to this unwanted background illumination, the IrDA Protocol defines the characteristics and the minimum levels of background illumination at which its transceivers must be able to operate. The maximum illumination levels generated by different sources that can be tolerated at the optical port without risk of interruption of the communication link are [224]:

- Incandescent light: 1 kilolux maximum
- Fluorescent light: 1 kilolux maximum
- Sunlight: 10 kilolux maximum
- Electromagnetic field: 3 volts/meter maximum

The IrDA also specifies a maximum bit error rate (BER). This BER represents the number of errors occurring in the system per number of transmitted bits. The maximum BER defined by the IrDA is 10^{-8}. This implies that for the maximum distance under typical illumination conditions, the link must not transmit more than one error bit every hundred million bits, regardless of the encoding scheme used.

There are also a number of parameters defined by the IrDA standard for its receivers. Two of these parameters are the minimum and maximum irradiances required at the receiver to achieve good-quality communication and avoid the disruption of the link. The minimum irradiance has been established to ensure that the signal is still received at the maximum distance defined by the protocol even when transmitted with minimum intensity. The maximum irradiance has been defined to ensure that the detector does not saturate when operating at short distances from the transmitter or when the emitter transmits with maximum intensity. The maximum irradiance defined by the IrDA Protocol is 500 mW/cm^2, while the minimum irradiance value depends on the transmission speed. The values set by the protocol for minimum irradiance are 4 mW/cm^2 for standard IR receivers and 10 µW/cm^2 for high-speed IR receivers. For the low-power versions of the receiver, these values are 9 mW/cm^2 for standard infrared and 22.5 W/cm^2 for high-speed infrared receivers. It must be taken into account that the maximum distance defined for low-power transceivers is very small compared to that of the standard version (20 cm compared to 1 m), which explains why standard power transceivers have a minimum irradiance limit lower than their low-power counterparts. The minimum and maximum signal irradiance values for the receiver are summarized in Table 9.3.

With regard to the angular capability of the receiver, the minimum field-of-view (FOV) value that a receiver based on the IrDA standard must offer is ±15°. This minimum value has been established to guarantee that the link does not suffer from alignment problems. In practice, IrDA-based transceivers have a larger FOV due to the fact that they employ semi-hemispherical concentrators, which, as explained in Chapter 5, have a wide FOV.

The IrDA standard also defines a number of characteristics with regard to the encoding schemes used by its transceivers (and by devices based on the IrDA standard) at different speeds. These characteristics include data rates and modulation schemes. In the case of standard IR, the data rates available are 2.4, 9.6, 19.2, 38.4, 57.6, and 115.2 kbps. The modulation scheme used at these low speeds consists of routing bytes of information obtained from a 16550-compatible universal asynchronous receiver transmitter (UART) through an IR transmit encoder that inverts the data bits, generates a return-to-zero (RZ) signal of specific width (in this case, 3/16 of the data bit time width), and adds start and end bits to the 8-bit string ("0" to indicate start and "1" to indicate stop). The information is sent from there to the transmitter driver and to the emitter, where pulses of light are generated using IM.

For medium infrared speeds, which as discussed in Section 9.3 are 0.576 Mbps and 1.152 Mbps, the modulation scheme is slightly different. In this case too, bytes are obtained from a UART and converted by an IR transmit encoder that, instead of generating bits of $^3/_{16}$ data bit time width, generates encoded bits of ¼ the time width of the data bits.

The data rate corresponding to fast infrared is 4 Mbps. For this speed, the modulation scheme described by the IrDA Protocol is pulse position modulation (PPM), which works in a different way and has a higher complexity than the modulation schemes employed by the SIR and MIR versions. PPM, as described in Chapter 8, is a modulation scheme in which the temporal position of the pulses is varied according to the characteristic of the modulating signal. In the case of the IrDA standard, the width of these pulses is specified as 125 ns. This narrow pulse width is required to provide the necessary data rate through PPM. The position of the pulse can be assigned to a number of time slots within a small frame corresponding to a data symbol. Each of the symbols represents a bit. Depending on the number of time slots "L" (also known as *chips*) per data symbol, the modulation scheme is called *L*-PPM. For example, if there are eight time slots per data symbol, the modulation scheme is called 8-PPM. The IrDA Protocol defines 4-PPM as the modulation scheme required for FIR, which means that there are four time slots per data symbol. If each time slot corresponds to 125 ns, this means that in this scheme each data symbol consists of 500 ns. Each of these symbols represents a data bit pair, where 0001 represents the pair 11, 0010 10, 0100 corresponds to 01, and 1000 represents 00. It is worth noting that only one of the time slots is activated each time. This corresponds to the time that the LED is switched on.

As explained in Chapter 8, the two main advantages of this modulation scheme are (1) its low power consumption (as the time the LED is switched on is only one time slice of the data symbol) and (2) the relative simplicity of the receiver (because the number of pulses that can appear next to each other is reduced to two). Unfortunately, this modulation scheme is not without drawbacks, the main limitation being that the clock rate cannot be used completely due to the fact that the data symbols represent 2 bits. This implies that the system can only operate at 50 percent of the clock rate.

The highest data rate defined by the IrDA Protocol is the one corresponding to very fast infrared (VFIR), which is specified as 16 Mbps. At such speeds, it becomes increasingly difficult for the receiver to detect pulses of very narrow time widths, which appear to the receiver as ambient illumination noise (this problem can potentially disrupt the communication). The fact that detecting the edges of the pulses becomes increasingly difficult at high bit rates makes it necessary for VFIR to use something called run-length limited (RLL) encoding, which is a predictive method employed by the receiver to assume bit patterns. The specific RLL encoding used by VFIR is HHH(1,19), where 1 corresponds to the minimum number of empty time slots that must follow a pulse, and 19 corresponds to the number of empty slots before the next pulse occurs.

Despite the fact that the packet frame format used at this data rate is based on the FIR framing, the modulation is not only different because of the use of run-length limited coding, but also because instead of having a chipping rate of 8 Mchips per second, the chipping rate is 24 Mchips per second.

The way in which the IrDA Protocol deals with the problem of having to detect very narrow pulse widths at high speeds is by using a long pulse called a serial infrared interaction pulse (SIP), the time width of which is wide enough to be detected by any IrDA receiver. This pulse is generated at regular intervals of 500 ms (minimum) in order to be detected by a device before discovery is activated or before the link turns around. The minimum length of the SIP pulse defined by the IrDA standard is 8.7 μs.

In terms of certification, the IrDA has its own certification process to ensure device compliance. This self-certification process is described in the IrDA Physical Layer Measurement Guidelines.

9.5 Framer and Driver

The operation and interaction of the components described in the physical layer of the protocol (with the upper layers of the model) is based on the software and hardware described by the IrDA standard in the framer/driver layer. This software and hardware includes framers, controllers, and drivers that allow the conversion of data from the upper layers of the model into defined groups of consecutive bits prior to transmission. They also drive the LED to convert the data from the protocol stack into optical pulses and convert the signal received by the photodetector back into an appropriate set of data that can be used by the upper layers of the model. Some of these functions are performed by the same module.

9.5.1 Framers

The function of the framer is to perform the conversion (discussed above) of the data received from the IrLAP layer into structured "frames" prior to transmission. These frames consist of, in addition to the information packets received from the IrLAP layer, boundary bytes (at the beginning and the end of the data packets), synchronization bytes, and error detection bytes.

The format of the frames, which can be synchronous or asynchronous, depends on the data rate. SIR, for example, uses asynchronous framing, while MIR employs synchronous HDLC (High-level Data Link Control) framing. FIR uses synchronous 4-PPM framing and VFIR employs HHH (1,3) framing. The framing work of MIR, FIR, and VFIR is performed within dedicated controllers designed for IrDA transmission at the corresponding speed. The framing work of SIR, on the other hand, is not built within a controller specially designed for the IrDA, which means that writing a program for the framer to interact with the UART controller is more challenging [224].

The format of an SIR frame also consists of extra Beginning of Frame bytes (XBOFs), which alert a remote receiver and allow it to react to the relative intensity of the received signal; a Beginning of Frame byte (BOF), which marks the start of the frame; a Frame Check Sequence (FCS), a 2-bit value calculated in SIR from the content of the IrLAP packet; and an End of Frame byte (or EOF), which indicates the end of the frame.

9.5.2 Controllers

A controller performs two functions: (1) as a receiver, it accepts the signals from the transceiver's detector and transforms them into data that can be understood by the protocol layer; and (2) as a transmitter, it uses the data received from the protocol stack to drive the transceiver's optical source.

The IrDA defines two different controllers, depending on the data rate. SIR transmission, for example, requires a universal asynchronous receiver-transmitter (UART) compatible controller, while faster data rates employ dedicated controllers.

The UART creates an intermediary queue of bytes between the driver and the transmitter that allows asynchronous transmission. Here, bits of information are accessed in a first-in-first-out (FIFO) fashion. The capacity of this queue of data depends on the UART standard. The 16550 UART, for example, has a 16-byte queue that supports the same data rates as the IrDA SIR, while further versions of UART have up to 128-byte queues. The driver puts bytes into this queue and the transmitter takes bytes from it. Once the driver puts a byte into the queue, it can go on to other tasks without waiting for the transmitter to generate the signal knowing that it will eventually perform this operation. A similar queue is generated when the transceiver is receiving information. In this case, the bytes decoded by the receiver are placed into the queue from where the driver retrieves them. The driver performs a periodic check on the queue to know if new bytes have arrived [224].

9.5.3 Drivers

The driver (or device driver) is the software that interacts with the controller. This software is responsible for initializing the controller (which includes setting its data rate in the normal response mode [NRM]), transferring the information obtained from the controller to the protocol stack, and transferring the information obtained from the protocol stack to the controller prior to transmission. All these functions vary depending on the specific controller used.

9.6 IrLAP

The Link Access Protocol (IrLAP) is the layer of the IrDA protocol stack located between the framer/driver and the IrLMP (cf. Figure 9.2). It is responsible for interacting with the framer/driver explained in the previous section to establish and maintain an effective point-to-point half-duplex communication. IrLAP, for example, is responsible for managing the half-duplex access to the link in such a way that it avoids conflicts of transmission. It also allows discovery of devices, exchange of link characteristics, connection, data exchange, and link disconnection.

Because the communication of IrDA devices is half-duplex, only one device can transmit at a time. In an IrLAP connection, the device that transmits first is called the *primary device*, while the device that receives (and that transmits in the next turn) is called the *secondary device*. The tasks of the primary device include [224]:

- Device discovery
- Negotiation and connection
- Maintaining the link

The secondary device, on the other hand, is responsible for responding to the actions of the primary device. Thus, it must respond to device discovery, negotiation, and connection requests..

The potential flow of communication between IrLAP and the higher layers of the protocol may occur in two ways, depending on who initiates the transaction. If the transaction is initiated by the upper layers of the protocol, the flow of communication occurs through a *request* and *confirm* pair. If, on the other hand, the transaction is initiated by IrLAP, the flow occurs through an *indication* and *response* pair. In the first case, the upper layers of the protocol send a request to IrLAP when they require a service from it. Once this service has been provided, a confirmation is sent from IrLAP to the upper layer that created the request. In the second case, an indication is sent by IrLAP to the upper layers when it receives an event directed to them. The upper layers reply to this indication by sending a response back to IrLAP.

IrLAP can operate in two different states: (1) normal state mode (NRM) and (2) normal disconnect mode (NDM). The first mode (also called a *connected state*) refers to the condition in which IrLAP sets to once discovery, negotiation, and connection have taken place. NDM (also called a *contention state*), on the other hand, refers to the state of IrLAP when no connection exists. The services provided by IrLAP can be grouped into classes depending on its state. Thus, *connectionless* services, which include device discovery and broadcast data, correspond to NDM; and *connection-oriented* services, which include establishment of connection and data exchange, correspond to NRM.

9.6.1 The IrLAP Frame

The IrLAP frame is the basic unit of information exchange in IrDA. It consists of one address byte, one control byte, and from zero to 2047 bytes of information. The address byte is an 8-bit field composed of a connection address (which is generated by the primary device and identifies the connection between this and the desired secondary) of 7 bits and a C/R bit, which is used to indicate if the packet is sent by the primary device (Command frame) or by the secondary (Response frame).

The control byte is an 8-bit field used to identify the type of the frame, which can be [224]:

■ Information (I-frame): used to move the data between devices. The format of this frame consists of 3 Nr bits (which identify the number of the next I frame to be received), 1 P/F (Poll/Final) bit at bit 4, 3 Ns bits (which identify the number of the I-frame) in which it is found, and a "0" at the end of the frame.

■ Supervisory (S-frame): used for flow control in IrLAP and to reject distorted frames.

■ Unnumbered (U-frame): used for device discovery, negotiation, and connection establishment in NDM.

9.6.2 *Discovery and Device Selection*

During initialization, IrLAP generates a device address of 32 bits (which can be changed in case of conflict) and sets its operation parameters (which include data rate, packet size, window size, etc.) according to the parameters of the default connection. Once initialization has taken place, IrLAP is ready to proceed with device discovery, which is done as follows. First, the device waits for at least 500 ms prior to initiating device discovery in an effort to know if transmission by other devices is taking place and avoid interference. When the device has detected that no other transmission is taking place, it then proceeds to initiate device discovery using XID commands corresponding to the U-frame. The slotted discovery process consists of announcing the presence of the primary device and requesting the secondary devices to select a random number between 0 and $n-1$, where n can take the value of 1, 6, 8, or 16, with 16 being assumed as the maximum number of devices that can be within the 30° cone FOV of the receiver. The primary device then moves through all the slots and waits for the response of the secondary devices at each slot. When the secondary device replies to the first one, it provides its 32-bit physical address, its nickname (the character string displayed to the user), and service hint bits (which give an idea of the services provided by the secondary device). In situations when multiple devices are detected, the primary device may inform the user that multiple devices are in range and request that only the desired device is kept within the device's FOV. Another possibility is that the device informs the user of the nicknames of each device within range and asks that user to select one of them. Alternatively, the primary device may make use of the hint bits of the secondary devices and make a selection accordingly [224].

Note that when secondary devices detect the XID of the primary, it is possible that more than one choose the same slot number. In these cases, the secondary devices respond to the primary on the same time slot, creating a garbled packet that the primary device can interpret either as interference or as simultaneous transmission from secondary devices within range, which prompts it to reinitiate the device discovery process.

9.6.3 *Link Negotiation and Connection*

Once the discovery and device selection has taken place, the primary device can initiate the link negotiation stage (which allows devices with different capabilities to choose the optimal link capabilities). This is done by sending a connect request packet to the secondary. This packet is called *set normal response mode* (SNRM) and contains negotiation parameters that indicate the desired settings and the features of the primary device.

These negotiation parameters can be of two types: Type 0 and Type 1. If the parameters need to be agreed upon by both devices, they are classified as Type 0;

while if they do not need to be agreed on, they are classified as Type 1. It must be mentioned that even if Type 1 parameters do not have to be agreed on, they need to be respected by the other device. The negotiation parameters include [224]:

- *Data size.* This is the maximum size of a packet that a device can receive. This size can be 64, 128, 256, 512, 1024, or 2048 bytes.
- *Baud rate.* This represents the transmission speed. It corresponds to the different speeds supported by the IrDA Protocol.
- *Minimum turnaround time.* This refers to the minimum time that a device must wait for another device to be ready and begin transmission. The possible minimum turnaround time sizes are 10, 5, 1, 0.5, 0.1, 0.05, and 0.01 ms.
- *Maximum turnaround time.* This corresponds to the maximum time that a device can wait before responding. Its maximum value is 500 ms.
- *Window size.* This is the largest number of packets that can be transmitted by a device before the link is turned around. Slow speeds generally have a window size of 1, while faster speeds have windows of up to 127 (for the extended window size).
- *XBOF.* XBOFs lengthen the time a device has to get ready to receive information. This is necessary due to the fact that sometimes hardware presents poor latency.
- *Link disconnect/threshold time.* This corresponds to the period of time agreed upon by both devices in a connection to see if a communication (which was interrupted) resumes. The link disconnect threshold potential values are 3, 8, 12, 16, 20, 25, 30, and 40 seconds.

From these parameters, the baud rate and the link disconnect/threshold time are Type 0, while the rest of the parameters are Type 1. When the secondary device receives this packet, it proceeds to generate an Unnumbered Acknowledgment (UA) packet (indicating its preferences and capabilities) and sends it to the primary device. Each device can then proceed to determine the characteristics of the link by following exactly the same algorithm.

Because the IrLAP layer is used prior to the link parameters exchange, these parameters have a default value. These values, which are used during NDM, are data size = 64 bytes, baud rate = 9600 bps, window size = 1, minimum turnaround time = 10 ms, maximum turnaround time = 500 ms, XBOF = 10, and link disconnect time = 40 s.

9.6.4 Connection Establishment

Once the link negotiation stage has taken place, the devices are ready to establish a connection, which is done in the following way. After device discovery, the primary device sends an SNRM frame, which is used to invite the secondary device to establish a connection. The secondary device can accept the invitation by issuing a UA frame with its negotiation parameters or reject the invitation by sending a DM (Disconnect Mode) frame. If the invitation is accepted, both check the negotiation parameters of each other and compare them to their own to determine the common parameters to use for the connection. Once this is done, they change their parameters (data rate and link disconnect threshold) to the appropriate values; and the primary device sends an RR (Receive Ready) frame to the secondary device. The secondary device replies with its own RR frame, which completes the connection process [224].

9.6.5 Information Exchange and Flow Control

The devices are ready to exchange information once a connection has been established (that is, when they have changed to NRM). At this stage, data is sent in I-frames (although S- and U-frames can also be exchanged) which, as noted in Subsection 9.6.1, present a format that includes three Nr bits, one P/F bit, three Ns bits, and a "0". The fields Nr and Ns are used to accommodate the information and ensure that it is exchanged correctly. The Ns field, for example, contains the number of the frame being sent, while the field Nr contains the number of the next frame to arrive.

The flow control in IrLAP refers to the process by which the receiver requests the sender to stop sending data while it processes the information stored in the buffers, when information is being sent faster than that with which the receiver can cope. This is achieved through a Receive Not Ready (RNR) frame, which informs the sender that the receiver is saturated and requests it to stop sending data while it processes the information. The sender acknowledges this request through a Receive Ready (RR) frame. Communication can resume once the buffer has space again. In this situation, the receiver informs the sender that it is ready to receive again by sending an RR frame, which allows the transmission of information to resume.

Once the transfer of information has finished, the primary device emits a Disconnect frame (DISC). The secondary device acknowledges this frame by sending a UA, which allows the devices to enter NDM.

9.7 Link Management Protocol (IrLMP)

The Link Management Protocol (IrLMP) is the layer that provides multiple access to several applications (or services) through an IrLAP connection (although it also offers an application programming interface (API), a Tiny Transport Protocol (TinyTP), and an Information Access Service (IAS).

The services that can establish connections through IrLMP are the applications that communicate with the IrDA stack via the API. These services have a

Table 9.4 Required IrLMP Primitives and Their Use

Primitive	Use	Specific Function
DiscoverDevices	Device discovery	To invoke the IrLAP device discovery process
Data	Data transfer	To make IrLAP send an I-frame with IrLMP data as a payload
Udata	Data transfer	To make IrLAP send an unnumbered Information frame (UI-frame)
Connect	Link control	To establish an IrLMP connection over IrLAP
Disconnect	Link control	To terminate an existing IrLMP connection
Status	Link control	To determine if it is safe to terminate a connection

unique 7-bit logical address (called *Logical Service Access Point Selector* [LSAP-Sel]), assigned to them by IrLMP, which is available to remote devices via the IAS discovery mechanism. LSAP-Sels, which are used to identify the source and the destination applications, are used by IrLMP to establish connections and to exchange information. The source and destination information is contained in the first 2 bytes of an IrLAP packet. The source byte is known as SLSAP-Sel and the destination packet as DLSAP-Sel.

Device discovery, data transfer, and link control are functions provided by IrLMP through its service primitives, which may be required or optional (a list of the required properties and their characteristics are presented in Table 9.4; and a list of the optional properties is shown in Table 9.5 [224]). This is done through calls from the IrLMP to IrLAP, after which the latter can perform a function that generates a packet to be transmitted by IrLAP, act as a carrier for a control or data packet from IrLMP to be transmitted through IrLAP, or create data that is not transmitted.

IrLMP exchanges information by putting data as the payload of its exchange frames. These frames, which consist of a DLSAP-Sel byte, a SLSAP-Sel byte, and the data, become the data section of an IrLAP frame.

In addition to data exchange, IrLMP can exchange link control frames to establish and to terminate IrLMP connections. The format of the IrLMP link control frame consists of a DLSAP-Sel byte, a SLSAP-Sel byte, an opcode byte (that indicates the packet type), and parameters. The opcode byte can be of three types: (1) connect (to establish a connection), (2) disconnect (to terminate a connection), and (3) accessmode (to change between the multiplexed and exclusive connection modes).

Table 9.5   Optional IrLMP Primitives and Their Use

Primitive	Use	Specific Function
Sniff	Device discovery	To invoke the IrLAP sniff capabilities
ConnectionlessData	Data transfer	To make IrLAP broadcast data even if a connection does not exist
Idle	Link control	To mark a local LSAP-Sel as active or idle
AccessMode	Link control	To toggle between connection modes (exclusive or multiplex)

The multiplexed connection mode refers to the normal state of IrLMP in which multiple services can be connected over an IrLAP physical link. The exclusive connection mode, on the other hand, refers to the use of the IrLAP link by a single service.

9.8 Information Access Service and Protocol (IAS)

The Information Access Service (IAS) enables the registration, discovery, and access of services. IAS, for example, assigns LSAP-Sel addresses to the IrLMP services. The IAS data store includes the class name and the attributes of the registered services. Class names may be generic or proprietary, but the use of generic class names is encouraged to facilitate interoperability between devices. All services need to register the class *Device* to provide information about the device.

Service classes include a number of attributes that can be standard or specific. One of these attributes is the LSAP-Sel described above. An IAS query, therefore, can ask for a service class and for its LSAP-Sel through a *GetValueByClass* query. This way, a remote device can locate the desired service and establish a connection.

Attribute data types are used to define the data that can be used to describe IAS attributes. The attributes data types include Class name, Object Identifier, Attributes (Attribute Name, Attribute Value: missing, integer, octet sequence, and user string), and List. All the primitives of the IAS service are related to remote querying. Table 9.6 presents the primitives that facilitate IAS queries by remote devices [224].

The Information Access Protocol (IAP) is defined in the IAS specification. It specifies the rules of use of the IrLMP frames for access between remote IAS servers and applications. The format of the IAP frame consists of a control field (which is contained in one byte and specifies an opcode related to one of the service primitives) and a data field of variable size. The IAP frame is appended to the IrLMP frame (after the IrLMP header); and the IrLMP is part of the IrLAP frame.

Table 9.6   IAS Service Primitives and Their Function

Primitive	Required or Optional	Specific Function
GetValueByClass	Required	To specify a particular class and ask for a specific attribute
GetInfoBaseDetails	Optional	To check the number of services in a remote IAS database as well as the highest object identifier
GetObjects	Optional	To check the services in a remote IAS database.
GetValue	Optional	To query the value of a specific attribute
GetAttributesNames	Optional	To enumerate the attributes for a specific object
GetObject Info	Optional	To provide a remote IAS with the name of a specific object

9.9 Tiny Transport Protocol (TinyTP)

The IrDA Tiny Transport Protocol (TinyTP) is responsible for the flow control at IrLMP level and for segmentation and reassembly (SAR). This layer of the IrDA Protocol is generally implemented as part of IrLMP.

TinyTP presents the same addressing scheme as IrLMP. It includes 2 bytes of 7 bits (called TinyTP Service Access Points [TTPSAPs]) with source and destination addresses.

The primitives of TinyTP include [234]:

- *Data*: used to exchange information over a TinyTP connection.
- *Udata*: makes IrLAP to send an UI-frame.
- *Connect*: used to create a TinyTP/IrLMP connection.
- *Disconnect*: used to terminate a TinyTP/IrLMP connection.
- *Local flow*: controls data flow on a particular channel.

The TinyTP of the IrDA also defines data units called *TTP* Data Units (TTP-PDUs), which are used to establish and maintain TTP connections. The TTP-PDUs, can be used for data or for connection maintenance. The Connect TTP-PDU may consist of a 1-byte TTP header (containing flow control information) and of user data contained in 0 to 59 bytes (when no parameters are assigned). Another possibility is that the TTP-PDU may be formed by the TTP header, the user data, and parameter bytes between the header and the user data (when parameters are present). The first configuration is called the "parameterless TTP-PDU format"; and the second, which is adopted when parameters are present, is the "TTP-PDU format with parameters." The existence or lack of parameters is indicated by the P in the TTP header. The parameters field that follows the header when parameters are present consists of a length byte and up to 255 parameter bytes. The parameter defined by TTP is called the MaxSDUSize and indicates the maximum packet size that it can reassemble. Its maximum size is $2^{32} - 2$; and if this parameter is not provided, UserData becomes limited by the IrLAP frame [224].

The format of the Data TTP-PDU frame is similar to that of the Connect TTP-PDU. It consists of a byte containing an M bit (which is cleared if no more segmented data follows or if the TTP-PDU is smaller than an IrLAP packet, and is set if more segmented data follows this set of user data) and 7 bits of DeltaCredit (which convey flow control credit) followed by the user data.

The TTP layer can break large files (or objects) into individual packets of information for easier management and transmission. The information can be reassembled at the other end, which allows the original file to be recovered. The functions of segmentation and reassembly are possible thanks to SAR (Segmentation and Reassembly).

Another important capability of the TTP layer is flow control, which is responsible for the efficient use of the channel and for avoiding unnecessary retransmission of information. Credit-based flow control is used in situations where the processing system is not capable of emptying the receive buffers (being filled in by TinyTP packets) fast enough. Here, the connection endpoint issues credit (according to the number of packets that can be received without saturating the buffers) to the remote endpoint at the beginning of the session. When the buffers are empty, new credit can be issued to the endpoint. No credit can be issued until the buffer data is used by the processing system.

9.10 Session and Application Layer Protocols

The Session and Application layers of the protocol is a comunication protocol that defines the rules to exchange. Some of the session layers include IrOBEX, IrComm, IrWW, IrTran-P, and IrLAN. IrDA Object Exchange (IrOBEX), for example, provides the base model for *point-and-shoot* communication and allows interoperability between devices for the exchange of objects (which are complete logical units of data and include data files, music files, digital pictures, business cards, etc.). IrComm, on the other hand, allows applications that would traditionally communicate via serial and parallel data links (such as printers and modems) to communicate via an IrDA link.

IrOBEX follows the model of the HyperText Transport Protocol (http), which provides functionality for services with different requirements and characteristics of operation. These layers are located above the fundamental IrDA protocol stack layers illustrated in Figure 9.2 objects such as the ones discussed above. Thus, the function of IrOBEX also is to allow object transfers in an efficient and flexible manner, regardless of their type. The main services of IrOBEX are to provide a common protocol for data exchange and to offer a common application program interface for data exchange. These services facilitate the very important feature of interoperability of devices. To do this, it is important that different devices (and applications) understand not only the type of objects that are to be transferred, but also the type of data. The three logical components of IrOBEX include [234, 224]:

■ 1. *The object model.* This component optimizes the flexibility of object exchange by organizing them in a specific way. The objects exchanged consist of two parts: one that describes it (called *meta-information*) and another that contains the data. The format of the object consists of a Header ID (contained in 1 byte, of which the first 2 bits describe the format) and a Header value or HV (note that the term "header" in this context refers to a section of the object and not to extra bits to which the rest of the information is appended). The data of the object is sometimes sent over different headers due to the limitations of the packet size. Table 9.7 presents the four header types. The complete list of header IDs of IrOBEX can be found in [224].

Table 9.7   IrOBEX Header Types

Header	Type	Characteristics
0b00000000 (0 x 00)	Null-terminated Unicode	Its 1 byte Header ID includes 2 bits at the beginning (0,0). It also contains a Header Size of 2 bytes, followed by the header value.
0b01000000 (0 x 40)	Byte Sequence	Its 1 byte Header ID includes 2 bits at the beginning (0,1). It also contains a Header Size of 2 bytes, followed by the header value.
0b10000000 (0 x 80)	1-byte quantity	The header consists of only 2 bytes. The first one corresponds to the Header ID, where the first 2 bits are (1,0). The second byte contains the Header Value.
0b11000000 (0 x C0)	4-byte quantity	The header consists of 5 bytes. The first one corresponds to the Header ID, where the first 2 bits are (1,1). The rest of the bytes contain the Header Value.

- 2. *The session protocol.* This component specifies the commands of the protocol and their functions. The Request and Response packet format employed by OBEX consists of an Opcode of 1 byte (used to specify the packet type), a packet length of 2 bytes (used to define the length of the OBEX packet), and an unspecified number of object headers. The types of opcodes include request (used to define the packets sent — from the requester to the OBEX served — to control connection management and to control data movements), response (used for acknowledgments for single-packet and multi-packet operation), folder browsing (used to browse the file system of a remote device), and authentication (used in secure applications that require a PIN or a password for authentication purposes).
- 3. *The application framework.* This component is a client/server architecture that defines the way in which services interact with each other.

IrComm provides support to the 232 protocol (which defines serial transmission between devices) and to the Centronics protocol (which defines parallel communication between computers and other devices). With regard to the 232 serial service, IrComm support is provided for the three-wire raw frame (where no wire is used as a control channel). Here, IrComm provides flow control through IrLAP), three-wire and nine-wire versions (where no wires are used for flow and are called "cooked"). The three-wire 232 serial service, for example, is used to implement three RS-232 circuits and flow control is provided through Tiny-TP.

The nine-wire 232 serial service, on the other hand, is used to implement nine RS-232 circuits, but flow control is also provided through IrLAP. In these two cases, it is important that the flow control provided by IrComm is consistent with the software and hardware flow control mechanisms used for wired communication (that is, that the IrComm appears as the port driver after the legacy application). One way of doing this is by manipulating the XON/XOFF characters used by the software mechanism of wired communication services in such a way that IrComm can simulate the behavior of the wired application.

IrComm also provides the following services: connect (used to create an IrComm connection by exchanging connect packets), disconnect (used to refuse an IrComm connection or to terminate a connection), and data and control (used, among other things, for link control). A list of the control services is presented in Table 9.8 [224].

Finally, the Centronics service is designed to emulate the Centronics parallel port. Here too, flow control is provided through TinyTP.

Note that while IrDA provides half-duplex communication, some of the applications that traditionally communicate via the serial or the parallel port support full-duplex connection. Therefore, in some of these cases, it may not always be possible for an IrDA link to support the application requirements.

Table 9.8 List of Control Services Provided for Different IrComm Communication Types

Communication Type	Control Parameters
3-Wire	Data format, flow control, data rate, XON/XOFF flow control characters, ENQ/ACK flow control characters, break and line status
9-Wire	Data format, flow control, data rate, XON/XOFF flow control characters, ENQ/ACK flow control characters, break, line status, CTE line settings and changes, DTE line settings and changes, and poll for line settings
Centronics	Status query, status query response, set busy timeout, set busy timeout response, request IEEE 1284 Mode support, IEEE 1284 Mode support response, request IEEE 1284 Mode Device ID, IEEE 1284 Mode Device ID response, select IEEE 1284 Mode, select IEEE 1284 Mode response, IEEE 1284 ECP/EPP Data transfer, IEEE 1284 ECP/EPP Data transfer response

The two frame formats used for IrComm are (1) the three-wire raw frame and (2) the cooked frame. The first consists of 2 bytes of data in a single packet, while the second consists of 4 bytes and contains a TinyTP byte (for flow control), a control sector, and user data. In addition, IrComm relies on hint bits used for device discovery.

9.11 Summary and Conclusions

The IrDA is an organization that provides interoperability standards for the software and hardware used in wireless infrared communication. This standard serves as the basis for the creation of simple, low-cost, and low-power transceivers that enable point-to-point wireless IR communication in a number of devices such as PDAs, laptops, printers, and digital cameras.

The IrDA consists of three committees — Technical, Marketing, and Test and Interoperability — and an Architecture Council.

The core specifications of the IrDA Protocol include a Serial Infrared physical layer (SIR), a Link Access Protocol (IrLAP), a Link Management Protocol (IrLMP), and a Fast Infrared physical layer extension (FIR). Additional specifications include the Infrared Object Exchange Protocol (IrOBEX) used for generic object exchange between remote devices.

To provide interoperability between low-cost and low-power devices of all types, the IrDA standard defines the following parameters for the physical layer:

- Continuous point-to-point operation over 1 m (although it can reach 2 m in practice)
- Wavelength of operation = is 880 nm
- Data rates from 9.6 kbps to 16 Mbps (with 100 Mbps under development)
- Bi-directional operation in any system (half duplex)
- A ±15 to ±30° cone from the transmitter that equates to a 0.5- 1-m beam diameter at a 1-m range
- A very wide receiver FOV

The physical layer must also be able to operate according to the following constraints:

- BER no worse than 10^{-8}
- LED minimum intensity
  - 40 mW/sr (SIR standard)
  - 3.6 mW/sr (SIR LP)
- LED minimum intensity = 100 mW/sr (MIR/FIR/VFIR)
- LED maximum intensity = 500 mW/sr
- Rise and fall times:
  - 600 ns (SIR)
  - 40 ns (MIR/FIR)
  - 19 ns (VFIR)

The data rates supported by the IrDA include 2.4, 9.6, 19.2, 38.4, 57.6, and 115.2 (for SIR); 0.576 and 1.152 Mbps (for MIR); 4 Mbps (for FIR); and 16 Mbps for VFIR. Support for a further speed of 100 Mbps (denoted Ultra Fast Infrared [UFIR]) is currently under development.

An IrDA connection can be in two modes: (1) Normal Disconnect Mode (NDM) or (2) Normal Response Mode (NRM). The NDM refers to the disconnected state when two remote devices are trying to establish a connection. When a device initiates a connection, the devices enter the NDM where they can start transferring data.

In addition to the physical components, it is important to have hardware or software that facilitates the communication between IrLAP and the physical layer. This intermediate layer consists of the controllers, framers, and drivers used to receive the data from the upper layers of the stack and drive the LED. The framer and the driver functions are generally contained in the same software module.

The IrLAP layer is responsible for connection establishment (which involves device discovery and connection negotiation) between remote devices and for maintaining the link once a connection has been established. Packet delivery is also supported by the IrLAP layer.

The IrLMP layer is responsible for providing multiple service connection through the same IrLAP connection. It functions include service registry and service discovery, service level connection, and data exchange.

The Tiny Transport Protocol (TinyTP) is responsible for providing flow control at the IrLMP layer level and for providing segmentation and reassembly, which allows an application to submit arbitrarily large files that are broken by SAR into individual packets and reconstructed at the other end.

The IrDA Protocol includes a description of the Minimal IrDA Protocol Implementation or IrDA Lite, which is used for small devices that have limited memory and computing resources. This description focuses on IrLAP and IrLMP, and contains strategies to provide maximum functionality when limited resources are available. They can save data (RAM), code (ROM), or both.

Chapter 10

Wireless Infrared Networking

10.1 Introduction

The proliferation of wireless communication technologies and the evolution of optical fiber techniques suggest that the future network will primarily consist of an optical-fiber backbone as the primary system, with short-range wireless communication links providing additional mobility, either as stand-alone systems or as part of the larger networking infrastructure. Users of future mobile systems should have easy access to networks via reliable, high-speed wireless links anywhere in the world.

As one of the current dominant technologies in the wireless indoor arena and with many implicit technologies already being used in different scenarios, infrared (IR) has put itself in a strong position to be an essential part of the future global network. The six link configurations presented in Chapter 1 suggest its usage for different scenarios. Among them, only the diffuse and the non-directed LOS topologies can form a real network naturally, and could exist as networks on their own. This chapter focuses on networks formed by these two link configurations.

As discussed in Chapter 1, due to the characteristics of the technology itself, infrared can be considered as complementary to RF for the support of wireless local area networks. This principle has been followed in the specification and standard designs. IEEE 802.11 [236] networks based on infrared technology are well suited for low-cost, low-range applications, such as small area networks that are set up only for a short period of time. IEEE 802.11 networks based on radio, on the other hand, tend to be favored for applications in which user mobility must be maximized or where transmission through walls and over long distances is required.

This chapter discusses the issues and challenges faced when designing an optical wireless network. It describes the existing forms of infrared networks — network architecture. Current standards and specifications for infrared networking, IEEE 802.11 and IrDA Advanced Infrared (AIr) protocols, are reviewed. Several technology aspects on mobile ad hoc networking, in which IR communication seems to have a promising future, are discussed and illustrated. Finally, a forecast of the future infrared network is presented.

10.2 Network Architecture

As for many other wireless networks, there are two basic optical wireless network architectures: (1) ad hoc and (2) infrastructure. The current IEEE 802.11 specification allows network interface cards to be set to work in either of these modes, but not in both simultaneously. An ad hoc network as illustrated in Figure 10.1 is the most basic form of a network, in which stations or mobile nodes cooperatively and spontaneously form the connections, independently of any fixed infrastructure or centralized administration such as a base station (BS) in a cellular wireless network or an access point (AP) in a wireless local area network. In general, these networks, which are characteristically limited both temporally and spatially, can be set up rapidly. In an ad hoc network, the nodes are dynamically and arbitrarily located, and the use of a multi-hop route to deliver data through other nodes in the network is possible.

Figure 10.2 illustrates the basic form of an infrastructure-based architecture. In contrast to an ad hoc network, an infrastructure network requires the existence of a centralized controller, which may be an AP or a BS. In this network, mobile nodes communicate with each other through the AP/BS. The IEEE 802.11 specification, for example, requests the implementation of a join network function to associate the station with the access point before other operations can take place. The station can then communicate with other stations through the access point. The access point/base station can also be treated as a path used by mobile nodes within the cell to access other networks, such as those connected to the wired network for Internet or virtual private network (VPN) services.

Ad hoc wireless networks without infrastructure are highly appealing for many applications due to their flexibility and robustness, which explains why they have become increasingly popular in recent years. This type of network is used, for example, in situations where infrastructure is either not available or temporary network connectivity is needed. Other examples include battlefield communications, disaster relief situations, conferencing, temporary offices, electronic classrooms, and sensor networks for various research purposes, as well as for personal area networks. Infrastructure wireless networks are normally used for other network accesses, such as Internet service and VPN access. In VPN, for example, while the WLAN (wireless local area network) user is roaming somewhere else rather than in his or her company network, it is possible for him or her to obtain access to the company's network through an Internet connection with certain security protection.

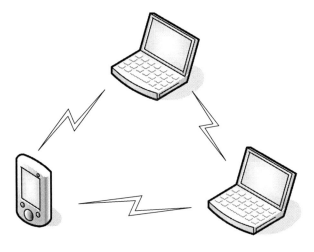

Figure 10.1   Ad hoc network.

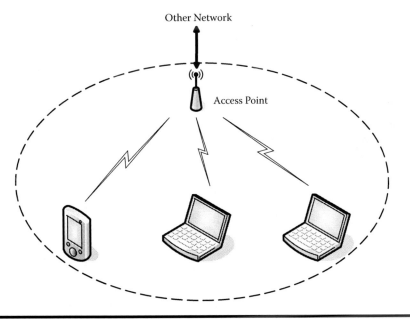

Figure 10.2   Infrastructure network.

10.3 Optical Wireless Network Specifications

One of the elements necessary for the success of any networking technology is the availability of appropriate networking standards. Along with the tremendous growth in the wireless market during the past few years, both the IEEE and the

IrDA have become actively involved in producing standards and specifications for infrared networking — namely, the IEEE 802.11 standard and the Advanced Infrared (AIr), respectively [226]. An overview on both specifications is given in this section.

10.3.1 IEEE 802.11 Specification

The IEEE 802.11 group was founded in July 1990 to work on the specifications of a WLAN for different technologies, including radio and infrared [236]. The standard was approved in June 1997. An essential characteristic of the IEEE 802.11 specification is the existence of a single Medium Access Control (MAC) sub-layer common to all physical (PHY) layers. This feature allows for easier interoperability among many physical layers (some of which may be defined in the future), driven by the fast technological progress in this field. Currently, three different technologies are specified in the physical layer: (1) Diffuse Infrared (DFIR), (2) Frequency Hopping Spread Spectrum (FHSS), and (3) Direct Sequence Spread Spectrum (DSSS). The MAC layer and the physical layer defined in the IEEE 802.11 standard are shown in Figure 10.3.

The aim of the IEEE 802.11 standard is to enable a mobile host to communicate with any other mobile or wired host in a transparent manner. That is, an IEEE 802.11 WLAN appears to layers above the MAC layer like any other IEEE 802.X LAN (for example, Ethernet or Token Ring). This requires several additional functions, when compared to wired networks, to be able to adapt to the particular characteristics of the transmission channel and to the mobility of the stations. Thus, in addition to the basic medium access control function, wireless networks require the IEEE 802.11 MAC sub-layer to include additional functions [272]:

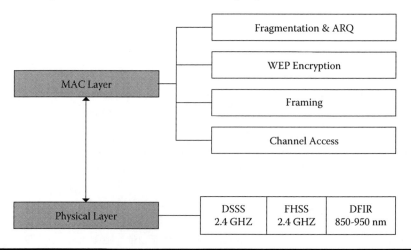

Figure 10.3   Physical and MAC layers in IEEE 802.11.

- Fragmentation and reassembly of frames when the quality of the transmission channel is poor
- Association and re-association of stations with access points when presented
- Temporal synchronization for the support of delay-sensitive applications
- Power management for battery operated stations
- Data rate adaptation

Before transmitting a frame, the MAC coordination function must first gain channel access to the network using a defined medium access method. In IEEE 802.11, the basic channel access function is Carrier Sense Multiple Access with Collision Avoidance (CSMA/CA). More details on this method are discussed below. Because of the open broadcast nature of WLANs, designers need to implement appropriate levels of security. The IEEE 802.11 standard describes two types of authentication services: (1) open system and (2) shared key. Open system authentication is the IEEE 802.11 default authentication method. It is a Null authentication method. Any station that requests authentication with this algorithm may become authenticated if dot11 AuthenticationType at the recipient station is set to Open System authentication. Shared key authentication assumes that each station has received a secret shared key through a secure channel independent of the 802.11 network. Stations authenticate through shared knowledge of the secret key. The use of shared key authentication requires implementation of the encryption method known as Wired Equivalent Privacy (WEP). Note that the security level provided by WEP is not satisfactory in some cases. IEEE 802.1X [237] has recently offered an effective framework for authenticating and controlling user traffic to a protected network, as well as dynamically varying encryption keys. Note also that because the access point is involved in the key exchange process when using IEEE 802.1X, in theory, this framework only applies to infrastructure networks.

The fragmentation and Automatic Repeat Request (ARQ) block provide the functions of retransmission and error correction. The ARQ scheme is integrated with the MAC layer, which determines the ARQ relevant round-trip delay by specifying the inter-frame space, the short inter-frame space (SIFS). The fragmentation method is used here to adapt the ARQ scheme to the channel conditions. The degree of fragmentation is determined by a parameter called fragmentation threshold. If the packet size is larger than this threshold, it is transmitted as fragments.

The physical layer properties were described in the previous chapter (cf. Chapter 9), and this section thus focuses on the MAC mechanism in the IEEE 802.11 standard.

10.3.1.1 Medium Access Control

The IEEE 802.11 legacy Medium Access Control is based on coordination functions, which determine when an operating station is permitted to transmit and is able to receive frames via the wireless medium. Two coordination functions are defined: (1) the mandatory distributed coordination function (DCF) that provides a contention-based service, and (2) the optional point coordination function (PCF) that implements contention-free service based on a poll-and-response mechanism. The DCF provides the basic access method of the 802.11 MAC Protocol and is based on a CSMA/CA scheme. The PCF is implemented on top of the DCF and is based on a polling scheme. It uses a Point Coordinator that cyclically polls stations, giving them the opportunity to transmit. Due to the fact that the PCF specification is not well defined, PCF has not been implemented in real WLAN cards. It has been set only as an optional part within the whole IEEE 802.11 standard. The following subsection focuses on the widely used DCF.

10.3.1.2 Distributed Coordination Function (DCF)

DCF was designed for asynchronous data transport, which refers to traffic that is relatively insensitive to time delay. Examples of asynchronous data traffic are electronic mail (e-mail) and file transfer. The functions in the MAC protocol layer are provided independently of the characteristics of the underlying physical layers and data rates. In IEEE 802.11 WLANs, the DCF is implemented in all stations, for use within both ad hoc and infrastructure networks.

10.3.1.3 CSMA/CA

Figure 10.4 illustrates how CSMA/CA works in DCF. For a station that intends to transmit, it starts by sensing (carrier sense) if the channel is busy. If this is the case, it defers until the end of the ongoing transmission. When the last frame detected on the medium is received correctly, it then waits for a period of time equivalent to the length of the distributed coordination inter-frame space (DIFS). The DIFS includes carrier sensing time, and is computed as DIFS = SIFS + (2 × slot time), where SIFS is the short inter-frame space. After that, it randomly selects a time slot within the backoff window (contention window). At the first transmission attempt,

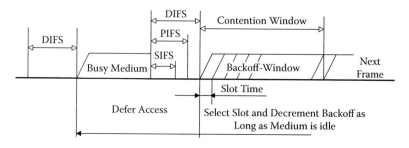

Figure 10.4   DCF basic access method. (*Source:* From [236].)

the contention window (CW) is equal to the minimum contention window (defined as CWmin), and the value doubles at each retransmission, up to the maximum contention window (CWmax). If no other station has started transmitting before reaching the slot, the station starts its own transmission. Collisions can now only occur in the case when two or more stations have selected the same slot. If another station has selected an earlier slot, the station freezes its backoff counter, waits for the end of this transmission, and then only waits for the slots remaining from the previous competition.

A backoff timer is set with a random backoff integer drawn from a uniform distribution over the interval [0 to 1]. The backoff integer is the number of idle "slots" the station must wait until it is allowed to transmit. The value decreases by one for each idle slot detected. The backoff timer suspends when the medium becomes busy, before the backoff integer reaches zero. The timer resumes only after the medium has been idle longer than the designated inter-frame space interval. The station starts transmitting the frame when the backoff timer reaches zero. For an ith successive retry for access to the medium, the contention window becomes $2^i \times$ CWmin. Upon a successful transmission, the contention window returns to CWmin. The maximum contention window CWmax is given by $2^m \times$ CWmin, where m is the maximum backoff stage. For each successive retransmission, the values of CW increase exponentially (that is, $CW_{new} = CW_{old} \times 2 - 1$), until it reaches and then stays at CWmax. CW is restored to CWmin after a successful transmission.

It is argued that, due to the fact that the backoff algorithm has to pay the cost of collisions to increase the backoff time when the network is congested, this algorithm may significantly degrade the channel utilization in conditions of high contention. For this reason, there are several proposals to extend the standard backoff protocol in IEEE 802.11 by adjusting the backoff time according to the network contention level [238–240].

10.3.1.4 Acknowledgment and Reservation: RTS/CTS

Because a source station in a wireless network cannot hear its own transmissions, when a collision occurs, the source continues transmitting the complete data frame. If the data frame is large, a lot of channel bandwidth is wasted due to a corrupted data frame. The special short request to send (RTS) and clear to send (CTS) control frame can be used by a station to reserve channel bandwidth prior to the transmission of an actual data frame and to minimize the amount of bandwidth wasted when a collision occurs. Figure 10.5 illustrates the RTS/CTS exchange process.

In the RTS/CTS scheme, after gaining access to the medium, and before starting transmitting the data packet itself, a potential transmitter sends an RTS frame to the receiver announcing the upcoming transmission. All stations in the service set, hearing the RTS packet, read the *duration field* in the frame head and set their

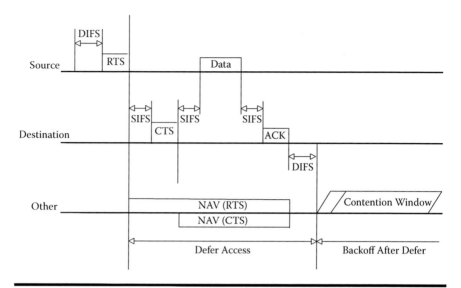

Figure 10.5 RTS/CTS message exchange. (*Source:* From [236].)

network allocation vector (NAV) accordingly. The *duration field* of the frame indicates the projected length of the transmission and thus informs all stations within range of both stations how long the channel will be used. The NAV, which is an indicator in each station, maintains a prediction of future traffic on the medium based on the duration information announced in the RTS/CTS frame prior to the actual exchange of data. A CTS frame is sent from the receiver in response to the received RTS frame after waiting for a time interval SIFS. Stations hearing the CTS packet read the *duration field* of the CTS frame and update the NAV period. If the CTS frame is not received within a predefined time interval, the RTS frame is retransmitted by executing the backoff algorithm. After successful exchange of the RTS and CTS frames, the data frame can be sent by the transmitter after waiting for a SIFS. Because the stations that can hear either the transmitter or the receiver refrain from transmitting until their NAV has expired, the probability of a collision occurring due to a hidden station decreases [273].

The RTS/CTS scheme could increase bandwidth efficiency by reserving the channel for ongoing transmission, and reduces collision for the long data packets using transmission of the small control frame beforehand. It is reasonable to think that the RTS/CTS mechanism improves the performance when data frame sizes are large when compared to the size of the RTS frame. Consequently, the RTS/CTS mechanism relies on a threshold, the *RTS threshold*. The mechanism is enabled for data frame sizes over the threshold and disabled for data frame sizes under the threshold. Alternatively, the overhead introduced by the RTS/CTS exchange, and the reservation, could also decrease the bandwidth efficiency when it might not be actually needed. For these reasons, the RTS/CTS scheme in IEEE 802.11 is suggested as optional.

10.3.2 IrDA AIr

The IrDA 1.x protocols have been widely adopted by many manufactures for a broad range of devices, such as laptop computers, PDAs, palmtops, printers, calculators, mobile phones, etc. As explained in Chapter 9, these protocols are mainly designed for point-to-point link architectures, in which the optical interface has a maximum emission cone angle of ±15° [227]. The point-to-point communication nature of the IrDA standard means that only one pair of devices can be connected at a time, and multiplexing must be on an application basis only. Although the architecture may be able to accommodate a point-to-multipoint operation in some way, the protocol has never been properly defined.

To expand the traditional point-to-point mode and enable true multipoint operation, the IrDA has proposed the Advanced Infrared (AIr) Protocol for indoor local area networking connectivity [226]. The technology was developed within IBM, including the IR group of their labs in Zurich and other members of the IrDA. The rights to AIr were sold later by IBM to Infinion Technologies Inc. [241].

In the AIr Protocol, the investment in upper layer applications and services is preserved by ensuring that the semantics of the service definitions at the upper layers of the platform are maintained. However, a new physical layer (PHY) and Medium Access Control (MAC) protocol [226] have been proposed. The AIr protocol stack is presented in Figure 10.6.

The new AIr physical layer (AIr-PHY) utilizes wide-angle infrared ports capable of operating at angles of up to ±60°. These wide communication angles have increased spatial coverage, which means that transceivers do not have to aim at each other. This has given the IrDA devices a certain degree of freedom for roaming. An efficient variable repetition 4-Pulse Position Modulation (4-PPM) variable repetition (VR) encoding scheme has also been adopted for AIr. The number of repetitions is called the repetition rate (RR) and ranges from 1 to 16 (1, 2, 4, or 16), corresponding to bit rates between 4 Mbps and 250 kbps. This provides a dynamic

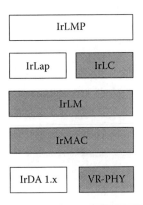

Figure 10.6   AIr protocol stack.

trade-off between bit rate and transmission range or link quality. In AIr-IrPHY, the receiver estimates the signal-to-noise ratio (SNR) and informs the transmitter of the bit rate that can be supported. In addition, AIr offers the possibility of scaling the transmission range and emission angle with variable levels of transceiver power to suit different communication scenarios. It has been claimed that a number of devices within 8 meters of each other can communicate without interference.

To introduce multi-access communication, AIr has further split the IrLAP Protocol of IrDA 1.x into three sub-layers: (1) AIr Medium Access Control (AIr-MAC), (2) AIr Link Manager (AIrLM), and (3) AIr Link Control (AIrLC) sub-layers [226, 228, 242, 243]. The original IrLAP Protocol has been kept in the protocol stack to provide legacy connectivity to IrDA 1.x devices. IrLC and IrLAP provide identical services to the IrLMP layer. Thus, from an upper layer protocol point of view, the IrLAP, IrLC, and IrLM protocol entities can be regarded as a single logical entity. The AIr-MAC sub-layer is responsible for establishing access to the IR medium and avoiding packet collision. AIr-MAC employs a burst reservation CSMA/CA-based protocol with RTS/CTS medium reservation that operates in a manner very similar to the IEEE 802.11 DCF access method. The AIrLM sub-layer provides multiplexing for client protocols, and the AIrLC sub-layer is a High Level Data Link Control Asynchronous Balanced Mode (HDLC-ABM)-based data link layer. To improve link quality, the AIr Protocol proposes a Go-back-N (GBN) automatic repeat request (ARQ) retransmission scheme at the LC layer; and an optional stop-and-wait (SW) ARQ in the AIr-MAC sub-layer [274].

The Air-MAC Protocol is a four-way handshake (RTS, CTS, data, ACK) with multiple data-packet transmission in each reservation. This is referred as burst transmission. The AIr-MAC Protocol can be summarized as shown in Figure 10.7.

Before sending an RTS frame, the station waits for a randomly chosen set of collision avoidance slots (CAS). Each CAS is 800 milliseconds long. If another contending station with a collision avoidance timer running receives the start of

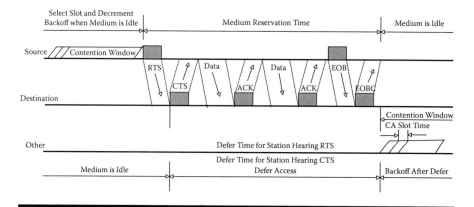

Figure 10.7   AIr channel access. (*Source:* From: [226].)

an RTS frame, the collision avoidance (CA) timer is paused, and resumed when the media becomes free again. The intended destination replies to the RTS frame with a CTS packet. The channel reservation is done through the short RTS and CTS frame exchange. Upon receiving the CTS, the source station sends its data immediately. The data is sent in bursts, and the number of packets in each burst time is restricted by the reservation time and the upper limit, which is 500 milliseconds. After the transmission of multiple data and their ACK packets, the source station sends an end-of-burst (EOB) packet, and waits for an end-of-burst-confirmation (EOBC) packet from the intended receiver. In this system, all other stations overhearing the RTS/CTS exchange or data transmission defer their transmissions until EOB/EOBC exchange.

The choice of a CAS slot number is random. The duration of CAS is relatively long. Air-MAC employs linear adjustment of the CW size to minimize delays emerging from the long CAS duration. Collision occurs when two stations choose the same CAS value. This is detected by the expiration of a Wait-for-CTS timer, after which a new CAS value is chosen and the contention is restarted. To improve contention, Air-MAC proposes a CAS window adjustment algorithm. When RTS collision occurs, the retry CAS window is increased and after every successful transmission, the CAS window decreases.

The revolutionary AIr Protocol extends the use of IR from traditional remote control to a real network connection. With AIr, infrared has gained the functionality needed to beam it into mainstream communications.

10.3.2.1 AIr and IEEE 802.11

Despite the fact that a similar access method to IEEE 802.11 DCF was adopted by AIr-MAC, there are significant differences in their collision avoidance and media access methods. This subsection explains the features of AIr compared to IEEE 802.11.

One of the main differences between IEEE 802.11 and AIr is the way in which they deal with the "hidden nodes" problem. "Hidden nodes" in wireless communication refer to those nodes that are out of the range of other nodes. Due to node mobility and to various communication features, the bounds of a given network segment are difficult to define. The IEEE 802 Medium Access Control (MAC) service defines a best-effort, ordered delivery service and assumes a transitive communications relationship. At the MAC layer:

- IF A can communicate with B
- AND B can communicate with C,
- THEN A can communicate with C.

Unfortunately, in wireless communications, this is not the case. It is very normal to have nodes that cannot hear each other. In this topology, node C may not be able to hear node *A*'s transmissions, in which case node *C* is the "hidden node" for node *A*. Other possible hidden node phenomena could happen between node *A* and node *B* even if they are within each other's communication range if the communication relationship is asymmetric: node *A* may be "heard" by node *B*, but node *B* may not be "heard" by node *A*. In a real network, this may present a more complex communication relationship. It is known that in CSMA, a node is allowed to transmit a packet only if it senses that the channel is idle. Packet collisions are rare as long as the nodes can hear each other and the propagation delay is small [275]. The "hidden nodes" problem has presented a big challenge for the implementation of the CSMA scheme. For this reason it has generated a great deal of interest among different researchers on multiple medium access control design for wireless networks.

In IEEE 802.11, the RTS/CTS exchange and handshake packets scheme can partly overcome the "hidden nodes" problem by reserving the channel beforehand. However, the scheme itself and its variants assume that all the "hidden nodes" are within the transmission range of the receivers, such that they can correctly receive the CTS packet. This assumption is not always true in wireless communications, as discussed above. A number of recent simulations conducted by different researchers further indicate that some stations can be out of the transmission range of both the transmitting and the receiving nodes, but still capable of interfering with the frames' reception. The fact that the interference range may be larger than the transmission range has been identified as one of the major reasons of inefficiency for both the RTS/CTS mechanism [244] and the TCP unfairness/capture problems [245]. A further discussion of the "hidden nodes" problem in wireless networks follows. RTS/CTS is not a complete solution and it may even further decrease the throughput when not being used properly.

AIr presents special considerations with regard to the "hidden nodes" problem. The solution is taken both from the physical layer design and the MAC access control. The AIr access protocol employs extended CTS range and channel symmetry to avoid hidden terminal situations. The CTS frame is transmitted to announce channel reservation, and its transmission range should cover all the potentially interfering terminals, in such a way that individual terminals can rely on a symmetrical channel. To support this, AIr PHY has been designed to derive the conditions for achieving symmetry of the IR channel. This is where the VR coding technique comes into play. Repetition coding trades SNR for transmission rate by repetition of the physical layer symbol. It provides a means of robustly coding the media reservation message so that they reach more potential sources of interference.

In addition to finding a solution in protocol design, in practice, there are other ways that can be used to avoid the "hidden nodes" problem. They include increasing the power to the nodes if possible, the use of omni-directional antennae, the removal of obstacles, and moving the node and using protocol enhancement software.

There are also other significant differences between IEEE 802.11 and AIr. Two of them are the bandwidth and the transmission range. While the maximum available data rate is 2 Mbps for IEEE 802.11, the data rate of AIr is 4 Mbps. The number of packets sent after medium reservation is 1 in IEEE 802.11, but it is a burst of packets in AIr. IEEE 802.11 employs exponential backoff contention windows adjustment. AIr, on the contrary, uses a linear mechanism.

10.4 The Ad Hoc Network

As discussed in Chapter 1, infrared and radio can be considered complementary technologies for the support of WLANs. The characteristics of infrared make it more useful for low-cost and short-range applications, such as ad hoc networks. As a matter of fact, infrared has been considered a primary candidate for many ad hoc network applications, such as sensor networks, battlefield communication, etc. This section looks at the challenges in ad hoc network design and gives a detailed description of network issues such as ad hoc routing, security, and quality of service (QoS).

10.4.1 Issues in Ad Hoc Network Design

Ad hoc networking has become one of the most popular fields of study in recent years, both in academic research and commercial environments, due to the availability of inexpensive, widely used wireless devices and the development of mobile computing. This type of network can be seen as either an autonomous system with its own routing and management method, or as a multi-hop wireless extension to the Internet that requires a flexible and seamless interface for Internet access. The design of these systems is not straightforward, and the following characteristics of ad hoc networks have posed a number of challenges from different aspects:

- *Power constraints.* In many cases, wireless devices must rely on batteries, which make energy efficiency one of the most critical problems in ad hoc networks. In addition, because of the limited range of each mobile node's wireless transmission, to communicate with nodes outside its transmission range, it requires another node to act as a relay station to forward packets to the destination. This route relay increases energy consumption, especially in large-scale ad hoc networks. In addition to hardware improvements, the solutions from the networking point of view have focused on saving power from unnecessary uses, such as selectively putting the receiver into a sleep mode in the MAC layer, and introducing energy-awareness routing protocols.
- *Dynamic network topology.* In an ad hoc network, nodes might join, move around, or leave at any time. The network topology changes with the movement of the nodes and is very unpredictable. Well-designed routing protocols and mobility management schemes are required to handle these problems if one is to achieve continuous and reliable communication.

- *Limited communication bandwidth and capacity.* Despite the fact that the IR medium potentially offers very high-speed data rates, the actual bandwidth available is very limited due to technology obstacles. Meanwhile, open space communication has introduced high BERs, especially when multi-hop is introduced. There are several proposals for solving these problems, such as optimizing the network structure and improving the transmission techniques.

- *Distributed operation.* Without a central administration, a node in an ad hoc network must rely on other nodes to support security and routing functions. In such an environment, passive or active link attacks (for example, eavesdropping and denial-of-service) are possible. Meanwhile, the intermediate nodes may refuse to cooperate in an effort to save power. The cooperation issue has attracted considerable attention in recent years. Several solutions (such as introducing an appropriate acceptance algorithm [226], "Virtual Currency" distribution [237], etc.) are proposed.

- *Open space communications.* The high error rate in wireless channels has always been a major challenge for QoS, especially for some applications where accurate transmission is in high demand. Improvements in this area mainly focus on physical designs, such as increasing transmission power, introducing efficient error correction methods, etc.

The following subsections detail the different issues related to ad hoc networking, namely, routing, security and QoS. Because it is very difficult to discuss each of these issues independently of the others, the relationship between each of them also is discussed to some extent.

10.4.2 Routing in Infrared Ad Hoc Networking

Routing in ad hoc networking faces a number of challenges, including bandwidth restrictions, dynamic network topology, and uncertainty of path quality. The route establishment procedure in ad hoc networks should be dynamic and adaptive. Most of the approaches currently proposed are derived from extensions to existing routing protocols in IP-based wired networks. The routing protocols can be roughly classified as proactive, reactive, or hybrid.

10.4.2.1 Proactive

Proactive routing attempts to maintain consistent, up-to-date routing information of the entire network. Each node must maintain one or more tables to store up-to-date routing information and to propagate updates throughout the network. The advantage of this method is that the routing information is not only current, but also immediately available when required. Like many traditional protocols, they are typically table-driven and distance-vector protocols. Examples of these protocols include Dynamic Destination Sequenced Distance-Vector Routing (DSDV) [246], Optimized Link State Routing (OLSR) [247, 248], Fisheye State Routing (FSR) [249], Wireless Routing Protocol (WRP) [250], Clustered Gateway Switch Routing (CGSR) [251], and Global State Routing (GSR) [252]. These protocols employ different mechanisms on route change broadcast and the number of routing-related tables used. These proactive protocols try to maintain valid routes to all communication mobile nodes all the time, regardless of whether or not they are needed. They also need to update periodically the route changes, which may introduce heavy overheads to the network.

10.4.2.2 Reactive

To eliminate the heavy overheads introduced by conventional routing tables and periodical routing information update in proactive protocols, reactive routes are created only when desired by the source node, and the routing information update is only propagated when necessary. To assist the non-constant route updates, a route discovery process is required in the network. For optimization purposes, many of the reactive routing protocols are on-demand driven. Examples include source-initiated ad hoc routing protocols such as Dynamic Source Routing (DSR) [253, 254], Ad Hoc On-Demand Distance Vector Routing (AODV) [255, 256], Temporally-Ordered Routing Algorithm (TORA) [257], and Location Aided Routing (LAR). The disadvantage of the reactive protocols is that they create a lot of overhead when the route is being determined, because the routes are not necessarily up-to-date when required.

10.4.2.3 Hybrid

Hybrid protocols try to incorporate the good properties of both the reactive and the proactive approaches and offer means to switch dynamically between the parts of the two protocols. For example, table-driven protocols could be used between networks and on-demand protocols inside the networks, or vice versa. Most hybrid schemes are proposed to solve the scalability problem in ad hoc networking. A good example is the Zone Routing Protocol (ZRP) [258, 259]. The ZRP separates the network into different zones. Each node inside a zone employs a proactive routing algorithm for any nodes inside the same zone, but uses a reactive protocol for any nodes outside the zone. Details on ZRP can be found in [260].

Although large-scale ad hoc networking using IR is not considered a major scenario in current applications, it should not be totally excluded. In situations where radio cannot be used, a group of IR cells clustered together provides an attractive alternative.

Despite the fact that each of the approaches presented above brings particular advantages, none of them suits all the possible conditions in ad hoc environments. The choice of a suitable routing scheme depends on several factors.

10.4.3 Security in Infrared Ad Hoc Networking

The salient features of open space transmission and dynamically changed network structure in wireless ad hoc networking pose several challenges to security for this type of network. In some working scenarios, especially for security-critical applications, security is a highly demanding requirement in ad hoc network design. In a fully self-organized ad hoc network, the establishment of security association is purely based on mutual agreement between users. Due to the wide availability of IrDA ports in today's portable equipment, there are currently a number of proposals to use infrared wireless links to bypass the encryption and authentication stages necessary for securing radio frequency (RF) communication. Because it can benefit from the high data rate of LOS infrared links, IR provides a broadband channel for secure data exchange. Two IR devices pointing to each other within a "secure range" can be used to form a "side channel." This "side channel" is released after the end of the exchange of "security" information, which could be a secure key, a password, etc. For example, in an open café, mobile users may declare their friends by pointing to each other through the IrDA port to exchange security codes. They could then switch to radio for other information exchange, such as file sharing, messenger service, etc. Users can also authenticate their own devices through the IrDA port to share services among them.

With regard to the security issues in a pure IR ad hoc network, which is formed via non-LOS or diffuse links, there are more challenges in these topologies than in a simple point-to-point secure link. Compared to its radio counterpart, IR presents better security performance, due to the fact that IR radiation cannot pass through walls and that IR cells provide low coverage. Therefore, malicious attacking from outside the network is preventable. However, in a wide field-of-view configuration, there is no guarantee that eavesdropping will not occur, even if the signal is weak. Danger may arise from within the network by placing compromised nodes in some specific situations. Mobile nodes roaming in a hostile environment with relatively poor physical protection, for example, have a non-negligible probability of being compromised in a battlefield or sharing information through an electronic device in a public place, such as a café or an airport.

A traditional method of protecting data is via cryptographic methods, in which a secret key is encrypted and exchanged between communicating nodes. The optional WEP encryption method in the IEEE 802.11 standard is a good example. This method can be used in both ad hoc networks and infrastructure networks. In WEP encryption, each mobile node must use the same encryption key; the frame is encrypted with a WEP key before transmission, and the received frame is decrypted with an identical key. If the WEP keys are not the same string or strength, etc., the station is not able to communicate. The WEP key must be updated regularly; otherwise, an unauthorized person can easily decrypt the package by sniffing the network for a few hours (due to the simple encryption algorithm). To improve security, a more complicated encryption method is necessary. A station may be able to support several encryption methods and sharing different keys with its peers in the network. Recently there have been some proposals to increase data protection with the help of routing. One approach is to divide the data into N parts and to send them along different routes. The message is reconstructed at the destination. The major concern regarding the encryption key in this case is that the key distribution and the frequent refresh introduce high overheads to wireless ad hoc networks. The more complicated the key, the heavier the overhead that is potentially introduced. This represents a further restriction to the already-constrained wireless bandwidth.

Another security issue in ad hoc networking concerns the cooperation among the nodes. Most ad hoc routing algorithms assume that each routing node accepts the packets it receives. However, for an ad hoc network that consists of unknown (or untrusted) nodes, the misbehavior caused by these nodes can introduce lots of problems to the network operation. A malicious node might jam the communication channel by sending bursts of data or by trying to deprive a device of battery power, by keeping it awake and engaging it in communication all the time. A "selfish" node may also break down the routing route reactively by falling into "sleep" to save its power. Even in a cooperative network, there is the risk of a node being captured and compromised. This node might then send false routing information or key update information that could paralyze the entire network very quickly. That is why the cooperation issue in security has started to attract more attention these days. However, due to its complexity, it remains an open issue to this day.

10.5 Quality of Service (QoS)

Quality-of-Service (QoS) is a measure of network performance that reflects the network's transmission quality and service availability. It can come in different forms to guarantee the performance of network flow in terms of delay, jitter, bandwidth, packet loss probability, error rate, and other characteristics. To ensure QoS, these measuring characteristics must be improved, and to some extent, guaranteed in advance. QoS provision in wireless communication is a well-known problem due to the poor quality, the limited bandwidth, and the highly dynamically changing network topology of the links. Recently, the rising popularity of multimedia streaming applications, the continuous transmission of high-bandwidth video, and many other potential commercial applications are demanding that more effort be put on QoS support in wireless networks.

There are numerous attempts to improve the QoS from the physical to the application layer, which reflect the multi-layer nature of ad hoc networks. The methods proposed to improve QoS from lower layers are based on the argument that the weakness of QoS in wireless is due to the lack of reliability of the links. The physical layer is required to adapt to the rapid changes in the link characteristics.

To deal with the dynamic network architecture, individual nodes are responsible for dynamically discovering the other nodes with which they can directly communicate [276]. To achieve energy efficiency in ad hoc networks, all processing and control must be performed by the network nodes in a distributed fashion. "The MAC layer needs to minimize collisions, allow fair access, and semi-reliably transport data over the shared wireless links in the presence of rapid changes and hidden or exposed terminals. The network layer needs to determine and distribute the information used to calculate paths in a way that maintains efficiency when links change frequently and bandwidth is at a premium. It also needs to integrate smoothly with conventional networks, which are normally more reliable and faster. In addition, ad hoc networks need to perform functions such as auto-configuration in changing environments. The transport layer must be able to handle delay and packet loss statistics that are very different than that of wired networks. Finally, applications must be designed to adapt to frequent disconnection and reconnection with peer applications, as well as to widely varying the delay and the packet loss characteristics."

The improvement in the physical layer depends primarily on the transceiver design, which was presented in previous chapters. In this section, two other major issues on QoS support in mobile ad hoc networks are discussed: (1) QoS in Medium Access Control and (2) QoS routing. The issue of QoS in applications is out of the scope of this chapter and is not described further.

10.5.1 QoS in the MAC Layer: IEEE 802.11e

To support bandwidth-sensitive applications such as voice and video, the IEEE approved the IEEE 802.11e standard in November 2005 [261, 262]. This standard presents a MAC layer enhancement over the traditional IEEE 802.11 standard. It is known that the DCF and PCF defined in IEEE 802.11 standard do not differentiate between the traffic types of the sources. DCF only supports best-effort service, and it does not have any QoS guarantee. PCF is also an inefficient central polling scheme. In IEEE 802.11e, both coordination modes are enhanced to facilitate QoS. These changes allow for the fulfillment of critical service requirements while maintaining backward-compatibility with current 802.11 standards.

In IEEE 802.11e, the DCF remains the fundamental access method used by non-QoS STAs (nQSTAs). Additionally, a new Hybrid Coordination Function (HCF) protocol has been created. Within HCF, two priority access variants have been used — (1) Enhanced Distributed Channel Access (EDCA) and (2) HCF Controlled Channel Access (HCCA) — which derive from their earlier versions Enhanced Distributed Channel Function (EDCF) and Hybrid Coordination Function (HCF). The relationships between these MAC methods are shown in Figure 10.8. The DCF is the basis for the other three methods: PCF is optional, non-QSTA will keep on using DCF and PCF, and QSTA will utilize either EDCA or HCCA.

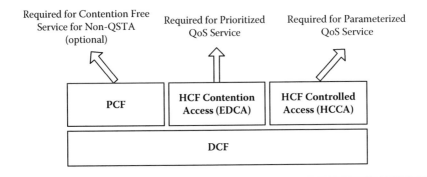

Figure 10.8 IEEE 802.11e MAC architecture. (*Source:* From [262].)

HCCA is a contention-free based mechanism, as is PCF in the IEEE 802.11 standard. The main innovation introduced in HCCA is the ability to poll the stations in both the contention and contention-free periods (CFP). The HCCA mechanism manages access to the wireless medium, using a hybrid coordinator (HC) that has higher medium access priority than non-AP STAs. This allows it to transfer MSDUs to non-AP QSTAs, and to allocate transmission opportunities (TXOPs) to non-AP QSTAs. Due to the involvement of AP in HCCA, it is not discussed further in this section.

10.5.1.1 EDCA in IEEE 802.11e

The EDCA is the prioritized CSMA/CA access mechanism used by QSTAs. It provides differentiated, distributed access to the wireless medium for QSTAs using eight different user priorities. The EDCA mechanism defines four access categories (ACs) that provide support for the delivery of traffic with user priorities (UPs) at the QSTAs. The four ACs include AC_BK (background traffic), AC_BE (best effort traffic), AC_VI (video traffic), and AC_VO (voice traffic), among which AC_BK has the lowest priority and AC_VO has the highest priority. Each AC contains two different UPs. The values a UP can take are the integer values from 0 to 7 and are identical to the IEEE 802.1D priority tags [263]. Differentiation among the user priorities is achieved by varying the following values:

- Amount of time a station senses the channel as idle before backoff or transmission
- Length of the contention window to be used for the backoff
- Duration a station may transmit after it acquires the channel

For example, the Arbitration Inter-Frame Space (AIFS) is used by QSTAs to transmit all data type frames and other management or control type frames.

Stations try to send data after detecting the medium is idle and after waiting a corresponding AIFS according to the traffic category. "A higher-priority traffic category has a shorter AIFS than a lower-priority traffic category. Thus, stations with lower-priority traffic must wait longer than those with high-priority traffic before trying to access the medium. To avoid collisions within a traffic category, the station counts down an additional random number of time slots, known as a *collision window* (CW), before attempting to transmit data. If another station transmits before the countdown has ended, the station waits for the next idle period, after which it continues the countdown where it left off [277]." The traffic categories can be configured during the station initiation period or broadcast by access points in an infrastructure network. In ad hoc networking, it depends on the agreement among mobile nodes, and this sort of information can be broadcast in a distributed manner.

IEEE 802.11e has enhanced the RTS/CTS exchange scheme in IEEE 802.11 by introducing piggybacking — the possibility to attach the ACK of a previously received frame or a poll to a data frame for a specific QSTA. This procedure decreases the overhead by reducing the number of frames to be exchanged. In addition, the network allocation vector (NAV) approach to keep track of the periods that a station does not try to access the wireless medium is also used in IEEE 802.11e.

EDCA does not guarantee the service provided. However, it establishes a probabilistic priority mechanism to allocate bandwidth based on traffic categories.

10.5.2 QoS in Routing

Most available routing protocols in ad hoc networking aim to achieve the shortest and most reachable path, but with little concern for QoS. The services provided by these selected routing paths are normally best-effort. In ad hoc networks, dynamically changed network structures and relatively slow and unreliable physical links have high demands in service assurance that best-effort-only service would find very difficult to meet. Routing with QoS looks for a path with enough resources and high reliability to meet the requirements of different applications. Thus, the QoS routing protocols should work together with resource management to establish paths through the network that meet end-to-end QoS requirements. In most cases, the QoS routing does not reserve resources, but leaves it as a task for QoS signaling.

QoS in routing is a very difficult issue, especially for multi-hop ad hoc networks. The routing algorithms for wired networks cannot be applied directly to mobile wireless ad hoc networks. This is because the performance of most wired routing algorithms depends on the precise state of the information, which may not be attainable in a wireless ad hoc network due to its dynamic nature, which results in the imprecision of the link state information. In recent years, QoS routing has become a very popular research area. However, despite the number of proposals and solutions coming up, until today, QoS routing is still an open question due to the complexity it brings.

One good example of a routing algorithm among the recent influential work presented by different researchers is the ticket-based probing (TBP) routing proposed by Chen and Nahrstedt [264], which was designed for imprecise state information in the dynamic multi-hop mobile network environment. This routing algorithm intends to select a network path with sufficient resources to satisfy the delay or bandwidth requirements of certain applications to achieve a near-optimal performance with modest overhead using a limited number of tickets and making intelligent hop-by-hop path selection. When a source wants to find QoS paths to a destination, it issues probe messages with some tickets. The number of tickets is based on the available state information. One ticket corresponds to one path searching; and one probe message should carry at least one ticket. Thus, the maximum number of searched paths is bounded by the tickets issued from the source. When an intermediate host receives a probe message with n tickets, then based on its local state information, it decides whether and how to split the n tickets and where to forward the probe (or probes). When the destination host receives a probe message, a possible path from the source to the destination is found. The ticket distribution in this QoS routing scheme may introduce a heavy communication overhead when looking for an appropriate route path.

Recently, there have been some proposals to improve QoS with the aid of mobile node location information — and GPSR [265] is a very good example. In this algorithm, packets are forwarded greedily to a neighbor closer to the physical location of the destination, which repeats at each intermediate node until the destination is reached. Due to the complexity of QoS routing, there are many more complicated routing algorithms proposed these days. As in the use of GPSR, when choosing an algorithm, the QoS overhead must be taken into account due to the constrained bandwidth in wireless links.

A number of researchers working on QoS tend to discuss a certain aspect of networking (such as routing, bandwidth, or mobility) without considering the entire picture. Unfortunately, it is difficult or even impossible to judge and evaluate an algorithm without doing so. A promising method for satisfying QoS requirements that has been proposed is cross-layer or vertical-layer integration [266]. The idea in this method is to violate many of the transitional layering styles to allow different parts of the stack to adapt to the environment in a way that takes into account the adaptation and available information of other layers. This method intends to cover all the important design issues — medium access control, routing, multicasting, and transport layer, security, quality-of-service provisioning, and energy management — in ad hoc wireless networking. The design of the algorithm for each one of these issues is based on the knowledge of different layers in the protocol stack. Each protocol layer can access the common information by sharing crossing layers and implement the proper algorithms according to the information it acquired.

10.6 Future Infrared Networking

Infrared offers a flexible, secure, and economic option for short-range wireless communications. The standards and specifications from the IEEE and IrDA have accelerated the adoption of this technology for small-scale networks. The widespread IR devices in different application categories these days have also greatly reduced the risks and costs of introducing this technology to new devices.

Infrared is a very promising technology for the future wireless network. With potential high speeds to exploit, it provides an ideal platform for short-range, indoor communication. As discussed in Chapter 1, infrared will not replace radio but will coexist as a complementary option; it will also play essential roles, especially in applications where radio interference from the surroundings must be avoided and in applications that require highly secure communication.

There still remain many challenges to be addressed in infrared networking, such as the development of mechanisms for efficient use of limited bandwidth and communication capacity, further exploration of potential bandwidth, mechanisms for reducing power consumption and network overhead in networking management, development of algorithms for enhancing information security and providing efficient routing, etc. Optical wireless networks are also facing difficulties relating to integrating with reliable and fast wired links. The vast performance difference between wired and wireless links is a well-known concern in wireless network design. Seamless interfacing and smooth convergence with wired links are significant challenges.

Despite the fact that there are still lots of obstacles to overcome, the development of IR technology in networking gets stronger every day. The rapid merges of other communication technologies at the same time also has sped up the utilization of IR in networking. It is certainly influencing the way we see the world of communication and networking today.

References

1. Ofcom, The Office of Communications, http://www.ofcom.org.uk/
2. RA, Radiocommunications Agency Web page, http://www.radio.gov.uk
3. FCC, Federal Communications Commission Web page, http://www.fcc.gov
4. S. Hagihira, M. Takashina, T. Mori, N. Taenaka, T. Mashimo, and I. Yoshiya, Infrared Transmission of Electromagnetic Information via LAN in the Operating Room, *Journal of Clinical Monitoring and Computing*, 16, 171–175, 2000.
5. E. Hanada, K. Kodama, K. Takano, Y. Watanabe, and Y. Nose, Possible Electromagnetic Intereference with Electronic Medical Equipment by Radio Waves Coming from Outside the Hospital, *Journal of Medical Systems*, 25, 4, 257–267, 2001.
6. A.A. Klein and G.N. Djaiani, Mobile Phones in the Hospital — Past, Present and Future, *Anaesthesia*, 58, 353–357, 2003.
7. I. Pavlosoglou, R. Ramirez-Iniguez, M.S. Leeson, and R.J. Green, A Security Application of the Warwick Optical Antenna in Wireless Local and Personal Area Networks, presented at *London Communications Symposium*, 2002, pp. 225–228.
8. R. Ramirez-Iniguez and R.J. Green, Indoor Optical Wireless Communications, presented at *Optical Wireless Communications*, IEE Savoy Place, London, U.K., 1999, pp. 14/1–14/7.
9. F.R. Gfeller, H.R. Muller, and P. Vettiger, Infrared Communication for In-House Applications, presented at *IEEE COMPCON '78*, Washington, D.C., 1978, pp. 132–138.
10. F.R. Gfeller and U.H. Bapst, Wireless In-House Data Communication via Diffuse Infrared Radiation, *Proceedings of the IEEE*, 67, 1474–1486, 1979.
11. F.R. Gfeller, Infranet: Infrared Microbroadcasting Network for In-House Data Communication, IBM Zurich Research Laboratory, Ruschlikon RZ 1068 No. 38619, April 27, 1981.
12. Cablefree, Cablefree Solutions Web page, http://cablefreesolutions.com
13. Lightpointe, Lightpointe Web page, http://www.lightpointe.com
14. SONA, SONA Optical Wireless Web page, http://www.fsona.com
15. PAV, PAV Web page, http://www.pavdata.com
16. iRLan, iRLan Web page, http://www.irlan.co.il
17. IrDA, Infrared Data Association Web page, http://www.irda.org
18. Bluetooth, Bluetooth Web page, http://www.bluetooth.com
19. J.R. Barry, *Wireless Communication Using Non-Directed Infrared Radiation*. PhD dissertation: University of California, 1992.

20. A.M. Street, P.N. Stavrinou, D.C. O'Brien, and D.J. Edwards, Indoor Optical Wireless Systems — A Review, in *Optical and Quantum Electronics*, 29, 349–378, 1997.
21. J.M. Kahn and J.R. Barry, Wireless Infrared Communications, *Proceedings of the IEEE*, 85(2), 265–298, 1997.
22. G.W. Marsh and J.M. Kahn, Channel Reuse Strategies for Indoor Infrared Wireless Communication, *IEEE Transactions on Communications*, 45(10) 1280–1290, 1997.
23. P.P. Smyth, M. McCullagh, D.R. Wisely, D. Wood, S. Ritchie, P.L. Eardley, and S. Cassidy, Optical Wireless Local Area Networks — Enabling Technologies, *BT Technology Journal*, 11(2), 56–64, 1993.
24. C. Singh, J. John, Y.N. Singh, and K.K. Tripathi, Indoor Optical Wireless Systems: Design Challenges, Mitigating Techniques and Future Prospects, *IETE Technical Review*, 21(2), 101–117, 2004.
25. P.P. Smyth, P.L. Eardley, K.T. Dalton, D.R. Wisely, P. McKee, and D. Wood, Optical Wireless — A Prognosis, presented at *SPIE*, 1995, pp. 212–225.
26. K.S. Natarajan, K.C. Chen, and P.D. Hortensius, Considerations for the Design of High Speed Wireless Optical Networks, presented at *IEEE Workshop on Wireless LANs*, 1991, pp. 140–143.
27. D.R. Pauluzzi, P.R.H. McConnell, and R.L. Poulin, Free-Space, Undirected Infrared (IR) Voice and Data Communications with a Comparison to RF Systems, presented at *IEEE ICWC '92*, 1992, pp. 279–285.
28. I.A. Parkin and J. Zic, An Application of Infra-red Communications, *Journal of Electrical and Electronics Engineering*, 4(4), 331–336, 1984.
29. J.R. Barry, J.M. Kahn, E.A. Lee, and D.G. Messerschmitt, High-Speed Nondirective Optical Communication for Wireless Networks, in *IEEE Network Magazine*, 5(6), 44–54, 1991.
30. Canobeam, http://www.canon.com/bctv/canobeam/index.html
31. Terescope, http://www.pulsewan.com/wireless/laser_pp_menu.htm
32. OPTICOMM, http://www.optel.com/en/prod_index.htmOPTEL
33. VIPSLAN-100, JVC-Victor Web page, http://babelfish.av.com/babelfish/trurl_pagecontent?lp=ja_en&trurl=http://www.jvc-victor.co.jp%2Fpro%2Flan%2Flineup%2Findex.html
34. T. Minami, K. Jano, T. Touge, H. Morikawa, and O. Takahashi, Optical Wireless Model for Office Communication, presented at *National Computer Conference*, 1983, pp. 721–728.
35. O. Takahashi and T. Touge, Optical Wireless Network for Office Communication, presented at *JARECT*, 1985, pp. 217–228.
36. Y. Nakata, J. Casio, T. Kojima, and T. Noguchi, In-House Wireless Communication Systems Using Infrared Radiation, presented at *International Computer and Communication Conference*, 1985.
37. C. Yen and R.D. Crawford, The Use of Directed Optical Beams in Wireless Computer Communications, presented at *IEEE GLOBECOM*, 1985, pp. 1181–1184.
38. M.D. Kotzin and A.P. Van den Heuvel, A Duplex Infra-red System for In-Building Communications, presented at *IEEE VTC*, 1986, pp. 179–185.
39. T.S. Chu and M.J. Gans, High Speed Infrared Local Wireless Communication, in *IEEE Communications Magazine*, 25(8), 4–10, 1987.

40. T. Fuji and Y. Kikkawa, Optical Space Transmission Module, *National Tech. Report*, 34(1), 101–106, 1988.

41. R.L. Poulin, D.R. Pauluzzi, and M.R. Walker, A Multi-Channel Infrared Telephony Demonstration System for Public Access Applications, presented at *IEEE International Conference on Wireless Communication*, 1992, pp. 286–291.

42. G.W. Marsh and J.M. Kahn, 50 Mb/s Diffuse Infrared Free-Space Link Using On-Off Keying with Decision Feedback Equalization, *IEEE Photonics Technology Letters*, 6(10), 1268–1270, 1994.

43. G.W. Marsh and J.M. Kahn, Performance Evaluation of Experimental 50 Mbps Diffuse Infrared Wireless Link Using On-Off Keying with Decision-Feedback Equalization, *IEEE Transactions on Communications*, 44(11), 1496–1504, 1996.

44. G. Faulkner, K. Jim, E. Zyambo, F. Parand, D.C. O'Brien, and D.J. Edwards, High Speed Integrated Optical Wireless Transceivers for In-Building Optical LANs, www.eng.ox.ac.uk/~eoswww/research/Cellular/cellular%20opical%20network.htm

45. Showa, 100 Mbps Short-Range Infrared Wireless Transceiver, *Showa Electric Wire and Cable Review*, 110, 61, 2004.

46. H. Park and J.R. Barry, Performance of Multiple Pulse Position Modulation on Multipath Channels, *IEE Proceedings in Optoelectronics*, 143(6), 360–364, 1996.

47. J.R. Barry, *Wireless Infrared Communications*. Kluwer Academic Publishers, 1994.

48. H. Park and J.R. Barry, Modulation Analysis for Wireless Infrared Communications, presented at *IEEE International Conference on Communications*, ICC 95, Seattle, 1995, pp. 1182–1186.

49. B. Yu, J.M. Kahn, and R. You, Power-Efficient Multiple-Subcarrier Modulation Scheme for Optical Wireless Communications, *Proceedings of SPIE*, 4873, 41–53, 2002.

50. R. Otte, L.P. De Jong, and A.H.M. Van Roermund, *Low-Power Wireless Infrared Communications*. Kluwer Academic Publishers, 1999.

51. M. Achour, Simulating Atmospheric Free-Space Optical Propagation. II. Haze, Fog and Low Clouds Attenuation, *Optical Wireless Communications V, Proceedings of SPIE*, 4873, 1–12, 2002.

52. N.J. Veck, Athmospheric Tranmission and Natural Illumination (visible to microwave regions), *GEC Journal of Research*, 3(4), 209–223, 1985.

53. C. C. f. R. S. CCRS, *Definition of Atmospheric Attenuation*, http://ccrs.nrcan.gc.ca/glossary/index_e.php?id=34

54. M. Gebhart, E. Leitgeb, and J. Bregenzer, Atmospheric Effects on Optical Wireless Links, presented at *7th International Conference on Telecommunications (ConTEL)*, Zagreb, Croatia, 2003, pp. 395–401.

55. A.R. Tebo, Those Pesky Aerosols: The Bane of Good Seeing through the Atmosphere, *Electro-optical Systems Design*, 1982, pp. 23–24.

56. P.L. Eardley and D.R. Wisely, 1 Gbit/s Optical Free Space Link Operating over 40 m Systems and Applications, *IEE Proceedings in Optoelectronics*, 143(6), 330–333, 1996.

57. M.A. Bramson, in *Infrared Radiation, A Handbook for Applications*. Plenum Press, 1969, p. 602.

58. D.B. Rensch and R.K. Long, Comparative Studies of Extinction and Backscattering by Aerosols, Fog and Rain at 10.6 Micrometers and 630 nm, *Applied Optics*, 9(7), 1563–1573, 1970.

59. A. Al-habasch, K.W. Fischer, C.S. Cornish, K.N. Desmet, and J. Nash, Comparison between Experimental and Theoretical Probability of Fade for Free Space Optical Communications, *Optical Wireless Communications V, Proceedings of SPIE*, 4873, 79–89, 2002.

60. R.J. Hill, R.G. Frehlich, and W.D. Otto, The Probability Distribution of Irradiance Scintillation, NOAA Environmental Research Laboratories, Colorado, ERL ETL-274, 1996.

61. A. Al-habasch and L.C. Andrews, Mathematical Model for the Irradiance PDF of a Laser Beam Propagating through Turbulent Media, *Optical Engineering*, 40, 1554–1562, 2001.

62. L.C. Andrews and R.L. Phillips, I-K Distribution as Universal Propagation Model of Laser Beams in Atmospheric Turbulence, *Optical Society of America*, A 2, 160–163, 1985.

63. P.L. Eardley, D.R. Wisely, D. Wood, and P. McKee, Holograms for Optical Wireless LANs, *IEE Proceedings in Optoelectronics*, 134(6), 365–369, 1996.

64. M.R. Clay and A.P. Lenham, Transmission of Electromagnetic Radiation in Fogs in the 0.53–10.1 Micrometer Wavelength Range, *Applied Optics*, 20(22), 3831, 1981.

65. R. Narasimhan, M.D. Audeh, and J.M. Kahn, Effect of Electronic-Ballast Fluorescent Lighting on Wireless Infrared Links, *IEE Proceedings in Optoelectronics*, 143(6), 347–354, 1996.

66. K. Pahlavan, Wireless Communications for Office Information Networks, in *IEEE Communications Magazine*, 23(6), 19–27, 1985.

67. D. Rollins, J. Baars, D. Bajorins, C. Cornish, K. Fischer, and T. Wiltsey, Background Light Environment for Free-Space Terrestrial Communications Links, *Proceedings of SPIE*, 4873, 99–110, 2002.

68. N.J.J. Bunnik, *The Multispectral Reflectance of Short Wave Radiation by Agricultural Crops*, Veenman and Zn, 1978, pp. 9.

69. V.G. Sidorovich, Solar Background Effects in Wireless Optical Communications, *Proceedings of SPIE. Optical Wireless Communications*, 4873, 133–142, 2002.

70. J.M. Kahn, W.J. Krause, and J.B. Carruthers, Experimental Characterization of Non-Directed Indoor Infrared Channels, *IEE Transactions on Communications*, 43(2/3/4), 1613–1623, 1995.

71. J.J.G. Fernandes, P.A. Watson, and J.C. Neves, Wireless LANs: Physical Properties of Infra-Red Systems vs. Mmw Systems, *IEEE Communications Magazine*, 1994, pp. 68–73.

72. L. Matthews and G. Garcia, *Laser and Eye Safety in the Laboratory.* in SPIE Press, IEEE Press,1995.

73. P. Benitez, J.C. Minano, F.J. Lopez-Hernandez, D. Biosca, R. Mohedano, M. Labrador, F. Munoz, K. Hirohashi, and M. Sakai, Eye-Safe Collimated Laser Emitter for Optical Wireless Communications, *Proceedings of SPIE*, Optical Wireless Communications V, 2002, pp. 30–40.

74. D.J.T. Heatley, D.R. Wisely, I. Neild, and P. Cochrane, Optical Wireless: The Story So Far, *IEEE Communications Magazine*, 36(12), 72–82, 1998.

75. IEC, International Electrotechnical Commission Web page, http://www.iec.ch

76. M.P. Dames, R.J. Dowling, P. McKee, and D. Wood, Efficient Optical Elements to Generate Intensity Weighted Spot Arrays: Design and Fabrication, in *Applied Optics*, 30(19), 2685–2691, 1991.

77. J. Yao and G.C.K. Chen, Holographic Diffuser for Diffuse Infrared Wireless Home Networking, *Optical Engineering*, 42(2), 317–324, 2003.

78. M. Wen, J. Yao, D.W.K. Wong, and G.C.K. Chen, Holographic Diffuser Design Using a Modified Genetic Algorithm, *Optical Engineering*, 44(8), 1–8, 2005.

79. J.R. Brown and A.W. Lohmann, Complex Spatial Filtering with Binary Masks, *Applied Optics*, 5, 969–976, 1966.

80. Photonics.com, Definition of kinoform filter, http://www.photonics.com/dictionary/lookup/XQ/ASP/url.lookup/entrynum.2778/letter.k/pu./QX/lookup.htm

81. S. Kirkpatrick, C.D. Gelatt, and M.P. Vecchi, Optimization by Simulated Annealing, *Science*, 220(4598), 671–680, 1983.

82. NIST, National Institute of Standards and Technology, http://www.nist.gov/dads/HTML/simulatedAnnealing.html

83. F.J. Mendieta, E. Mitrani, H. Cervantes, J.P. Martínez, J.A. Santiago, H. Martínez, and M.A. Martínez, Atmospheric Near Infrared Communications at CEPT-1 Rate, presented at *International Conference in Wireless Communications, IEE*, 1992.

84. W.T. Welford and R. Winston, *High Collection Nonimaging Optics*. Academic Press, 1989.

85. E. Hetch, *Optics*, 4th ed. Addison-Wesley, 2002.

86. M.J.N. Sibley, *Optical Communications*. Macmillan New Electronics, 1995.

87. J. Gowar, *Optical Communication Systems*. Prentice Hall, 1993.

88. W. Grisé and C. Patrick, Passive Solar Lighting Using Fibre Optics, *Journal of Industrial Technology*, 19(1), 2–7, 2002.

89. X. Ning, R. Winston, and J. O'Gallagher, Dielectric Totally Internally Reflecting Concentrator, *Applied Optics*, 26(2), 300–305, 1987.

90. K. Ho and J.M. Kahn, Compound Parabolic Concentrators for Narrowband Wireless Infrared Receivers, *Optical Engineering*, 34(5), 1385–1395, 1995.

91. J.R. Barry and J.M. Kahn, Link Design for Nondirected Wireless Infrared Communications, *Applied Optics*, 34(19), 3764–3776, 1995.

92. M.E. Marhic, M.D. Kotzin, and A.P. Van den Heuvel, Reflectors and Immersion Lenses for Detectors of Diffuse Radiation, *Optical Society of America*, 72(3), 352–355, 1982.

93. J.P. Savicki and S.P. Morgan, Hemispherical Concentrators and Spectral Filters for Planar Sensors in Diffuse Radiation Fields, *Applied Optics*, 33(34), 8057–8061, 1994.

94. I. Fresnel Technologies, http://www.fresneltech.com/

95. J.M. Kahn, J.R. Barry, W.J. Krause, M.D. Audeh, J.B. Carruthers, G.W. Marsh, E.A. Lee, and D.G. Messerschmitt, High-Speed Non-Directional Infrared Communication for Wireless Local-Area Networks, presented at *26th Asilomar Conference on Signals & Computers*, 1992, pp. 1–5.

96. R. Ramirez-Iniguez and R.J. Green, Totally Internally Reflecting Optical Antennas for Wireless IR Communication, presented at *IEEE Wireless Design Conference*, London, U.K., 2002, pp. 129–132.

97. R. Ramirez-Iniguez and R.J. Green, Optical Antenna Design for Indoor Wireless Communication Systems, *International Journal of Communication Systems*, 18, 229–245, 2005.

98. R. Ramirez-Iniguez and R.J. Green, Elliptical and Parabolic Totally Internally Reflecting Optical Antennas for Wireless IR Communications, presented at *IEEE/IEE IrDA Optical Wireless Communications Symposium*, Warwick University, 2003, electronic version.

99. Zeon, http://www.zeonchemicals.com/zeon/default.asp

100. Lightpointe, *FlightExpress 100 Datasheet*, www.lightpoint.com, July 2006.

101. J.C. Palais, *Fiber Optic Communications*, 5th ed. Pearson Prentice Hall, 2005.

102. J.M. Senior, *Optical Fiber Communications: Principles and Practice*, 2nd ed. Prentice Hall International, 1992.

103. H.T. Lin and Y.H. Kao, Nonlinear Distortions and Compensations of DFB Laser Diode in AM-VSB Lightwave CATV Applications, *IEEE Journal on Lightwave Technology*, 14(11), 2567–2574, 1996.

104. H. Kressel (Ed.), *Semiconductor Devices for Optical Communications*. Springer-Verlag, 1980.

105. D.C. O'Brien, G. Faulkner, K. Jim, E. Zyambo, D.J. Edwards, M. Whitehead, P.N. Stavrinou, G. Parry, J. Bellon, M.J.N. Sibley, V.A. Lalithambika, V.M. Joyner, R.J. Samsudin, D.M. Holburn, and R.J. Mears, High-Speed Integrated Transceivers for Optical Wireless, *IEEE Communications Magazine*, 41(3), 58–62, 2003.

106. D.J.T. Heatley and I. Neild, Optical Wireless — The Promise and the Reality, presented at *IEE Colloquium in Optical Wireless Communications*, London, 1999, pp. 1/1–1/6.

107. P.L. Eardley, D.R. Wisely, D. Wood, and P. McKee, Holograms for Optical Wireless LANs, *IEE Proceedings in Optoelectronics*, 41, 899–911, 2002.

108. C. Sinah, J. John, Y.N. Singh, and K.K. Tripathi, Design Aspects of High-Performance Indoor Optical Wireless Transceivers, presented at *IEEE ICPWC2005*, New Delhi, 2005, pp. 14–18.

109. C. Yan, N. Yongqiang, L. Qin, Q. Wang, L. Zhao, Z. Jin, Y. Sun, G. Tao, Y. Liu, G. Chu, L. Wang, and H. Jiang, A High Power InGaAs/GaAsP Vertical Cavity Surface-Emitting Laser and Its Temperature Characteristics, *Semiconductor Science Technology*, 19, 685–689, 2004.

110. M. Yoshikawa, A. Murakami, J. Sakurai, H. Nakayama, and T. Nakamura, High Power VCSEL Devices for Free Space Optical Communications, presented at *55th Electronic Components and Technology Conference*, 2005, pp. 1353–1358.

111. S. Hranilovic and F.R. Kscischang, Short-Range Wireless Optical Communication Using Pixilated Transmitters and Imaging Receivers, presented at *IEEE International Conference*, 2004, pp. 891–895.

112. M. Stach, F. Rinaldi, M. Chandran, S. Lorch, and R. Michalzik, Bidirectional Optical Interconnection at 100 Gbps Data Rates with Monolitically Integrated VCSEL-MSM Transceiver Chips, *IEEE Photonic Technology Letters*, 18(22), 2286–2388, 2006.

113. P.J. Winzer and A. Kalmar, Sensitivity Enhancement of Optical Receivers by Impulsive RZ Coding, *IEEE Journal on Lightwave Technology*, 12, 171–177, 1999.

114. C. Bobbert, J. Kreissl, L. Molle, F. Raub, M. Rohde, B. Sartorius, A. Umbach, and G. Jacumeit, Compact 40 GHz RZ-Transmitter Design Applying Self-Pulsating Lasers, *Electronic Letters*, 40(2), 134–135, 2004.

115. C. Bornholdt, J. Slovak, M. Mohrle, and B. Sartorius, Jitter Analysis of All-Optical Clock Recovery at 40 GHz, presented at *Optical Fiber Communication Conference*, 2003, pp. 120–121.

116. E. Sackinger, *Broadband Circuits for Optical Fiber Communication*. John Wiley & Sons, 2005.

117. D.A. Fishman and B.S. Jackson, *Transmitter and Receiver Design for Amplified Lightwave System*, Optical Fibre Telecommunications IIIB. Academic Press, 1997.

118. K. Yonenaga, S. Kuwano, S. Norimatsu, and N. Shibata, Optical Duobinary Transmission System with No Receiver Sensitivity Degradation, *Electronic Letters*, 34(4), 302–304, 1995.

119. M. Uhle, The Influence of Source Impedance on the Electro-optical Switching Behaviour of LEDs, *IEEE Transactions on Electronic Devices*, 23(4), 438–441, 1976.

120. C. Lin (Ed.), *Optoelectronic Technology and Lightwave Communication Systems*. Van Nostrand Reinhold, 1989.

121. R. Olshanky and D. Fye, Reduction of Dynamic Linewidth in Single-Frequency Semiconductor Lasers, Electronic Letters, 24, 928–929, 1984.

122. L. Bickers and L.D. Westbrook, Reduction of Laser Chirp in 1.5 Micrometer DFB Lasers by Modulation Pulse Shaping, *Electronic Letters,* 21, 103–104, 1985.

123. A. Maxim, A 10 Gbps SiGe Compact Laser Diode Driver Using Push-Pull Emitter Followers and Miller Compensated Output Switch, presented at *Conference on European Solid-State Circuits*, 2003, pp. 557–560.

124. T. Wang, T. Chen, C. Tsai, L. Huang, and D. Li, A 10 Gbps Laser Diode Driver in 0–35 pm BiCMOS Technology, presented at *IEEE International Symposium on VLSI Design, Automation and Test*, 2005, pp. 253–256.

125. S.K. Kang, T. Lee, H. Park, and D.V. Plant, A Novel Automatic Power Control Method for Multichannel VCSEL Driver, presented at *CLEO Pacific Rim*, 2005, pp. 836–838.

126. A.R. Shah and B. Jalali, Adaptive Equalization for Broadband Predistortion Linearisation of Optical Transmitters, *IEE Proceedings in Optoelectronics*, 152(1), 16–32, 2005.

127. J.S. Kenney, Y. Park, and W. Woo, Advanced Architectures for Predistortion Linearization of RF Power Amplifiers, presented at *IEEE Topical Workshop on Power Amplifiers*, La Jolla, CA, 2002, pp. 122–129.

128. J. Sills, *Improving PA Performance with Digital Predistortion*, www.commsdesign. com/story/OEG20021002S00012, July 2006.

129. S.P. Stapleton, Amplifier Linearization Using Adaptive RF Predistortion, *Applied Microwave Wireless*, 2001, pp. 40–46.

130. K.C.E. Wang, High-speed Circuits for Lightwave Communications, *World Scientific, Singapore*, 1999.

131. A. Yariv, *Optical Electronics*, 4th ed. Sounders College Publishing, 1991.

132. S.M. Sze, *Photodetector in Physics of Semiconductor Devices*, Chapter 13. John Wiley & Sons, 1981.

133. K. Kishino, S. Uhlu, J.I. Chyi, J. Reed, L. Arsenault, and H. Markoc, Resonant Cavity-Enhanced (RCE) Photodetectors, *IEEE Journal of Quantum Electronics*, 27(8), 2025–2034, 1991.

134. I.H. Tan, C.K. Sun, K.S. Giboney, J.E. Bowers, E.L. Hu, B.I. Miller, U. Koren, and M.G. Young, 108-GHz GaInAs/InP PIN Photodiodes with Integrated Bias Tees and Matched Resistors, *IEEE Photonic Technology Letters*, 5(11), 1310–1312, 1994.

135. K. Kato, Ultrawide-Band/High-Frequency Photodetectors, *IEEE Transactions on Microwave Theory Technology*, 47(7), 1265–1281, 1999.

136. K. Kato, S. Hata, A. Kozen, J. Yoshida, and K. Kawano, High-Efficiency Waveguide InGaAs PIN Photodiode with Bandwidth of Over 40 GHz, *IEEE Photonic Technology Letters*, 3(5), 473–474, 1991.

137. H. Ito, T. Furuta, S. Kodama, and T. Ishibashi, InP/InGaAs Uni-Travelling-Carrier Photodiode with 310 GHz Bandwidth, *Electronic Letters*, 36(21), 1809–1810, 2000.

138. C.C. Barron, C.J. Mahon, B.J. Thibeault, G. Wang, W. Jiang, L.A. Coldren, and J.E. Bowers, Resonant-Cavity-Enhanced PIN Photodetector with 17 GHz Bandwidth-Efficiency Product, *Electronic Letters*, 30(21), 1796–1797, 1994.

139. I.H. Tan, C.K. Sun, K.S. Giboney, J.E. Bowers, E.L. Hu, B.I. Miller, and R.J. Kapik, 120 GHz Long-Wavelength Low-Capacitance Photodetector with an Air-Bridged Coplanar Metal Waveguide, *IEEE Photonic Technology Letters*, 7(8), 1477–1479, 1995.

140. F.M. Madani and K. Kikuchi, Design Theory of Long-Distance WDM Dispersion-Managed Transmission Systems, *Journal of Lightwave Technology*, 17(8), 1326–1335, 1999.

141. K. Shiba, T. Nakata, T. Takeuchi, T. Sasaki, and K. Makita, 10 Gbps Asymmetric Waveguide APD with High Sensitivity of –30dBm, *Electronic Letters*, 42(20), 1177–1178, 2006.

142. T. Otani, K. Goto, H. Abe, M. Tanaka, H. Yamamoto, and H. Wakabashi, 5.3 Gbps 11300 Km Data Transmission Using Actual Submarine Cables and Repeaters, *Electronic Letters*, 31(5), 380–381, 1995.

143. B.J. Van Zeghbroeck, W. Patrick, H.J.M., and P. Vettiger, 105-GHz Bandwidth Metal-Semiconductor-Metal Photodiode, *IEEE Electron Device Letters*, 9(10), 527–529, 1988.

144. M. Stach, L. Stoferle, F. Rinaldi, and R. Michalzik, Monolithically Integrated Optical Transceiver Chips at 850 nm Wavelength Consisting of VCSELs and MSM Photodiodes, presented at *Conference on Lasers and Electro-Optics Europe*, 2005, pp. 482–488.

145. M. Lang, W. Bronner, W. Benz, M. Ludwig, V. Hurm, G. Kaufel, A. Leuther, and J. Rosenzweig, Complete Monolithic Integrated 2.5 Gbit/s Optoelectronic Receiver with Large Area MSM Photodiode for 850 nm Wavelength, *IEEE Electronics Letters*, 37(20), 1247–1249, 2001.

146. F. Niblers, K. Kupfer, W. Janssen, N. Krausse, M. Lang, P. Pauli, A. Rupp, and F. Schmerhr, *High Frequency Circuit Engineering*. IEE, London, 1996.

147. G. Jacobsen, *Noise in Digital Optical Transmission System*. Artech House, 1994.

148. S.B. Alexander, *Optical Communication Receiver Design*, IEE Telecommunication Series, IEEE/SPIE, Vol. 37, 1997.

149. W. Ciciora, J. Farmer, and D. Large, *Modern Television Technology: Video, Voice and Data Communication*. Morgan Kauffmann, 1999.

150. R.J. Green and M.G. McNeill, Bootstrap Transimpedance Amplifier: A New Configuration, *IEE Proceedings*, 136, 57–61, 1989.

151. R.J. Green, Experimental Performance of a Bandwidth Enhancement Technique for Photodiodes, *Electronic Letters*, 22(3), 153–155, 1986.

152. S. M. Idrus, S.S. Rais, and A.S. Supaat, Analysis of Shunt Bootstrap Transimpedance Amplifier for Lange Windows Optical Wireless Receiver, presented at *International Joint Conference TSSA & WSSA*, Bandung, 2006, pp. 62–66.

153. C. Hoyle and A. Peyton, Bootstrapping Techniques to Improve the Bandwidth of Transimpedance Amplifiers, presented at *IEE Colloquium on Analog Signal Processing*, 1998, pp. 7/1–7/6.

154. A.M. Street, P.N. Stavrinou, D.J. Edwards, and G. Parry, Optical Preamplifier Designs for IR-LAN Applications, presented at *IEE Colloquium on Optical Free Space Communication Links*, 1996, pp. 8/1–8/6.

155. M.J. Teare, Low Noise Detector Amplifier, U.K. Patent 3801933, 2 April 1974.

156. P. Dueme and M. Schaller, MMIC GaAs Transimpedance Amplifiers for Optoelectronic Applications, *IEEE MTT-S Digest*, , 1997, pp. 13–16.

157. M. McCullagh and D.R. Wisely, 155 Mbit/s Optical Wireless Link using a Bootstrapped Silicon APD Receiver, *Electronic Letters*, 30[5], 430–432, 1994.

158. M. McCullagh, D.R. Wisely, and P.P. Smyth, A Low Noise Optical Receiver for a 155 Mbps 4km Optical Free Space Link, presented at *IEEE LEOS*, 1993, pp. 365–367.

159. W. Tomasi, *Electronics Communication System*. Pearson, Singapore, 2004.

160. J.S. Beasley and G.M. Miller, *Modern Electronics Communication*. Pearson and Prentice Hall, 2005.

161. C. Schick, C. Feger, E. Soenmez, K.B. Schad, A. Trasser, and H. Schumacher, Attenuation Compensation Techniques in Distributed SiGe HBT Amplifiers Using Highly Lossy Thin Film Microstrip Lines, *IEEE MTT-S International Digest*, 2005, pp. 85–88.

162. L.W. Couch, *Digital and Analog Communication Systems*. Prentice Hall, 2001.

163. E.A. Lee and D.G. Messerschmitt, *Digital Communication*. Kluwer Academic, 1994.

164. Telcodia-Technologies, SONET Transport System: Common Criteria, *GR-252-CORE, Telcodia, New Jersey*, Issue 3, 2000.

165. B. Razavi, *Design of Integrated Circuits for Optical Communications*. McGraw-Hill, 2003.

166. K.K. Ng, *Complete Guide to Semiconductor Devices*. John Wiley & Sons, 2002.

167. M. Miyashita, N. Yoshida, Y. Kojima, T. Kitano, N. Higashisaka, J. Nakagawa, T. Takagi, and M. Otsubo, An AlGaAs/InGaAs Pseudomorphic HEMT Modulator Driver IC with Low Power Disipation for 10-Gb/s Optical Transmission Systems, *IEEE Transactions on Microwave Theory Technology*, 45(7), 1058–1064, 1997.

168. J. Jeong and Y. Kwon, 10 Gbps Modulator Driver IC with Ultra High Gain and Compact Size Using Composite Lumped-Distributed Amplifier Approach, presented at *Gallium Arsenide Integrated Circuit Conference*, 2003, pp. 149–152.

169. R. Baucknecht, H.P. Schneibel, J. Schmid, and H. Melchior, 12 Gbps Laser Diode and Optical Modulator Drivers with InP/InGaAs Double HBTs, *Electronic Letters*, 32(23), 2156–2157, 1996.

170. Z. Lao, M. Yu, V. Ho, K. Guinn, M. Xu, S. Lee, V. Radisic, and K.C. Wang, High-Performance 10–12.5 Gbps Modulator Driver in InP SHBT Technology, *Electronic Letters*, 39(13), 983–985, 2003.

171. H.M. Rein, R. Schmid, P. Weger, T. Smith, T. Herzog, and R. Lachner, A Versatile Si-Bipolar Driver Circuit with High Output Voltage Swing for External and Direct Laser Modulation in 10 Gbps Optical-Fiber Links, *IEEE Journal of Solid-State Circuits*, 29(9), 1014–1021, 1994.

172. S. Galal and B. Razavi, 10 Gbps Limiting Amplifier and Laser/Modulator Driver in 0.18 Micrometer CMOS Technology, *IEEE International Solid-State Circuits Conference (ISSCC)*, 2003, pp. 188–189.

173. Y. Umeda, A. Kanda, K. Sano, K. Murata, and H. Sugahara, 10 Gbps Series-Connected Voltage-Balancing Pulse Driver with High-Speed Input Buffer, *Electronic Letters*, 40(15) , 934–935, 2004.

174. S. Mandegaran and A. Hajimiri, A Breakdown Voltage Doubler for High Voltage Swing Drivers, presented at *IEEE Custom Integrated Circuits Conference (CICC)*, 2004, pp. 103–106.

175. D. Li and C. Tsai, 10-Gb/s Modulator Drivers with Local Feedback Networks, *IEEE Journal of Solid-State Circuits*, 41(5), 1025–1030, 2006.

176. D. Kucharski, Y. Kwark, D. Kuchta, D. Guckenberg, K. Kornehay, M. Tan, C. Lin, and A. Tandon, A 20 Gb/s VCSEL Driver with Pre-Emphasis and Regulated Output Impedance in 0.13 Micrometer CMOS, presented at *IEEE International Solid-State Circuits Conference (ISSCC)*, 2005, pp. 222–223.

177. K.W. Kobayashi, J. Cowles, L.T. Tran, A. Gutierrez-Aitken, T.R. Block, A.K. Oki, and D.C. Streit, A 50-MHz–55-GHz Multi Decade InP-Based HBT Distributed Amplifier, *IEEE Microwave and Guided Wave Letter*, 7, 353–355, 1997.

178. D.C. Caruth, R.L. Shimon, M.S. Heins, H. Hsia, Z. Tang, S.C. Shen, D. Becher, J.J. Huang, and M. Feng, Low-Cost 38 and 77 GHz CPW ICs Using Ion-Implanted GaAs MESFETs, presented at *IEEE MTT-S Microwave Symposium*, 2000, pp. 11–16.

179. T. Nguyen, C. Kim, G. Ihm, M. Yang, and S. Lee, CMOS Low Noise Amplifier Design Optimization Techniques, *IEEE Transactions on Microwave Theory Technology*, 52(5), 1433–1422, 2004.

180. A. Bevilacqua and A.M. Nijad, An Ultra Wideband CMOS LNA for 3.1 to 10.6 GHz Wireless Receivers, *IEEE Journal of Solid-State Circuits*, 2004, pp. 382–383.

181. C.H. Wu, C. Lee, W. Chen, and S. Liu, CMOS Wideband Amplifiers Using Multiple Inductive-Series Peaking Technique, *IEEE Journal of Solid-State Circuits*, 40(2), 548–552, 2005.

182. A.K. Peterson, K. Kiziloglu, T. Yoon, F. Williams, and M.R. Sandor, Front-End CMOS Chipset for 10 Gbps Communications, *IEEE RFIC Symposium*, 2002, 93–96.

183. R. Chan and M. Feng, Low Noise and High Gain Wideband Amplifier Using SiGe HBT Technology, presented at *Microwave Symposium Digest, MTT-S International*, 2004, pp. 21–24.

184. D. Choudhury, M. Mokhtari, M. Sokolich, and J.F. Jensen, DC to 50 GHz Wideband Amplifier with Bessel Transfer Function, presented at *IEEE Radio Frequency Integrated Circuits Symposium*, 2004, pp. 329–332.

185. M. Sokolich, High Speed, Low Power, Optoelectronic In-P-based HBT Integrated Circuits, presented at *IEEE Custom Integrated Circuits Conference (CICC)*, 2002, pp. 483–490.

186. N. Bar-Chain, S. Margalit, A. Yariv, and I. Ury, Low Threshold Be Implanted (GaAl)As Laser on Semi-Insulating Substrate, *Journal of Quantum Electronics*, 16(4), 390–391, 1980.

187. A. Yariv, Recent Developments in Monolithic Integration of InGaAsP/InP Optoelectronic Devices, *Journal of Quantum Electronics*, 18(10), 1653–1662, 1982.

188. U. Hilleringmann and K. Goser, Optoelectronic System Integration on Silicon: Waveguides, Photodetectors and VLSI CMOS Circuits on One Chip, *IEEE Transactions on Electronic Devices*, 42(5), 841–846, 1982.

189. M. Yamamoto, M. Kubo, and K. Nakao, Si-OEIC with a Built-in PIN Photodiode, *IEEE Transactions on Electronic Devices*, 42(1), 58–63, 1995.

190. R. Li, J.D. Schaub, S.M. Csutak, and J.C. Campbell, A High-Speed Monolithic Silicon Photoreceiver Fabricated on SOI, *IEEE Photonic Technology Letters*, 12(8), 1046–1048, 2000.

191. S.M. Csutak, J.D. Schaub, W.E. Wu, R. Shimer, and J.C. Campbell, High-speed Monolithically Integrated Silicon Photoreceivers Fabricated in 130-nm CMOS Technology, *Journal of Lightwave Technology*, 20(9), 516–518, 2002.

192. S. Radovanovic, A.J. Annema, and B. Nauta, 3 Gbps Monolithically Integrated Photodiode and Pre-Amplifier in Standard 0.18 Micrometer CMOS, presented at *IEEE International Solid-State Circuits Conference (ISSCC)*, 2004, pp. 472–473.

193. R. Swoboda, J. Knorr, and H. Zimmermann, A 2.4 GHz-bandwidth OEIC with Voltage-Up-Converter, presented at *30th European Solid-State Circuit Conference*, 2004, pp. 223–226.

194. S. Brigati, F. Francesconi, D. Gardino, M. Poletti, and A. Maglione, A 20 Mbps Integrated Photoreceiver with Digital Outputs in 0.6 mm CMOS Technology, presented at *28th ESSCIRC*, 2002, pp. 503–506.

195. K. Phang and D.A. Johns, A CMOS Optical Preamplifier for Wireless Infrared Communications, *IEEE Transactions on Circuit Systems II*, 46, 852–859, 1999.

196. J. Nissinen, P. Palojarvi, and J. Kostamovaara, A CMOS Receiver for a Pulsed Time-of-Flight Lasser Range Finder, *29th ESSCIRC*, 2003, pp. 325–328.

197. H. Hein, M. Fortsch, and H. Zimmermann, Low-power 300 Mbps OEIC with Large-Area Photodiode, *Electronic Letters*, 41(7), 231–238, 2005.

198. M. Fortsch, H. Zimmermann, and H. Pless, 220 MHz Optical Receiver with Large-Area Integrated PIN Photodiode, presented at *IEEE International Conference on Sensors*, Toronto, Canada, 2003, pp. 1012–1015.

199. K. Murata, K. Sano, T. Enoki, H. Sughara, and H. Tokimitsu, InP-Based IC Technologies for 100 Gbps and Beyond, presented at *16th International Conference on Indium Phosphate and Related Materials (IPRM)*, Kagoshima, Japan, 2004, pp. 10–15.

200. H. Bulow, F. Buchali, W. Baumert, R. Ballentin, and T. Wehren, PMD Mitigation at 10 Gps Using Linear and Nonlinear Integrated Electronics Equalizer Circuits, *Electronic Letters*, 36(2), 163–164, 2000.

201. H.A. Ankerman, Transmission of Audio Signals by Infrared Light Carrier, *SMPTE Journal*, 89, 834–837, 1980.

202. J.M.H. Elmirghani and R.A. Cryan, New PPM-CDMA Hybrid for Indoor Diffuse Infrared Channels, *Electronic Letters*, 3020), 1646–1647, 1994.

203. J.M.H. Elmirghani, H.H. Chan, and R.A. Cryan, Sensitivity Evaluation of Optical Wireless PPM Systems Utilising PIN-BJT Receivers, *IEE Proceedings in Optoelectronics*, 143(6), 355–359, 1996.

204. J. Zhang, Modulation Analysis for Outdoors Applications of Optical Wireless Communications, *Communication Technology Proceedings*, 2, WCC – ICCT International Conference on Digital Object Identifier, 1483–1487, 2000.

205. E.D. Kaluarachchi, Z. Ghassemloy, and B. Wilson, Digital Pulse Interval Modulation for Optical Free Space Communication Links, presented at IEE Colloquium in Optical Free Space Communication Links, London, U.K., 1996, pp. 3/1–3/5.

206. L. Diana and J.M. Kahn, Rate Adaptive Modulation Techniques for Infrared Wireless Communications, presented at *International Conference on Communication*, Canada, 1999.

207. B. Wilson, Z. Ghassemloy, and I. Darwazeh, *Analogue Optical Fibre Communications*, IEE Telecommunications series, 1995.

208. D. Ho-Quang, D. Soo Lee, and M. Seop Lee, Relevant Modulation Schemes Combined with Coding Proposed for Indoor Wireless Infrared Communication Systems, presented at *International Conference on Telecommunications*, 2002, pp. 425–429.

209. A.J.C. Moreira, R.T. Valadas, and A.M. de Oliveira Duarte, Performance of Infrared Transmission Systems under Ambient Light Interference, *IEE Proceedings in Optoelectronics*, 143(6), 339–346, 1996.

210. H.H. Chan, K.L. Sterckx, M.H. Jaafar, J.M.H. Elmirghani, and R.A. Cryan, Performance of Optical Wireless OOK and PPM Systems under the Constraints of Ambient Noise and Multipath Dispersion, *IEEE Communications Magazine*, 1998, pp. 83–87.

211. A.J.C. Moreira, A.M.R. Tavares, R.T. Valadas, and A.M. de Oliveira Duarte, Modulation Methods for Wireless Infrared Transmission Systems — Performance under Ambient Light Noise and Interference, *SPIE*, 2601, 226–237, 1995.

212. C.R. Lomba, R.T. Valadas, and A.M. de Oliveira Duarte, Sectored Receivers to Combat the Multipath Dispersion of the Indoor Optical Channel, presented at *Proceedings of the 6th International Symposium on Personal, Indoor, and Mobile Radio Communications*, Toronto, Canada, 1995, pp. 321–325.

213. A.M.R. Tavares, R.T. Valadas, and A.M. de Oliveira Duarte, Performance of an Optical Sectored Receiver for Indoor Wireless Communication Systems in the Presence of Artificial and Natural Noise Sources, presented at *Conference on Wireless Data Transmission, Proceedings of SPIE*, Philadelphia, PA, 1995, pp. 264–273.

214. R.T. Valadas and A.M. de Oliveira Duarte, Sectored Receivers for Indoor Wireless Optical Communication Systems, presented at *5th IEEE International Symposium on Personal, Indoor, and Mobile Communications*, The Hague, The Netherlands, 1994, pp. 1090–1095.

215. F.R. Gfeller, W. Hirt, M. de Lange, and B. Weiss, Wireless Infrared Transmission: How to Reach All Office Space, presented at *Proceedings of the IEEE Vehicle Technology Conference*, Atlanta, GA, 1996, pp. 1535–1539.

216. A.P. Tang, J.M. Kahn, and K. Ho, Wireless IR Communication Links Using Multibeam Transmitters and Imaging Receivers, presented at *Proceedings of the Int. Conference on Communications*, Dallas, TX, 1996, pp. 180–186.

217. J.M. Kahn, R. You, P. Djahani, A.G. Weisbin, B.K. Teik, and A. Tang, Imaging Diversity Receivers for High-Speed Infrared Wireless Communication, *IEEE Communications Magazine*, 1998, pp. 88–94.

218. P. Djahani and J.M. Kahn, Analysis of Infrared Wireless Links Employing Multi-Beam Transmitters and Imaging Diversity Receivers, *IEEE Transactions on Communications*, 48, 2077–2088, 1999.

219. G. Yun and M. Kavehrad, Spot-Diffusing and Fly-eye Receivers for Indoor Infrared Wireless Communications, presented at *Proceeding of the IEEE Conference on Selected Topics in Wireless Communications*, Vancouver, Canada, 1992, pp. 286–292.

220. A.G. Al-Ghamdi and J.M.H. Elmirghani, Performance of an Optical Pyramidal Fly-eye Diversity Receiver for Indoor Communication Systems in the Presence of Background Noise and Multipath Dispersion, presented at *London Communications Symposium*, London, 2002, pp. 221–224.

221. A.G. Al-Ghamdi and J.M.H. Elmirghani, Analysis of Diffuse Optical Wireless, *IEEE Transactions on Communications*, 52(10), 1622–1631, 2004.

222. R.T. Valadas and A.M. de Oliveira Duarte, Sectored Receivers for Indoor Wireless Optical Communication Systems, presented at *IEEE International Conference in Communications*, 1994, pp. 1090–1095.

223. M. Karppinen, K. Kataja, J.T. Makinen, and S. Juuso, Wireless Infrared Data Links: Ray-Trace Simulations of Diffuse Channels and Demonstration of Diffractive Element for Multibeam Transmitters, *Optical Engineering*, 41(4), 899–910, 2002.

224. C.D. Knutson and J.M. Brown, *IrDA Principles and Protocols*, Vol. 1. MCL Press, Salem, UT, 2004.

225. IrDA, Advanced Infrared Physical Layer Specification, 1998.

226. IrDA, Advanced Infrared (AIr) MAC Draft Protocol, Specification Version 1.0, 1999.

227. IrDA, Serial Infrared Physical Layer Link Specification: Version 1.1, 1995.

228. IrDA, Serial Infrared Link Access Protocol (IrLAP), Version 1.1, 1996.

229. M. Watson, Foreword, in *IrDA Principles and Protocols*, Vol. 1. MCL PRess, 2004, pp. vii–viii.

230. D. Suvak, *IrDA and Bluetooth: A Complementary Comparison*, http://www.techon-line.com/pdf/pavillions/standards/extended_irda.pdf

231. M. Hassner and N. Heise, IrDA-VFIr (16 Mb/s): Modulation Code and System Design, *IEEE Personal Communications*, 2001, pp. 1–17.

232. Vishay_Telefunken, IrDA — Compatible Data Transmission, IrDA Standard — Physical Layer, Document No. 82500.

233. CENELEC, European Comittee for Electrotechnical Standardization Web page, http://www.cenelec.org/Cenelec/Homepage.htm

234. C.D. Knutson, Infrared Communications with IrDA, http://www.embedded.com/2000/0006/0006ia3.htm

235. Freescale, *AN2597/D: Using the MC9S12E128 to Implement an IrDA Interfarce*, www.freescale.com/files/microcontrollers/doc/app_note/AN2597.pdf

236. IEEE802.11, *Wireless LAN Medium Access Control (MAC) and Physical (PHY) Specifications*, 1999.

237. IEEE802.1X, IEEE Standard for Local and Metropolitan Area Networks — Port-Based Network Access Control, 2001.

238. C. Ware, J. Judge, J. Chicharo, and E. Dutkiewicz, Unfairness and Capture Behaviour in 802.11 Ad Hoc Networks, presented at *International Conference on Communications (ICC) 2000*, New Orleans, LA, 2000, pp. 159–163.

239. E. Dutkievicz, Impact of Transmit Range on Throughput Performance in Mobile Ad Hoc Networks, presented at *ICC 2001*, Los Angeles, LA, 2001.

240. K. Xu, M. Gerla, and S. Bae, How Effective Is the IEEE 802.11 RTS/CTS Handshake in Ad Hoc Networks, presented at *GLOBECOM 2002*, Taipei, Taiwan, 2002, pp. 72–76.

241. Z. IBM Research, Infrared Communication, http://www.zurich.ibm.com/sys/wireless/ircommunication.html#download. Accessed December 2006.

242. Corp. I., Advanced Ir Link Manager (AirLM) Specification, 1999.

243. Corp. I., Advanced Infrared Logical Link Control (AirLC) Specification, 1999.

244. K. Xu, M. Gerla, and S. Bae, Effectiveness of RTS/CTS Handshake in IEEE 802.11 Based Ad Hoc Networks, *Ad Hoc Networks*, 1, 107–123, 2003.

245. S. Xu and T. Saadawi, Revealing the Problems with 802.11 Medium Access Control Protocol in Multi-hop Wireless Ad Hoc Networks, *Computer Networks*, 38(1), 531–548, 2002.

246. C. Perkins and P. Bhagwat, Highly Dynamic Destination-Sequenced Distance-Vector Routing (DSDV) for Mobile Computers, presented at *SIGCOMM*, London, U.K., 1994.

247. P. Jacquet and P. Muehlethaler, Optimized Link State Routing Protocol in Internet Draft, 1998.

248. T. Clausen and P. Jacket, Optimized Link State Routing Protocol, Internet Engineering Task Force, RFC 3626, 2003.

249. A. Iwata, C. Chiang, G. Pei, M. Gerla, and T. Chen, Scalable Routing Strategies for Ad Hoc Wireless Networks, *IEEE Journal on Selected Areas in Communications*, 17(8), 1369–1379, 1999.

250. S. Murthy and J. Aceves, An Efficient Routing Protocol for Wireless Networks, ACM/Baltzer Journal on Mobile Networks and Applications Special Issue on Routing in Mobile Communication Networks, 1(2), 183–197, 1996.

251. C. Chiang and M. Gerla, Routing in Clustered Multihop, Mobile Wireless Networks with Fading Channel, presented at *IEEE Singapore International Conference on Networks*, Singapore, 1997, pp. 546–551.

252. T. Chen and M. Gerla, Global State Routing: A New Routing Scheme for Ad-Hoc Wireless Networks, presented at *IEEE International Communications Conference*, Atlanta, GA, 1998, pp. 171–175.

253. D.B. Johnson and D.A. Maltz, Dynamic Source Routing in Ad Hoc Wireless Networks, *Mobile Computing*, 353(3), 153–181, 1996.

254. D.B. Johnson, D.A. Maltz, and Y. Hu, The Dynamic Source Routing Protocol for Mobile Ad Hoc Networks (DSR), Internet Draft, 2004.

255. C.E. Perkins, Ad Hoc On-Demand Distance Vector (AODV) Routing, Internet Draft, 1997.

256. C.E. Perkins and E.M. Belding-Royer, Ad Hoc On-Demand Distance Vector (AODV) Routing, RFC3561, 2003.

257. V.D. Park and M.S. Corson, Highly Adaptive Distributed Routing Algorithm for Mobile Wireless Networks, presented at *IEEE Conference on Computer Communications*, Kobe, Japan, 1997, pp. 1405–1413.

258. Z.J. Haas, New Routing Protocol for the Reconfigurable Wireless Networks, presented at *IEEE International Conference on Universal Personal Communications*, San Diego, California, 1997, pp. 562–566.

259. Z.J. Haas, M.R. Pearlman, and P. Samar, The Zone Routing Protocol (ZRP) for Ad Hoc Networks, Internet Draft, 2002.

260. N. Beijar, Zone Routing Protocol (ZRP), Networking Laboratory, University of Technology: Finland, 2002.

261. IEEE 802.11e/D3.0, Medium Access Control (MAC) Enhancements for Quality of Service (QoS), 2002.

262. IEEE, IEEE Standard for Information Technology — Telecommunication and Information Exchange between Systems — LAN/MAN Specific Requirements. Part 11. Wireless Medium Access Control (MAC) and Physical Layer (PHY) Specifications: Medium Access Control (MAC) Quality of Service (QoS) Enhancements 2005.

263. IEEE, IEEE Standards for Local and Metropolitan Area Networks: Media Access Control (MAC) Bridges, IEEE Standard 802.1D, 2004.

264. S. Chen and K. Nahrstedt, Distributed QoS Routing with Imprecise State Information, presented at *IEEE ICCCN*, 1998, pp. 80–89.

265. B. Karp and H.T. Kung, GPSR: Greedy Perimeter Stateless Routing for Wireless Networks, presented at the *Sixth Annual ACM/IEEE International Conference on Mobile Computing and Networking*, Boston, MA, 2000, pp. 243–254.

266. M. Conti, G. Muselli, G. Turi, and S. Giordano, Cross-Layering in Mobile Ad Hoc Network Design Computer, 37(2), 48–51, 2004.

267. Scrambled telephone, http://www.scrambledtelephone.com.

268. T.N. Swe and K.G. Veo, An Accurate Photodiode Model for DC and High Frequency SPICE Circuit Simulation, http://www.nanotech2003.com.

269. http://www.imagineeringezine.com

270. M. Sokolich, "In P-HBTs fill high-speed, lower power optical transceiver requirements," http://www.cetimes.com.

271. Sona beam. http://www.Sonabeam.com.

272. R. Valadas, A. Moreira, C. Lomba, A. Tavares, A. Oliveira Duarte, "The Infrared Physical Layer of the IEEE 802.11 Standard for Wireless LANs," http://www.av.it. pt.

273. G. Anastasi and L. Lenzini, "Wireless Networks QoS provided by the IEEE 802.11 wireless LAN to advanced data applications: A simulation analysis,"P 6 (2000), 99–108.

274. P. Barker and A.C. Boucouvalas, "Performance comparison of the IEEE 802.11 and Air infrared wireless MAC protocols."

275. S. Raya, D. Stavabinskia, and J. Carruthers, "Performance of Wireless Networks with Hidden Notes: A Qeueuing-Theoretic Analysis," http://netlab1.bu.edu.

276. R. Ramanathan and J. Redi, "A Brief Overview of Ad Hoc Networks: Challenges and Directories," http://www.comsec.or/linepubx/ci1/public/anniv/romana.html

277. http://www.wirelessbrasil.org

Index